I0029482

SOCIOLOGY AND THE NEW SYSTEMS THEORY

SOCIOLOGY AND THE NEW SYSTEMS THEORY

Toward a Theoretical Synthesis

KENNETH D. BAILEY

State University
of New York
Press

I wish to thank the following authors and publishers for the use of their materials in the production of this book:

Chapter 5: Quotations from pp. 110–111 of James Grier Miller, *Living Systems*. Copyright © 1978 by McGraw-Hill, Incorporated. Reprinted by permission of the author and of McGraw-Hill, Incorporated. Also, Figure 5.1, 5.2, and 5.3, from James Grier Miller and Jessie L. Miller, "Greater than the Sum of its Parts I. Subsystems Which Process Both Matter-Energy and Information." *Behavioral Science* 37:1–38. Copyright © 1992 by *Behavioral Science*. Figure 5.1 is from page 4; Figure 5.2 is from pages 6 and 7; and Figure 5.3 is from page 2. Further, Figure 5.1 is an adaptation from Table 1.1 on page 3, and Figure 5.3 is an adaptation from Figure 1.2 on page 4 from James Grier Miller, *Living Systems*. Copyright © 1978 by McGraw-Hill, Incorporated. Reprinted by permission of the authors and of McGraw-Hill, Incorporated.

Chapter 8: Quotations from pages 161, 162, 163, and 174 of John Mingers, "An Introduction to Autopoiesis." *Systems Practice* 2:159–180. Copyright © 1989 by Plenum Publishing Corporation. Reprinted by permission of the author and of Plenum Publishing Corporation.

Published by
State University of New York Press, Albany

© 1994 State University of New York

All rights reserved

Production by Susan Geraghty
Marketing by Fran Keneston

Printed in the United States of America

No part of this book may be used or reproduced
in any manner whatsoever without written permission
except in the case of brief quotations embodied in
critical articles and reviews.

For information, address State University of New York
Press, State University Plaza, Albany, N.Y., 12246

Library of Congress Cataloging-in-Publication Data

Bailey, Kenneth D.
    Sociology and the new systems theory : toward a theoretical
synthesis / Kenneth D. Bailey.
        p.   cm.
    Includes bibliographical references and index.
    ISBN 0-7914-1743-3 (hard : alk. paper). — ISBN 0-7914-1744-1
(pbk. : alk. paper)
    1. Sociology—Methodology.   2. System theory.   3. Social systems.
I. Title.
HM24.B297   1992
301'.01—dc20   93-22318
                                                                        CIP

10 9 8 7 6 5 4 3 2 1

To my parents, Sherline S. Bailey and Kenneth R. Bailey

CONTENTS

LIST OF TABLES

LIST OF FIGURES

PREFACE

Whatever happened to social systems theory? Did this 1960s phenomenon simply disappear along with the rest? The answer may surprise some readers. Social systems theory is alive and well, and in fact its literature is growing rapidly. However, this contemporary variety is not characterized by the functionalism of the 1960s.

Rather, as this volume clearly documents, social systems theory of the 1990s and beyond deals with a large number of important sociological issues barely dreamt of in the 1960s, such as nonequilibrium analysis, social entropy theory, living systems theory, autopoiesis, and the new sociocybernetics. Living systems theory and autopoiesis were born during the late 1970s and early 1980s, and are reaching maturity (amid further development) in the 1990s. Social entropy theory was not published until 1990. Together, these developments represent an exciting new theoretical frontier that I refer to as the "new systems theory." More correctly, I should say the "new social systems theory," as this volume does not deal directly with exciting systems advances such as artificial intelligence, but only with those systems approaches that contribute directly to social science.

The goal of this theoretical monograph is to present, for the first time, a more integrated view of these interrelated approaches, chiefly (but not exclusively), living systems theory, social entropy theory, and autopoiesis. This book represents the first major attempt that I know of to synthesize these approaches in a fashion that maximizes their efficacy for social scientists of the 1990s and the twenty-first century. One of the guiding principles of the systems movement is the need for integration.

Fortuitously, integration is increasingly sought by social scientists, especially sociological theorists such as Alexander (Alexander and Colomy 1990). The age of increasing integration that Alexander foresees for the 1990s and beyond sets the stage for the second chief goal of this volume—the integration of systems theory and sociological theory. Lest "integration" seem too bold a goal for now, we can at least seek rapprochement, which is underway, as evidenced by increased recognition and acceptance of systems theory by sociologists. Thus, the present volume should alleviate much of the criticism that sociological theorists have leveled at system approaches in the past.

For one thing, 1990s systems theory is, as stated above, a very different entity. Much criticism was aimed at facets of functionalism (such as equilibrium) which are totally absent in the new systems theory, or of which only vestiges remain. Thus, if contemporary sociological theorists are to criticize the new systems theory, they will have to aim their guns in new directions, as their old targets have vanished, rendering their old critiques obsolete. In particular, the old "conflict theory versus functionalism" dichotomy now securely ensconced in introductory sociology texts has little relevance to the present volume.

The third goal of this volume is to simply make sociologists more aware of recent developments in social systems theory. It is safe to assume that many sociologists have not read systems theory much since Parsons. There are a number of reasons for this. One of course is that social theory, a la Sorokin, is simply subject to fads and foibles, and systems moved out of fashion for a while, as did many other approaches. Another is that systems theory is decidedly interdisciplinary, and those of us who are narrow specialists simply missed new developments. Yet another reason is that some of the literature discussed here was not written by Americans, and some of it is rather new (and admittedly jargon-filled, as in the case of autopoiesis) so that neophytes often find it relatively incomprehensible. Finally, perhaps the most troubling reason for recent sociological neglect of systems theory is that no one has yet dispelled the old stereotypical notions that systems theory is necessarily "functionalist," "positivistic," "tautological," particularly vulnerable to reification or hypostatization, etc. The variety of systems work discussed here, from the "soft systems" methodology of Checkland, to the curiously postmodern epistemology of Maturana and Varela, renders these old stereotypes sterile. At the very least, the wide variety of work presented insures that no one label or criticism can embrace systems theory. It certainly cannot all be labeled "functionalism" as before; the autopoiesis of Maturana and Varela takes an explicitly nonfunctional stance.

In summary, the three chief goals of this volume are to document the new systems theory and render it accessible to sociologists, to integrate it internally, and to integrate it with the larger body of sociological theory. The idea is to show that 1990s systems theory has a lot to offer sociology. After its contributions become evident, sociologists are free to criticize it, although they must be forewarned that the old 1960s criticisms will be generally obsolete.

How can these goals best be met? Chapter 1 begins with three salient approaches to sociological theory—those of Alexander, Giddens, and Collins. They are seen as points of comparison for systems theory. Chapter 2 makes the case for systems theory, while chapters 3 and 4 portray the historical foundations for contemporary analysis.

Presentation of the "new" systems theory begins in earnest with living systems theory in chapter 5. Chapter 6 discusses social entropy theory, while chapter 7 seeks congruence between the two. Chapter 8 presents autopoiesis. Chapter 9 discusses the "dual synthesis." While chapter 1 presents the theories of Alexander, Giddens and Collins as "points" for comparison, chapters 2 through 8 all conclude with a counterpoint, which discusses the interface between systems theory and sociological theory. Thus, after the new systems theory is synthesized in chapters 2 through 8, the counterpoints can be used as the basis for the dual synthesis of chapter 9 (that is, a synthesis between systems theory and sociological theory).

My view is that contemporary social systems theory will be a surprise to some readers. They will not always find it easy to read, but their extra effort will be rewarded. Social systems theory is not dead, but is vibrant, and seeks the very goals (such as integration) that many sociologists seek. In fact, careful reading of this volume will show that many of the goals espoused by sociological theorists (such as Giddens's call for increased emphasis on time and space) have in fact been earnestly addressed by contemporary systems theorists. I welcome you to the broad and exciting world of contemporary social systems theory, and I hope you see the same benefits in it that I do.

My intellectual debts are apparent in this volume, and are too numerous to mention them all. I especially benefited from my long association with James Grier Miller. I am thankful for the immersion in living systems theory that I gained through his tutelage. I also benefited greatly from discussion with a number of colleagues, including Niklas Luhmann, Gideon Sjoberg, and Jeffrey Alexander. I wish to particularly thank Melvin Pollner for suggesting the comparison of systems theory with sociological theory. I also wish to thank Leah Robin for editorial assistance and for compiling the index. Finally, special thanks go to Rosalie M. Robertson and Susan Geraghty of SUNY Press for their excellent editorial direction.

CHAPTER 1

Social Theory and Systems Theory

What is the "new" social systems theory? Does social systems theory even exist in the 1990s? Yes, social systems theory does exist, as this volume clearly documents. Further, it is alive and well, and bears little resemblance to the old structural functionalist macrotheory of two decades ago that most sociologists probably still identify as social systems theory. If this is so, how can it be that it is not more visible? One answer is that in a sense this "new" systems theory is still very new. Social entropy theory (Bailey 1990) is a product of the 1990s, while autopoiesis has, prior to publication of the present volume, not been presented in a comprehensive form to American sociologists. For whatever reason, the new systems theory has been developing on a path that is largely parallel to, but different from, the path followed by major "mainstream" sociological theorists such as Alexander, Giddens, and Collins. Thus, it has had until now somewhat the status of a theoretical yeti. Theorists may have seen its footprints in the snow, or perhaps even viewed it at a distance, but have had little actual proof of its existence.

The chief question seems to be why sociological theorists have not gotten a better glimpse of the new systems theory. Perhaps part of the answer is that the new systems theory has been quietly building upon its positivistic roots, while much of contemporary sociology has been looking in some other direction. Thus, to some extent, those who wanted to see systems theory could have seen it, except that they were looking South when they should have looked North. The parallel paths taken by systems theorists and mainstream theorists are well described by Turner, who writes:

> One possible scenario is that these more positivistic theorists will simply leave mainstream theory which, at present and into the foreseeable future, will have too many inhibitions and reservations about the prospects for scientific sociology. For increasingly, positivistic work simply ignores the way much theory is currently practiced in American sociology. (Turner 1990, p. 389)

This walking of parallel paths by social systems theorists and "mainstream" sociological theorists has already occurred during the last two decades. Clearly, some of it has to do with "positivism" (whatever that is). However, even though the new systems theory continues to be science based, it is certainly *not* a homogeneous entity, and variants of it differ greatly with respect to how positivistic they appear, as will be seen in this volume. It is safe to say that the new systems theory is decidedly less positivistic than the old.

Further, sociologists exhibit disagreement about the nature and role of positivism in American sociology. Turner (1990, p. 389) sees modern sociological theory as antipositivisitic, with positivistic theorists constituting "a relatively small group—fifty or sixty thinkers at most." In contrast, Alexander (1982, p. 5), while admitting that the formal methodological principles of classical positivism are eschewed by most sociologists, says, "Yet positivism in a more generic sense is, nonetheless, a persuasion that permeates contemporary social science." This lack of agreement extends to statements about Comte's dream. Pollner says that:

> Sociology has not of late dreamt Comte's dream. Few persons entertain the thought . . . that sociology is or could become the Queen of the Social Sciences. Indeed, for some, sociology's claim to presence in the court, let alone any sort of title, often seems precarious. (Pollner 1987, p. 1)

Turner (1990, p. 389) asserts that "Comte's original dream is alive and well."

Actually these positions are probably not as contradictory as they might at first appear. It is indisputable that American sociology is "permeated" with positivistic principles, and so I agree with Alexander. But Turner is also correct in being able to discern an antipositivistic attitude among many American social theorists. And again, Pollner and Turner are both correct. Relatively few still dream Comte's dream, but those few of us that exist do keep it "alive and well."

Interestingly and perhaps ironically, this mixed view of how American social theorists assess the degree of positivism in their midst extends as well to the new systems theory. Those who have a preconceived notion of systems theory as positivistic based on classical systems theory of twenty years ago will find that the new systems theory is considerably less positivistic (or a "gentler" positivism if you will). This could lead to a grudging acceptance even from antipositivists. On the other hand, strict positivists might find some parts of the new systems theory too nonpositivistic for their purist sentiments.

Alexander (1982, pp. 5–7) posits four postulates central to the "positivistic persuasion": (1) a distinction between the empirical and nonempirical; (2) exclusion of philosophical and nonempirical issues; (3) assumption of a scientific self-consciousness; and (4) theoretical issues dealt with only in relation to empirical observation. I will show that living systems theory (chapters 5 and 7) exhibits all of these characteristics of positivism to some degree. Even in this case the degree of positivism is more apparent than real. Miller (1978) advocates all four, but a reading of his book indicates a relative degree of nonempirical content. The other major approaches in the new systems theory—social entropy theory and autopoietic theory—are much less positivistic by Alexander's criteria.

In examining social entropy theory and autopoietic theory, the reader will be struck by the degree to which they *do not* distinguish the empirical from the philosophical. Even Miller grudgingly acknowledges the useful role of systems philosophy for a fledgling systems theory, but he expects a more mature systems theory to "grow out" of the philosophical phase into an empirical, formal, operationalized (and more clearly positivistic) phase.

In contrast, social entropy theory and autopoietic theory are both explicitly epistemological. Both autopoietic theory and the "new sociocybernetics" (see Geyer and van der Zouwen 1978) offer a sophisticated analysis of the relationship between observer and observed which transcends simplistic classical notions of "empirical observation" in positivism. Autopoietic theory, in fact, incorporates the notion of the observer with the model in a very sophisticated way. Rather than excluding philosophical issues, autopoietic theory often emphasizes them.

Social entropy theory also approaches Alexander's empirical-nonempirical distinction in a much more sophisticated (and I hope innovative) fashion than did classical approaches. I (Bailey 1990) recognize three levels—conceptual, empirical, and operational (indicator). Ironically, many of the fundamental principles of science, such as verification and replication, cannot easily be defined as strictly the empirical level in the three-level model. Rather, they involve a dialectical interaction between all three levels, and this is a diachronic process. That is, the scientist forms a perception (conceptual level) from observation (empirical level), then codes and recodes the data (indicator level) and repeats the whole process. Social entropy theory does not divide the empirical and nonempirical levels and exclude the latter, but rather emphasizes the dialectical interaction over time, among all three levels—the conceptual, empirical, and indicator. In this sense, social entropy theory is *not* positivistic, as it fails Alexander's initial fundamental criterion of positivism. However, as I share Comte's dream of a sociology which recognizes that humans are biological individuals in a physical world, I accept the label of positivist. I strive for a humanistic positivism (not as much of a contradiction as some think), or as I call it in this volume, a "positive positivism." It is certainly a positivism that transcends the simpler descriptions of some classical positivistic models, and should not be judged by them.

However, there is no question that the new systems theory is more methodological than traditional sociological theory. As Ritzer (1990b, p. 363) says, "There is a need for more methodologists and empirical researchers to address the micro-macro issue which to this time has been largely dominated by theorists. Some welcome signs in these areas are Bailey's (1987) work on micro-macro methods . . ." Thus, the issue of the degree of positivism aside, the new systems theory does add a needed methodological dimension to the analysis of the micro-macro issue as well as other issues. In fact, one of my long-range goals is to also further theory-method integration, a linkage that has been widely neglected by theorists. One can search almost in vain for equations or methodological discussion in theoretical treatises, and this volume on the new systems theory serves as a step toward theory-method integration by showing that social theory can be written in a methodologically informed fashion, just as scientific theory is. After all, we should never forget that although classical theorists such as Weber and Durkheim are known for their theoretical achievements, they both wrote books on methodology (Durkheim 1982; Weber 1949).

Why should sociologists study systems theory? There are many reasons, and the case is made in chapter 2. However, there is a need in this initial chapter for a discussion which frames systems theory in the larger context of extant sociological theory. I said earlier that social systems theory and mainstream sociological theory seem to be on parallel paths. This is fortuitous because parallel paths lead in the same direction. It is disquieting because one of the basic tenets of the systems movement is the need to integrate separate approaches, especially those which have some elements in common (as do systems theory and mainstream theory, as I will show shortly). Also, ironically, the new systems theory is sufficiently diverse that it is itself in need of synthesis, and that is one of the chief goals of this monograph.

One major reason that sociologists, and others, should study the new systems theory is because it deals with important societal processes more effectively than other approaches. These processes include entropy, autopoiesis, matter-energy processing, information processing, and control processes (sociocybernetics). While classical sociology and systems theory dealt extensively with equilibrium, the new systems theory emphasizes nonequilibrium approaches, framed largely in terms of entropy processes. All living systems must deal with entropy processes, including social systems, and sociologists can neglect study of this important topic only at their peril. Equally important is the notion of autopoiesis, or self-reproduction. The notion of autopoietic recursive systems is becoming increasingly emphasized in sociology, and systems theorists have taken the lead in its study. In addition to entropy and autopoiesis, energy and information processing are central processes in society that demand integrative theoretical attention. No sociological approach has accomplished this to the degree that the new systems theory, specifically living systems theory, has.

These important processes are crucial to complex society, as I will show in later chapters, yet they receive only scant attention in sociological theory. I believe that the best vehicle for their study is a newly synthesized systems approach. This new synthesis can most effectively deal with these processes, and for this reason alone deserves careful attention.

The goal of this volume is to construct a theoretical monograph which presents and synthesizes the new systems theory. This is done via an original conceptual framework combining the three-level model and the *Q-R* distinction (Bailey 1984c, 1990). This endeavor comprises chapters 2 through 9. Although this is in some sense a stand-alone endeavor, in order to affect the integration of systems theory and social theory just proposed, it is necessary now to engage in a brief discussion of some examples of extant sociological theory.

The examples I have chosen as criteria for my comparison of systems theory are Alexander's neofunctionalism (1984, 1985; Alexander and Colomy 1990), Giddens's structuration theory (1979, 1982, 1984), and Collins's conflict theory (1975, 1988). There is no claim that these exhaust the spectrum of contemporary theory, or that I can present all facets of each approach. However, these approaches are major and significant, and represent the macro-micro spectrum. They thus serve as good points of comparison. At the end of the volume I will make some brief comments about how systems theory relates to theoretical approaches in addition to these three.

By presenting these three approaches, I can serve a number of useful purposes. For one, I can document that social theory and social systems theory are indeed on parallel paths, and have a number of points in common. Secondly, I can use these commonalities as a starting point to initiate an integration of systems theory that is maximally compatible with a mainstream sociological theory. Third, I can demonstrate the major points addressed by systems theory that contemporary sociological theory does not address. I will thus show that by addressing the issues neglected by the mainstream, systems theory not only complements the mainstream, but actually adds breadth and richness to it. Fourth, the discussions of neofunctionalism, structuration, and conflict theory simply serve as a needed frame or criterion point, or stepping-off point, that provide some familiarity for the reader in his or her introduction to the less familiar world of social systems theory. I must stress that the discussion of neofunctionalism, structuration and conflict theory is not a critique, or even an analysis, but is merely the specification of a comparison point and focal point for the analysis of the new social systems theory.

The plan of the book is a simple one. I will sketch these three theories (neofunctionalism, structuration, and conflict) as the *point,* meaning the point of reference, or point at which social theory now stands (bearing in mind that other important perspectives are being omitted for lack of space). With these theories as the point, I will end each systems chapter (chapters 2–8) with a discussion of the *counterpoint* offered by that specific systems perspective. In some cases the counterpoint may be a literal counter, or statement of opposition or contradiction to the social theories in the point. Often, though, the counterpoint will be a complementary view, or parallel endorsement of the point made by the criteria theories.

The last chapter (chapter 9) will constitute a *point-counterpoint*. It will be a synthesis of the new systems theory. As a synthesis of the approaches such as the new sociocybernetics, living systems theory, social entropy theory, and autopoietic theory, it will be a synthesis of the counterpoint to the comparison major mainstream theories (neofunctionalism, structuration, and conflict). But chapter 9 will also strive for a subsequent synthesis of the integrated systems approach with the criteria theories. Thus, it will be a synthesis of a synthesis, or a dual synthesis, involving both point (neofunctionalism-structuration-conflict) and counterpoint (new sociocybernetics–living systems theory–social entropy–theory–autopoietic theory). While it may not be feasible at this point in time to totally bring the new systems theory into the "mainstream" of sociological theory, it is feasible to synthesize this new systems theory in a way that is maximally comparable to extant approaches, and thus is maximally accessible to sociologists. Analysis of the mainstream approaches will help to accomplish this. In other words, the strategy is to discover parallels between mainstream theory and systems theory, and to emphasize these parallels in the synthesis of systems theory. The end result should be a more methodologically informed contribution to the micro-macro issue as Ritzer (1990b, p. 363) has called for. It should also provide theorists an opportunity to examine systems theory "up close," and to see the contributions that it has to offer.

The selection of the three comparison theories is admittedly arbitrary. There are many others that could have been chosen. These three are useful because they are major theories that are visible and viable, and among them they cover a considerable degree of theoretical territory, as all three are to some degree micro-macro syntheses (see Ritzer 1990b). While Collins's approach is essentially micro and Alexander's is essentially macro (but deals with both action and order), Giddens's approach is intermediate, and deals extensively with the agency-structure issue. Since the reason for presenting these theories is not to critique or analyze them, but to provide comparison points from which systems theory can "bounce off," I will simply provide skeletal descriptions by using a lot of quotes, and sticking as closely as possible to the basic texts.

Alexander's Neofunctionalism

Alexander and Colomy (1990) identify three phases of postwar sociology: structural-functionalism, microsociology, and the emerging third phase, "marked by an effort to relink theorizing about action and order, conflict and stability, structure and culture" (Alexander and Colomy 1990, p. 36). This movement back to synthesis, directed largely but not entirely at bridging the micro-macro gap, is the third phase of postwar sociology. Neofunctionalism is part of this third phase. Alexander (1985) introduced the term "neofunctionalism" in order to emphasize the double element of continuity and internal critique in Parsonian thought (the analogy is to neo-Marxism). Neofunctionalism is a prototypically synthetic theory (in the sense of synthesis). Without the flaws in Parsonian thought that led to emphasis on separate macro and micro theories, neofunctionalism can once again return to the project of synthesis. Thus:

> It is not surprising, therefore, that as contemporary theorists have returned to the project of synthesis, they have often returned to some core element in Parsons' earlier thought. It is striking that this return is manifest in the work of theorists who have never had any previous association with Parsonian thought. The motive is theoretical logic, not personal desire. (Alexander and Colomy 1990, p. 39)

Further, functionalism was subject to elaboration and revision in the first theoretical phase, but was buffeted by shifts in disciplinary sensibilities in the second phase, and came near extinction. But now:

> In the emerging third phase, scientific sensibility has shifted once again. . . . In response, the functionalist tradition has entered a phase of reconstruction. Neofunctionalism is the result. (Alexander and Colomy 1990, p. 43)

The result of this reconstruction of the Parsonian core is that:

> A surprisingly large portion of earlier peripheral criticism has been accepted, just as the core itself is being reshaped in a responsive way. From this perspective, neofunctionalism is post-

Parsonian. Its aim is to go beyond both the first and second phase
of postwar sociology and to reconstruct a new synthesis on the
basis of the contributions of each. (Alexander and Colomy 1990,
p. 46)

Inasmuch as this new synthesis succeeds in combining the contributions of both the first phase (macro) and the second phase (micro), it is a true micro-macro synthesis. This reconstruction has resulted in a number of significant changes in the Parsonian core program. Not all of these are central to my concerns. I will list some specific reconstructions which have parallels in systems reconstruction or which are otherwise salient to the concerns of this volume.

One salient aspect addressed by Alexander is the problem of equilibrium in Parsons's work, which is dealt with in detail later in this volume. As Alexander and Colomy (1990, p. 45) say,

When Parsons converted this model into a cybernetic system,
however, he tilted toward one set of social system parts, the normative, raising it to a vertical position over another set, the material. He had great difficulty, moreover, in maintaining the analytical status of his model, often conflating the conceptualized ideal of equilibrium with the condition of an empirical society.

Alexander (1983) also attacked Parsons's idealist tendencies, and argued that they were responsible for many defects in his work, including the tendency to see change in teleological terms. The recognition of the problems with equilibrium and the emphasis on material rather than idealist factors are exceedingly important reconstructions for the new systems theory. It is crucial that neofunctionalism emphasize and maintain these features if neofunctionalism and the new social systems theory are to pursue parallel paths.

Another exceedingly important issue for systems theory, also addressed by Alexander (1988; Alexander and Colomy 1990) is the problem of order. Parsons's unnecessary conflation of equilibrium with the self-maintenance of order also is a major problem for systems theorists, as seen in detail in the following chapters of this volume. By sharply critiquing Parsons's positions on equilibrium and order, Alexander has made neofunctionalism exceedingly more attractive to the new systems theorists.

Still another reconstruction in neofunctionalism that is welcomed by systems theorists is Alexander's (1983) criticisms about the reification of functionalist and systems reasoning, and also the criticism of the conflict between the AGIL dimensions (see chapters 6 and 9) and the empirical differentiation in contemporary society (see Alexander and Colomy 1990, p. 47).

In summary, there fortuitously seem to be parallel programs of reconstruction in both neofunctionalism and the new systems theory. This work has not been planned or coordinated consciously between the two approaches, but the similarities have doubtlessly arisen largely because both approaches have been concerned with the elaboration, revision, and reconstruction of Parsons's work. The overlap between neofunctionalism and the new systems theory stems from the fact that both have analyzed Parsons's systems writings. The two approaches will probably never meet, as neofunctionalists continue to be more normative, cultural, and voluntaristic, and decidedly less methodological (and less "positivistic") than the new systems theorists. However, the paths are parallel, and without noticeable conflict or contradiction. In fact, the two approaches are distinctly complementary, as the new systems theory welcomes the neofunctionalist analysis of micro-macro links and such other concerns as material factors, idealist factors, culture, differentiation, etc. In return, I hope that neofunctionalists can appreciate some of the contributions of the new systems theory, such as the methodological approach to micro-macro analysis, the global-mutable-immutable distinction, allocation theory, the novel approach to power, and the critique of equilibrium.

Also important for the new systems theory is Alexander's (1982) emphasis on action and order as criteria for theoretical logic in sociology. These are extremely important concepts in the new systems theory, especially in social entropy theory (Bailey 1990). Thus, these important concepts will serve as important points of comparison in my point-counterpoint comparison of systems theory and mainstream sociological theory.

Alexander mentions the systems concept occasionally (see, for example, Alexander and Colomy 1990, p. 45; Alexander 1985, p. 8). It is not a main focus, and he is certainly (as he says) not a systems theorist. Nevertheless, his attention to the concept does provide an important and welcome point of reference for the subsequent comparison of systems theory and mainstream sociological theory.

An interesting parallel between neofunctionalism and social

entropy theory (Bailey 1990) is that while Alexander describes his neofunctional approach as "post-Parsonian" (Alexander and Colomy 1990, p. 46), Bailey has described his theory much earlier as "postfunctionalism" (Bailey 1983). This indicates the degree of parallel reconstruction in both neofunctionalism and the new systems theory.

Giddens's Structuration (and Agency/Structure)

It is interesting that while neofunctionalists such as Alexander have occasion to use the term "system" (see, for example, Alexander and Colomy 1990, pp. 45, 47), Giddens also uses the term "system" rather extensively. In fact, his discussion of the relation between systems and structure (Giddens 1979, pp. 59–81) is an important contribution to systems theory.

It has sometimes been implied that Giddens is "against" systems theory (see Archer 1985, p. 61). This probably stems from his critique of functionalism and the fact is that his structuration approach opposes morphogenetic theory in some ways (see Giddens 1979). However, it is also clear from perusing his work that Giddens accepts the notion of system as necessary, saying, "I want to suggest that *structure, system* and *structuration,* appropriately conceptualized, are all necessary terms in social theory" (Giddens 1979, p. 62, italics in the original). Giddens says (1979, pp. 61–62) that in functionalism, the notions of structure and system tend to dissolve into each other. This is because a structure is seen as synchronic, while a system is "functioning" over time. But when the social system ceases to function, it ceases to exist—thus structure and function dissolve into one another.

The culprit is time. "In functionalism and structuralism alike, an attempt is made to exclude time (or more accurately, time-space intersections) from social theory, by the application of the synchrony/diachrony distinction" (Giddens 1979, p. 62). But this distinction is unstable, as "time refuses to be eliminated" (Giddens 1979, p. 62). Giddens concludes that the term "social structure" thus includes two elements: (*a*) the patterning of interaction; and (*b*) the continuity of interaction in time. As Giddens (1979, p. 64) employs the term, "structure" refers to "structural property" or "structuring property," with structuring properties "providing the 'binding' of time and space in social systems" (Giddens 1979, p. 64).

As formally defined by Giddens (1979, p. 66), structure refers to "rules and resources, organised as properties of social systems. Structure only exists as 'structural properties.'" System refers to "reproduced relations between actors and collectivities, organised as regular social practices." Structuration refers to "conditions governing the continuity or transformation of structures, and therefore the reproduction of systems" (Giddens 1979, p. 66). Giddens also defines social systems, saying:

> Social systems involve regularised relations of interdependence between individuals or groups, that typically can be best analysed as *recurrent social practices*. Social systems are systems of social interaction; as such they involve the situated activities of human subjects, and exist syntagmatically in the flow of time. Systems, in this terminology, have structures, or more accurately, have structured properties; they are not structures in themselves. Structures are necessarily (logically) properties of systems or collectivities, and *are characterised by the 'absence of a subject'*. To study the structuration of a social system is to study the ways in which that system, via the application of generative rules and resources, and in the context of unintended outcomes, is produced and reproduced in interaction. (Giddens 1979, pp. 65–66, italics in the original)

Giddens (1979, p. 74) explicitly examines the work of Bertalanffy (1968) in *General System Theory*, specifically Bertalanffy's distinction between general systems theory, systems technology, and systems philosophy. While not finding systems philosophy of particular interest, Giddens finds systems technology to be crucial, saying:

> For, understood as a series of technological advances, systems theory has already had a great practical impact upon social life, an impact whose full implications will only be felt in the future. (Giddens 1979, pp. 74–75)

But Giddens also says that it is crucial to distinguish general systems theory from systems technology (he includes information theory and cybernetics in the latter category, saying that they were created in association with technological developments).
Giddens says further:

> Only by maintaining the distinction between the first and second categories is it possible to submit systems technology to ideology-critique. But sustaining this possibility, I think, also involves resisting the sort of claims that Bertalanffy and others have

made about the applicability of general systems theory to human conduct. (1979, p. 75)

Giddens goes on to say:

The reflexive monitoring of action among human actors cannot be adequately grasped in terms of principles of teleology applicable to mechanical systems. Purposive behavior is usually treated by systems theorists in terms of feed-back. I shall accept below Buckley's argument that systems involving feed-back processes are worthwhile distinguishing from the system mechanisms usually given prominence within functionalism, which are of a 'lower' kind. But I shall also want to differentiate feed-back system processes from a 'higher' order of reflexive self-regulation in social systems. (1979, p. 75)

Giddens says still further:

As employed by functionalist authors, the interdependence of system parts is usually interpreted as homeostasis. . . . But as critics of functionalism influenced by systems theory have pointed out, homeostasis is only one form or level of such interdependence: and one, borrowing from a physiological or mechanical model, where the forces involved operate most 'blindly'. It is not the same as self-regulation through feed-back, and is a more 'primitive' process. (1979, p. 78)

A few comments are in order concerning Giddens's views of systems theory. My basic conclusion is that virtually everything Giddens says is endorsed by comments independently conceived and written in subsequent chapters of this volume as well as in Bailey (1990). All of the comments referred to were written *before* reading Giddens's comments, so the parallels between the new systems theory and Giddens's approach (for example, the time-space intersections and the discussions of the relationships between structure and agency) are sometimes truly amazing. In fact, the comments that Giddens makes about homeostasis are reiterated frequently in this volume. Giddens's comments in essence form the foundation for reconstruction (in Alexander's terms), but without following through to a developed systems theory (in other words, if Giddens would follow his criticisms through with repair, he could have a systems theory). This book *does* follow through, and the exegesis of the new systems theory consists largely of these reconstructive efforts (chapters 2–9, but especially chapters 5–8 of this volume).

These latest systems efforts, in my opinion, are generally free from the charges that Giddens makes, as they represent a reconstruction and largely are sentiments parallel to his. Thus, the parallelism between structuration theory and the new systems theory, as between neofunctionalism and the new systems theory, is striking. This is true especially for social entropy theory. To the extent that living systems theory retains the homeostatic model (a very limited extent) it may be somewhat vulnerable to some of Giddens's criticisms, but bear in mind that these were directed specifically at functionalism and not at living systems theory. As we shall see, living systems theory has very different emphases from functionalism.

The only slight disagreement that I have with Giddens is his insistence on separating holistic general systems theory (Giddens's first category) from systems technology, with which Giddens includes information theory and cybernetics. Inasmuch as general systems theory rests soundly on cybernetics and information theory, to insist upon distinguishing these seems not only vastly unfair to general systems theory, but might be impossible, as it destroys the holistic quality that general systems theory strives for. Such piecemeal separation of systems theorists' holistic efforts by non-systems theorists is not uncommon, but is a bane of the systems movement. Holistic theories must be assessed as such, and cannot be fairly assessed by dissecting out some of their elements and leaving others. However, this is a relatively minor point, all things considered, especially since Giddens's critique is basically obsolete, dealing as it does with developments that are now some two decades old (the critique itself is now over a decade old).

To summarize, I share Giddens's critique of functionalism, and have written a reconstruction of functionalist systems theory (post-functional systems theory) that is based on a similar critique. This is social entropy theory (Bailey 1990). Social entropy theory is truthfully not so much a reconstruction of functionalism as it is a totally new effort. I simply took the same complex society that functionalism was interested in explaining, but built a new model from the ground up, with entropy and without equilibrium. It is only a reconstruction in the limited sense that I critiqued functionalism methodologically, and endeavored not to repeat its mistakes. Social entropy theory is thus based on the premise that the flaws of functionalist systems theory are flaws of *functionalism* and not of systems theory. In other words, the flaws are not found in the *general* systems model, but in the *specific application of it (functionalism)*. For further discussion of this point see Bailey (1990 and chapters 6 and 7 of this volume).

The only part of Giddens's critique of systems theory which has current relevance for the new systems theory is his discussion of autopoiesis (Giddens 1979, pp. 75–76). This is the topic of chapter 8. In speaking of autopoiesis, Giddens (1979, p. 75) says, ". . . it is probably too early to say just how close the parallels with social theory might be. The chief point of connection is undoubtedly recursiveness, taken to characterize autopoietic organization."

Turning more directly to structuration theory, Giddens says that:

> The concept of structuration involves that of the *duality of structure,* which relates to the *fundamentally recursive character of social life, and expresses the mutual dependence of structure and agency.* (1979, p. 69, italics in the original)

Thus the structural properties of social systems constitute both medium and outcome of the practices comprising those social systems. "The theory of structuration thus formulated, rejects any differentiation of synchrony and diachrony or statics and dynamics. The identification of structure with constraint is also rejected: structure is both enabling and constraining" (Giddens 1979, p. 69).

Explicitly rejecting the "snapshot" synchronicity of functionalism, Giddens (1979, p. 202, italics in the original), says *"any patterns of interaction that exist are situated in time;* only when examined over time do they form 'patterns' at all. This is most clear, perhaps, in the case of individuals in face-to-face encounters." He says further:

> To study the structuration of a social system is to study the ways in which that system, via the application of generative rules and resources, and in the context of unintended outcomes, is produced and reproduced in interaction. (Giddens 1979, p. 660)

Again, the degree of parallelism between Giddens's discussion of agency-structure relations and Bailey's (1990) discussion of the relationship between process and structure is truly amazing. These are not the same formulations by any means, but the parallels are striking. There are some clear differences. For one, Bailey's formulation is much more methodological, and his definition of structure is different from Giddens's. But essentially the two formulations reach the same conclusions: that action/process and structure are in a reciprocal relationship over time.

The difference is that Giddens has a different concept of structure. Bailey uses structure to indicate synchronic or static material components which can clearly be seen as products of diachronic action. These synchronic structures are symbolic structures in the indicator level X'' of the three-level model (see Bailey 1990, and chapter 2 of this volume). The symbolic structure can take a number of forms, such as sets of rules in etiquette books, books of laws, the Constitution, rule books for games, dictionaries, etc. These sets of symbols are symbolic-synchronic structure. Notice, however, that this is not "snapshot" synchronicity inasmuch as it is realized that this synchronic structure only exists in relation to, and as a result of, diachronic process. Thus, the symbolic-synchronic structure has a stand-alone quality, but not a snapshot quality.

What I mean by this is that it exists and can be viewed, but *cannot reproduce* itself. It can only be produced, reproduced, and changed via diachronic action. Thus, a true interaction or dialectic exists between the diachronic process and the synchronic structures. This will be discussed in more detail later. Consider synchronic static components such as dictionaries or rule books. Thus, an actor can use a dictionary (synchronic) to guide his or her writing (diachronic process), but continuing action can, over time, change the rules in the dictionary. Thus, structure (synchronic) guides process or action (diachronic), but diachronic action also produces a *product* in the form of symbolic synchronic structure, and can also alter this structure (at a later time). Again, while the formulations and certainly the language are different, the parallels are striking, especially inasmuch as Giddens uses the terminology of generative rules (1979, p. 66), but not in the context of a synchronic marker such as a rule book or etiquette book as does Bailey (1990).

The parallel formulations were also apparently conceived at about the same time. Bailey conceived his ideas beginning in 1978, with no knowledge of Giddens's writing on this topic. He did so by purposefully *not* reading the work of Giddens and others. He was essentially inspired to reconstruct the functionalist problem via systems theory, not by reviving or directly reconstructing functionalist tenets and principles, but by starting from the problem of how the complex social system functions, and building a new model (social entropy theory) from scratch or from the ground up. He was inspired primarily through reading Turner and Maryanski (1979) in draft form, as provided by an editor, in the winter of 1978.

The technique of constructing an approach without studying the work of others is common in art, and has clear advantages, especially if one is building an entirely new model from the ground up and wants to concentrate on internal consistency rather than on outside influences. However, it was supposedly practiced by Spencer, and is alleged to have been responsible for some of the flaws in his work (see Ritzer 1988). Thus, this insular practice obviously has pitfalls. I hope that in the case of social entropy theory this practice was justified. It led to some striking and I think valuable parallels between mainstream sociological theory and the methodological social theory of social entropy theory. The main flaw seems to be the self-imposed isolation of social entropy theory, which is being rectified in this volume. In other words, while I did not read mainstream theory while writing social entropy theory, I am correcting this now by making explicit connections between mainstream social theory and the new systems theory in this volume.

As one example, the process-structure model of Bailey (1990) is a model of diachronic-synchronic interaction. As such it is *not* subject to Giddens's criticisms of functionalism—that time is eliminated. Time and space are both used *explicitly* in both living systems theory and social entropy theory (see chapters 5 and 6). Thus, what has happened is that social systems theory has reconstructed itself in such a way as to avoid most all of the early criticisms of functionalism. It did so in relative isolation as a post-functional approach, but is now ready to reconnect to modern social theory as best it can.

The work of Giddens and Alexander provide clear anchoring points for this reconnection. The points of connection between these mainstream approaches and the new systems theory will become clearer throughout the volume. For example, Bailey's definition of structure may at this point seem much different from Giddens's, because Bailey uses the term synchronic structure, while

Giddens stresses the diachronic nature of structure. However, it will be seen that both are referring to rules. The difference is that Bailey is stressing the anchoring of symbol structure in physical markers (to be discussed later). While having synchronic qualities, these markers are used over time in a dialectic relationship with diachronic process, thus achieving the diachronic quality that Giddens stresses. Thus, the symbolic structure has a dualistic quality—it is static (until changed), but is utilized diachronically by human agents. This point is complex, and for now the reader will have to take my word that the agency-structure parallels between SET and structuration theory are very similar.

Archer (1985) has questioned Giddens's structuration approach, and explicitly compared it with morphogenetic theory (see chapter 4 of this volume). She says:

> Hence Giddens's whole approach turns on overcoming the dichotomies which the morphogenetic perspective retains and utilizes—between voluntarism and determinism, between synchrony and diachrony, and between individual and society. In 'place of these dualisms, as a single conceptual move, the theory of structuration substitutes the central notion of the *duality of structure*'. (Giddens 1979, p. 5)
>
> The body of this paper will: (a), question the capacity of this concept to transcend such dichotomies in a way which is sociologically useful; (b), defend the greater theoretical utility of *analytical dualism*, which underpins general systems theory, and, (c), seek to establish the greater theoretical utility of the morphogenetic perspective over the structuration approach. (Archer 1985, pp. 61–62, italics in the original)

My position is clearly intermediate between Archer and Giddens. I do not see any direct conflict between Giddens's structuration approach and systems theory. Rather, I view them as parallel and compatible. However, I do appreciate Archer's defense of systems theory, particularly of analytical dualism. I think this will be clear in Bailey's (1990 and chapters 6, 7, and 9 of this volume) discussion of synchronic-diachronic interaction. I agree with Archer that it is not necessary to get rid of these terms in order to meet Giddens's objective of explicating agency/structure relations. In other words, the answer is not to *eliminate* the distinction between synchrony and diachrony, but simply to show their dialectic intertwining. The flaw in functionalism was *not* that it used the synchronic/diachronic duality, but that it did not show that these are two sides of the same coin, and did not stress their interrelationships as SET does.

Collins's Conflict Sociology

The third mainstream work to be considered is Collins's (1975) conflict sociology. Collins also discusses functionalist theory (under the heading of "ideology"), saying:

> The functionalist effort to analyze human institutions as a system . . . has failed to pay off in genuine explanatory theory. This failing is due to a commitment to certain political values, which can be seen in system theorists from Comte and Durkheim through Pareto and Parsons. The commitment is to political unity. Systems theory is, in effect, a political (usually nationalist) utopia, hence the treatment of conflict is residual . . . (1975, pp. 20–21)

Collins goes on to say, "I believe that the only viable path to a comprehensive explanatory sociology is a conflict perspective" (1975, p. 21).

He says further:

> Conflict theory is intrinsically more detached from value judgments than is systems theory. To be able to recognize competing interests as a matter of *fact,* without trying to squeeze some of them out of existence as unrealistic, deviant, or just plain evil, is the essence of a detached position. It is for this reason that I argue for conflict theory as the basis of a scientific sociology, precisely because it moves farthest from the implicit value judgments that underlie most other approaches. Conflict theorists have come in a variety of political shades, ranging from anarchists and revolutionary socialists through welfare-state liberals to conservative nationalists. They have hardly been adverse to arguing for their political values, but it is not so difficult to separate their value judgments from their causal analysis, and it is to the best of them—Max Weber above all—that we owe the ideal of detachment from ideology in social science. (Collins 1975, pp. 21–22, italics in original)

Although I vowed not to analyze nor critique these three mainstream comparison theories, there are a number of surprising statements in this brief quote that cry for elaboration. The following comments on conflict theory apply *only* to Collins's (1975) book, and not to other approaches to conflict. One surprising assertion is that Max Weber is the "best" of the conflict theorists. I am not sure how Collins determined this, and whether he did so in a "value-free" manner. I would have guessed that in a poll, Marx might win the title of "best" classical conflict theorist, with Weber winning the "best" bureaucratic theorist award. More recently, Collins (1990, p. 68) seems to acknowledge this, attributing conflict theory to Marx and Engels, and "a little less obviously" to Weber. I freely admit, however, that I have no idea what constitutes the claim for "best" theorist, but I suspect it may be value laden (a cynic might suspect that Collins is basically trying to connect conflict theory to Weber's ideal of detachment from ideology). Another surprising statement is that it is "not so difficult" to separate conflict theorists' value judgments from their causal analysis. I for one find it very difficult, if not impossible, and think that Collins needs to elaborate more fully upon this claim.

Perhaps the reason I have difficulty understanding this latter claim is because conflict theory seems to me to be inherently value laden. When a person is lynched because of racial hatred, another is denied access to a living wage because of gender or sexual preference, and a nation declares war in the name of God, it seems obvious to me that entrenched value positions are at the base of the conflict in all cases. Yet Collins would have us believe that sociologists who study such visibly value-laden phenomena can be as detached and uninvolved as a physicist studying gas molecules. How can a sociologist keep from reacting to such value-laden issues when studying conflict? I do not understand it, and I need more evidence before I can accept the assertion that conflict theory "moves farthest from . . . implicit value judgments" (Collins 1975, p. 21).

The basic conclusion is that the subject matter of conflict theory is inherently ideological and value laden. As an example, in Collins 1975, table 1 (p. 238), the main heading is "Dominant ideology." Now I realize the distinction between a subject matter that is value laden, and the objectivity of causal statements about that subject matter. Still, a theorist venturing into the treacherous ideological currents of conflict and proposing to remain "objective" or "value free" may run the danger of saying that "I'll jump into the water but I won't get wet." I am not saying that conflict theory is *not* "farthest" from value judgments. I am simply saying that the case for that remains to be made, and is not so self-evident as Collins asserts.

Now as to the charges that concern systems theory. Again there are two, one that the systems model treats conflict as a residual, and that systems theorists as persons are committed to certain political values. I agree with Collins implicitly that *functional* systems theory, which is the *old* unreconstructed systems theory, emphasized return to the status quo and precluded or inhibited study of conflict and change. This is discussed in detail in chapters 2 and 3 and throughout this volume. This has been thoroughly reconstructed, and is absent in social entropy theory in particular and in the new systems theory in general. Thus, it is not correct to say that the new systems theory has this flaw, but only to say that the old functionalist form of systems theory (and not systems theory in general) has it. We must be careful not to equate functionalism and systems theory, for they are two different things. Thus, the conclusion, as with Alexander and Giddens, is that this comment is true for functionalism and the old systems theory, but applies only to it, and not to the new systems theory. From the standpoint of the new nonfunctionalist or postfunctionalist systems theory, these 1975 statements are obsolete.

Further, Collins's remarks concerning the commitment to certain political values on the part of Comte, Durkheim, Pareto, and Parsons is also clearly obsolete and generally irrelevant to the new systems theory. In chapter 2, I make a claim about the political commitments of systems theorists that is similar to the claim that Collins made about the political diversity of conflict theorists. Again, this claim about systems theorists was not copied from Collins's claim, but was made independently, offering yet another example of the parallelism between mainstream theory and the new social systems theory. Like conflict theorists, systems theorists come in all ideological shades. Contemporary systems theory has adherents in both capitalist and socialist countries. The International Institute for Applied Systems Analysis (IIASA) in Laxenburg, Austria, was supported by both the USSR and the United States, among other nations: Ironically, it was the conservative Reagan administration that stopped funding for the center, while the USSR continued to fund it. The reader can draw his or her own conclusions regarding the political commitment of systems theorists.

A look at chapter 6 and at Bailey 1990 will reveal that the discussion of race and gender in social entropy theory is "liberal." In fact, with the exception of conflict theory and explicit theories of race and gender, SET may consider these factors in greater detail than most general theories. I hope that a new generation can read the new systems theory with an open mind on this issue, and not be poisoned by the idea that systems theory is "conservative" as was a whole generation before them. The problem with labeling all of systems theory as conservative (see Lilienfeld 1978) when it is not is that it is an easy way to practice "contempt prior to investigation" (in Spencer's terms), and for "liberal" sociologists to neglect the approach entirely or to approach it in a biased fashion. And again, the basic response is that Collins's specific remarks are simply obsolete, as they extended only from Comte to Parsons, and do not include any of the new systems theorists such as Geyer, van der Zouwen, Miller, Bailey, Luhmann, Maturana, Varela, and so forth. Perhaps a similar case can be made for these theorists, but I doubt it. It will certainly be more difficult to do.

The whole situation can be rectified by simply making Collins's statement refer to functionalism, and not slipping, as he did, to applying these comments to systems theorists. I propose that these ideological comments are descriptive of functionalism only, but not of systems theory as a whole. Social entropy theory is a reconstruction, or more correctly a new model, which avoids these traps. It has the advantage of being aware of these criticisms, and so being able to avoid them. As Klir (1969) says, systems theory is so general that it has few ideological presuppositions. It is only when content is "loaded in" (in the form of functionalism, for example) that problems arise. The same can probably be said of conflict theory. As Collins says, the model of conflict between competing interests is relatively value free. But it is also not very useful for explanatory purposes until specific empirical content is loaded into the model. When the empirical content is added, so is the ideological component, and the value-free status of conflict theory is at this point obviously in doubt.

Still another very interesting feature of conflict theory as espoused by Collins is that while it clearly has a micro aspect (and is both micro and macro in some regards, see Collins 1975), it also has clear systemic qualities, not in terms of functionalism, but in terms of generic systems theory as defined in chapter 2 of this volume. Consider Collins's reliance on technology, as in "Technology and Military Organization" (Collins 1975, pp. 355–64). Technology is one of the six chief *PILOTS* (or *PISTOL*) macro variables of social entropy theory (see chapters 6 and 7 of this volume, and Bailey 1990). Further, Collins's (1975, p. 238) description of the role of social structure reveals a number of factors of interest to systems theory approaches such as SET. Not only is technology mentioned, but also stratification variables relating to level of living (component *L* in *PILOTS* or *PISTOL*). Further, by speaking of "tribal society," "stratified society," and "centralized state," Collins is not far from an explicit recognition of system boundary.

I think that Collins will find the new systems theory much more acceptable than functionalism. He says that compared to functionalist systems theory, "The conflict perspective, which grounds explanation in real people pursuing real interests, is a good deal more successful at realistic and testable explanation" (Collins 1975, p. 21). He will see in SET (Bailey 1990) real people pursuing real interests. He will also see real boundaries. As a systems theorist, one of the main problems that I have with micro conflict theory is that it often leaves the context for conflict unclear. While interpretive sociologists generally emphasize context as significant for the analysis of interaction, micro conflict theorists for some reason put less emphasis on the context for conflict as part of their general theory. It is instead introduced in specific cases (when empirical content is loaded into the general model) as when Collins (1975) writes of tribal society or American society.

The issue is important because to me, *conflict is by definition a systems problem, not an individual problem.* An individual cannot have conflict with himself or herself, except in the psychological sense of internal mental turmoil. Conflict is not defined for individuals, only for groups. Unless the boundaries for the conflict are specified, we cannot determine whether resources are scarce, what the distribution of power is, whether a zero-sum game exists, or any of the other group variables which are needed to analyze conflict. I would humbly propose that rather than rejecting the systems model as shown in SET, conflict theorists could benefit from the context it provides. It analyzes not only level of living, technology, and organization (three basic variables used by Collins) but also space, population size, and information (variables whose use has been advocated by Giddens and others). Careful analysis of

Collins's work will probably show that all of these are used to some degree (for example, "tribal society" implies lower population size than "stratified society"). The problem is that a comprehensive scheme for postulating these macro context variables is lacking, so their use can easily become inconsistent.

This leads us to one final point, the alleged "hypostatization" of systems theory. Collins (1975, p. 21) says, " 'Society' or 'system' is hypostatized, made the referent around which theory is to be constructed." This is in supposed contrast to the "reality" of conflict theory which deals with "real people pursuing real interests" (Collins 1975, p. 21). If we simply use a concrete systems model (defined in subsequent chapters) that includes real boundaries in time and space, we are not only using time and space as Giddens advocates, but we are removing the alleged hypostatization. Again, Parsonian functionalist theory, prior to reconstruction, did eschew grounding by insistence on the social role as the unit of systems analysis (an "abstracted" rather than a "concrete" system in Miller's [1978] terms). The concrete system used in living systems theory and social entropy theory is *not* hypostatized, but "real," and follows Giddens' suggestions to incorporate time and space.

If I have appeared to differ with Collins to this point, it is only from a desire to stress that the remarks he makes are accurate concerning functionalism, but do not apply to the new systems approaches such as social entropy theory that have been formulated since his volume was published in 1975. A more recent and more sympathetic treatment of systems theory is found in Collins 1988 (pp. 45–76), a full chapter of which discusses general systems theory, and even one "new systems theorist" (Luhmann is discussed as Parsons's student). I do appreciate the fact that he included systems as a full chapter, rather than simply neglecting the approach. The problem would be less acute if Collins always referred to functionalism, but unfortunately he occasionally lapses into statements about "systems theory," which if quoted out of context, could be assumed to apply to all systems theory.

Further, it seems to me that his conflict theory needs systemic variables such as space, organization, and technology as context variables. Still further, it seems to me that the most general models of *both* conflict and system are both relatively value free (and actually quite complementary), but that when content is added to either, problems of ideology and value commitment quickly arise.

To carry this one step further, I do not quite understand how consensus can be value laden and conflict be value free, when both are, from a systems perspective, two sides of the same coin, inasmuch as one can accept the notion that a dimension runs from total conflict at one end to total cooperation or consensus at the other. And lastly, as has been pointed out, both conflict theory and systems theory include both liberals and conservatives among their constituents. I hope that we can declare a truce on this issue and approach the new systems theory with an open mind instead of unfairly applying a clearly value-laden term such as "conservative" to it. At the very least, the term "conservative" masks the considerable diversity within systems theory. Collins's 1988 discussion of systems theory is a good step in this direction, as here he refrains from charges of hypostatization and conservatism that marked his more blatantly polemical 1975 discussion.

A last point to be made is that while Collins seeks a rather scientific explanatory theory, I feel that some of what passes for anticonservative attitude against systems theory is really antiscience attitude. The relatively positivistic stance of Miller's living systems theory can be readily contrasted (in chapters 5–7 of this volume) with the less positivistic stance of Bailey's social entropy theory, which prominently includes values within the analysis (which Miller is reluctant to do). If someone does not like systems theory because it is science based and has equations, then so be it, as this is not likely to change. Even then, this person may be more satisfied with certain variants of systems theory such as the "soft systems" approach of Checkland (1981).

Apart from this now obsolete criticism of systems theory (which was not obsolete with regard to the functionalism of 1975 when the book was published), I find some clear points of overlap between Collins's conflict sociology and social entropy theory. In addition to the discussion of technology and organization already mentioned, there are clear points of contiguity between SET and conflict sociology in the emphasis on sex. What should be another point of contiguity is SET's frequent allusion to race in its examples, which for some reason is relatively neglected in Collins's conflict sociology. Bailey's allocation theory (Bailey 1990, pp. 122–41) shows how persons are allocated into positions at all levels of society (group, organization, society). Such variables as race and gender as well as educational credentials are central to this allocation process, and this analysis should be of interest to conflict sociologists. Thus, I see SET and conflict sociology to be largely parallel and complementary, rather than contradictory. This is in keeping with the goals of SET to construct a general formulation which does not preclude approaches such as social change theory and conflict theory.

Another point of contiguity is the similarity of the explanatory hypotheses produced by both Collins (1975) and Bailey (1990). While Collins claims that conflict theory is more successful than systems theory in producing such hypotheses, once again this is seen to apply only to functionalism, and is obsolete with regard to new approaches such as living systems theory and social entropy theory. Both of these were formulated after the publication of Collins's 1975 book, and both have presented many testable hypotheses, as does Collins.

Now that the work of Alexander, Giddens, and Collins has been reviewed, and a number of interesting parallels and points of contiguity have been discovered, it is time to anticipate and analyze briefly what social systems theory of the 1990s looks like, before going on to its actual presentation.

Ritzer's Metatheorizing

Metatheory can be helpful in analyzing social systems theory. Ritzer (1990b, p. 347) says, "Metatheoretical tools provide us with the means to make sense of a wide range of seemingly disparate theoretical developments." Ritzer proposes use of the micro-macro instrument as a metatheoretical tool, saying "One of the things this paper will show is that . . . a wide range of theorists *are* converging on the micro-macro linkage issue (Ritzer 1990b, p. 348, italics in the original).

Ritzer writes further:

> Metatheorizing may be defined very broadly as the systematic study of the underlying structure of sociological theory. We can differentiate between three broad types of metatheorizing: metatheorizing as a prelude to the development of sociological theory [M_P]; metatheorizing in order to achieve a deeper understanding of sociological theory [M_U]; and metatheorizing in order to create an overarching theoretical perspective [M_O]. (1990a, p. 18)

Once again, there is a clear parallel development between mainstream social theory and systems theory. While explicit interest in metatheorizing in sociology is relatively recent, general systems theory has long been marked by widespread metatheorizing (for example, the interest in hierarchy theory or isomorphisms— see chapter 4 of this volume). This is because in their attempt to construct multidimensional general theory, systems theorists have been consistently faced with metatheoretical tasks such as how to formulate concepts and terms to represent multiple disciplines, how to deal with hierarchies and levels, how to construct boundaries, etc. In one sense, the whole part and parcel of the development of general systems theory is the task of metatheorizing.

Ritzer (1990a, pp. 18–19) presents four basic types of M_U. The first subtype focuses on internal-intellectual or cognitive issues. The identification of paradigms or approaches within systems theory (see Cavallo 1979) is an example of this effort. A variant of this internal-intellectual approach "involves the development of general metatheoretical tool[s] with which to analyze existing . . . theories and to develop new theories" (Ritzer 1990a, pp. 18–19). An example of this is Bailey's 1984a study of different descriptions and different theorists from the standpoint of equilibrium, homeostasis, and entropy. The second subtype of M_U focuses on social rather than cognitive factors, emphasizing "schools" of social thought (Ritzer 1990a, p. 19). This is also evident in the work of Cavallo (1979) and others.

The third variant of M_U (external-intellectual) involves turning to other academic disciplines for ideas, tools, concepts, theories, and the like that can be used in the analysis of sociological theory (Ritzer 1990a, p. 19). This subtype is endemic and almost definitional in systems theory. In a real sense, the whole task of general systems theory is to borrow from other academic disciplines. Examples of this include the "immigration" of such concepts as equilibrium, homeostatis, entropy, hierarchy, isomorphism, etc., from one discipline to another, or from one discipline to general systems theory.

The fourth example of M_U is the external-social, involving looking at the larger society and its impact on theorizing. This effort has been less prevalent in systems theory, although one example already mentioned is the case of the United States's national support for the International Institute for Applied Systems Analysis and its subsequent negative effect on the development of systems theory (when funding was suspended).

Returning to the first form, M_P is metatheorizing as a prelude to the development of sociological theory. The whole first section of this volume (chapters 2–4, as well as this chapter to some extent), is replete with examples of this type. The most visible of course is the extended critique of the functionalists' equilibrium concept as a prelude to the development of the new systems theory. The critique of equilibrium directly leads to formulation of nonequilibrium entropy theory.

The last of Ritzer's metatheoretical types is M_O—metatheorizing in order to create an overarching theoretical perspective. While not as prevalent in sociology as the other forms, M_O is quite evident in general systems theory (see chapter 4 of this volume). Collins (1988, p. 47) characterizes general systems theory (GST) as "the search for panscientific laws, for principles that apply to systems found in all areas of science."

The New Systems Theory

What summary description of the new systems theory—social systems theory of the 1990s and the twenty-first century—can be offered at this time? Some things are clear. The new systems theory has a number of striking parallels with mainstream sociological theory. These parallels were not copied from mainstream theory, but grew up independently. Fortuitously, the paths of mainstream social theory and of the new social systems theory seem to be generally proceeding in the same direction. Among these parallels are:

1. Both mainstream social theory and the new systems theory (NST) are characterized by an abundance of metatheory.

2. Both are concerned with the macro-micro link (see Bailey 1990, and chapters 5–7 of this volume).

3. Both deal with action and order.

4. Both are concerned with time and space.

5. Both are concerned with structure and system.

6. Both are engaged in programs of reconstruction (as well as elaboration and revision).

7. Both are in the third phase of postwar theoretical development, and are dealing with synthesis, not only of the micro-macro approaches, but other syntheses as well.

8. Both offer sets of explanatory hypotheses.
9. Both approaches have adherents who differ in terms of their political views (from liberal to conservative).

These parallel approaches are going in the same direction, and are prime targets for rapprochement. In addition to the similarities, there are some continuing differences between the new systems theory and mainstream sociological theory. New systems theory is demonstrably more methodological, and is more multidisciplinary.

It is also safe to say that systems theory is more "positivistic" or at least more science based. However, sociologists would likely be surprised at the number of humanists, Marxists, and other "nonconservatives" among the ranks of systems theorists (including a host of socialist scholars from Eastern nations). For example, the sociologist who gained notoriety as a leftist scholar and who exposed Project Camelot in Chile has also written rather widely on his entropy-based systems model of peace and conflict (Galtung 1975).

Consider this description from Ritzer:

> One of the key aspects . . . is its tendency to 'subvert' or 'explode' boundaries between disciplines and to create a multidisciplinary, multidimensional perspective that synthesizes ideas from a range of fields. . . (1990a, p. 14)

This might easily be a description of the systems movement, which has always been a multidisciplinary effort to synthesize and integrate. However, Ritzer is writing not about systems theory, but rather about postmodernism. To me, the postmodern movement at this point in time represents a humanistic parallel to the more science-based systems movement. I think that they are both going in the same direction. I am personally very interested in both approaches. I see both as examples of Giddens's (1987, pp. 6–32) predictions concerning the future of sociology. His first prediction is: *"Sociology will increasingly shed the residue of nineteenth- and early twentieth-century social thought."* (Giddens 1987, p. 26, emphasis in the original) His second prediction is: *"A theoretical synthesis will emerge giving a renewed coherence to sociological debates."* (Giddens 1987, p. 29, emphasis in the original) This volume shows ample evidence that the new systems theory has already taken great strides in "shedding the residue of nineteenth- and early twentieth-century" social systems thought.

"Bailey's dream" is that one day—not now, but perhaps in twenty or thirty years—this synthesis will reach the extent of combining parallel but now distant approaches such as postmodernism and the new systems theory. This involves elimination of a prevalent view which now sees humanistic sociology and positivism as opposites, and as generally incompatible. As a person who was a mathematics major and an English literature minor, I think that they *are* compatible. Further, verbal and quantitative skills *can* effectively be combined (although they are often not). There are, as has been noted, a number of humanists within systems theory. I see no inherent contradiction in a humanistic, yet quantitative systems approach (although I realize that others may).

For now, however, I will be content with a more modest and feasible rapprochement—the recognition of the myriad points of continuity between the parallel approaches of the new systems theory and mainstream sociological theory. It is entirely feasible for mainstream theorists to recognize the new systems theory as being, in Alexander's terms, a third-phase reconstruction of the old systems theory.

To this end, I propose to comment on systems theory from the standpoint of mainstream theory in a "counterpoint" statement at the end of each chapter, as noted above. Each counterpoint will utilize a list of points derived from this review of the work of Alexander, Giddens, and Collins. These points are not exhaustive, and were chosen somewhat arbitrarily. I also will not necessarily use them all in each chapter, and will not limit analysis to these points. Their main use is as a checklist of points of contiguity between mainstream social theory and the new systems theory. The final chapter (chapter 9) will present a synthesis of the new systems theory which will, I hope, demonstrate some degree of synthesis with mainstream sociological theory as well—thus a "dual synthesis." The checklist terms to be used, together with the author's work from which they were derived, are:

(a) Action, order, equilibrium, and idealism/ideational (Alexander);

(b) Agency, structure, system, structuration, system integration, and time and space (Giddens);

(*c*) Conservatism, conflict, age, sex, and hypostatization (Collins);

(*d*) Micro-macro (Alexander, Giddens, Collins, Ritzer);

(*e*) Forms of metatheory (M_P, M_U, M_O) (Ritzer).

A clarification is necessary at this point. When I say "counterpoint," I do not intend that the whole discussion deals with ways in which systems theory is "counter to," or contradicts, mainstream theory. On the contrary, I hope to show that points of contradiction are minimal or nonexistent. I use the term "counterpoint" in a more generic sense to point out the new approach of systems theory, and to be a general comparison with the mainstream. Thus, "counterpoint" here deals with comparison, continguity, similarity, and parallelism between systems theory and mainstream theory, as well as differences among the two approaches. Thus, as I use the term, "counterpoint" means "to show the other side," as in the other side of the coin. The other side of the coin in this case is the sorely neglected systems approach.

Chapter 2 deals in detail with the issue of why sociologists should study systems theory. In concluding the review of mainstream theories it is thus only necessary to point out, by way of comparison, what systems theory does that mainstream theory does not do. The first question is, what does mainstream theory do that systems theory does not do? Mainstream theory is superior to systems theory for dealing comprehensively with issues of structuration, conflict, normative and cultural studies, networks, ritualistic behavior, institutions, ethnomethodology, symbolic interactionism, and interpretive sociology in general. It is also superior to all the special theoretical areas such as feminist sociology, exchange theory, rational choice theory, Marxian sociology, stratification, demography, race and ethnic relations, etc.

The next question is, what do both mainstream and systems theory do well? These parallels have already been explored to some degree. A comprehensive comparison of mainstream theory and the new systems theory shows that both present explanatory hypotheses (Blau 1977; Collins 1975; Miller 1978; Bailey 1990), both use metatheory, both contribute to the analysis of micromacro linkages, both deal with order, action, structure and agency/structure relations, both are concerned with synthesis and integration.

The final (and crucial) question for this volume is, what does the new systems theory do (or do better) that mainstream social theory does not? Why should mainstream social theorists read the new systems theory? The answer is that there are a great number of gaps that mainstream theory has either neglected or is not well equipped to handle that systems theory is dealing with effectively. Thus, systems theory is valuable to mainstream theory because it is not contradictory nor antagonistic, but on the contrary fills gaps in mainstream theory.

What does the new social systems theory do that mainstream theory does not do? To be specific:

1. Systems theory lends needed methodological rigor. This includes methodological explication of such areas as the critique of equilibrium (chapters 2 and 3 of this volume and Bailey 1984a, 1990), a general methodological reconstruction of functionalism, and methodological analysis of the micro-macro link (see Ritzer 1990b, p. 363). Methodological tools of interest are the three-level model and the Q-R distinction (see chapters 6 and 7 of this volume, and Bailey 1990). Also of interest is the operationalization of entropy and operationalization of theory in general, a point consistently neglected by mainstream theorists.

2. It analyzes hierarchy and levels of analysis (such as Miller's 1978 levels of individual, group, organization, society, etc.). These mechanisms lend rigor and new analytical tools to micro-macro analysis, including a new perspective on the micro-macro link.

3. It provides operationalization and theoretical specification of the problem of order. Although mainstream theorists deal with order, systems theory gives a fresh and more methodological approach.

4. It provides a new approach to the relations between action/structure, process/structure, or agency/structure, including a new link between synchronic and diachronic analysis as advocated by Giddens (1979).

5. It provides specific inclusion of time and space in social systems theory (see Miller 1978; Bailey 1990; and chapters 5–7 of this volume) as advocated by Giddens (1979).

6. It provides a fresh analysis of action and order.
7. It presents the Q-R distinction and the three-level model.
8. It offers an analysis of systems philosophy and systems technology.
9. It presents an analysis of boundary theory.
10. It provides a context for the analysis of conflict, interaction, networks, etc.
11. It relates the ideational and empirical levels of analysis.
12. It presents the eight levels and twenty basic subsystems (Miller and Miller 1992).
13. It links energy and information (and not as a hierarchy as Parsons did).
14. It offers a comprehensive framework for comparison, both between groups at one point in time, and within a group over time.
15. It provides a comprehensive specification of salient macro variables (PILOTS).
16. It analyzes self-reproduction and self-regulation (autopoiesis).
17. It analyzes controlled systems, purposive behavior, self-steering and external steering, as well as morphogenetic behavior (cybernetics, the "second cybernetics," and the "new sociocybernetics").
18. It deals with issues of the relationship between the observer and the observed.
19. It deals with the notion of the reduction of complexity through systems (Luhmann 1982).
20. It emphasizes change through the vehicles of entropy and nonequilibrium analysis.
21. It provides a foundation for cultural and normative analysis.
22. It provides a comprehensive framework that does not preclude, exclude, or denigrate certain topics of lines of investigation.
23. It provides a framework for holistic analysis.
24. It provides a framework for macro analysis.
25. It provides a framework for multidisciplinary analysis.
26. It provides a framework for multidimensional analysis.

27. It provides a context and framework for the cumulation of knowledge.

28. It provides an inventory of concepts, so that one can tell what is missing from a theory.

29. It provides new concepts and vocabulary (input transducer, autopoiesis, structural coupling, three-level model, etc.).

Thus, the new systems theory is valuable to mainstream sociology because it contributes in at least twenty-nine important ways (and this is not an exhaustive list). Mainstream sociological theorists should wholeheartedly embrace an approach which, in twenty-nine ways, either deals with topics they have neglected entirely, or approaches problems that they are working on in a new and innovative way and from a fresh direction.

This assumes of course that the new systems theory has been sufficiently constructed and reconstructed so as to expunge the classical problems of functionalism, including conservatism, teleology, tautology, etc. I submit that this *has been done,* and that this is clearly shown throughout this volume, but particularly in chapters 2 through 4. Many of the classical problems have been removed simply by eschewing equilibrium. The neofunctionalists have also done this, but have not offered a replacement for equilibrium. The new systems theory fills this yawning gap by offering entropy and nonequilibrium analysis.

Apart from its specific contributions, the one thing that systems theory offers that mainstream sociological theory does not match is scope. Narrower theories have certain limitations that just cannot be rectified in any fashion except by extending their scope. For example, in social entropy theory, a central preoccupation is with the analysis of allocation of individuals into social positions. Allocation theory cannot even be defined in a narrower perspective. Only by sketching the boundaries of the macro society does the issue even become evident. Thus, there is no substitution for scope, and I am glad that mainstream theorists are becoming increasingly interested in scope (see Ritzer 1990b).

Some mainstream theories are simply too narrow in scope. As an analogy, these theorists are like persons building cars who do not add quite all of the parts. As a result, some of these theoretical cars run well, but do not look very good. Others look great, but do not run. Systems theory, by specifying a larger scope, at least deals with the whole car (although admittedly it is sometimes in need of a tune-up).

As another analogy, consider systems theory to be like a banquet where all of the silverware is set upon the table. It may be that a given diner (or a given user of theory) may not eat salad, and so may not need the salad fork. Many partial mainstream theories never lay out the full array of silver. Thus, these theorists often, in my opinion, find themselves trying to eat their soup with a fork, as their narrow theory simply does not provide the needed array of analytical tools.

There are some, like Collins (1975, p. 21) who charge that the "system" is usually a myth, and that system is hypostatized. As I said before, I agree with this as a criticism of unreconstructed functionalism, but I beg of you not to carelessly extend this to systems in general, as this quote, when taken out of context, can do.

Expansion of scope does not necessarily lead to hypostatization. As shown throughout this volume, the new systems theory (in contrast to Parsons) uses a *concrete system set in time and space* as Giddens advocates (see Miller 1978). Such a system uses "real people" just as does Collins's conflict sociology (see Collins 1975, p. 21) and is *not* an hypostatized system. While it is real, it is also comprehensive. It provides a needed context for the study of conflict, and it provides an inventory of concepts so that one can tell what is missing from the theory and what is not. Narrower theorists generally have no way to ascertain this. For example, Collins deals with technology, organization, and stratification, but less consistently with other important variables such as population size, information, and space. Social entropy theory, via the *PILOTS* model, provides a checklist of variables to guard against such neglect. A comprehensive theory can always be only partially used if necessary, but it is much more difficult, and usually impossible, to stretch a partial theory. One of the major reasons is that one simply does not know how to stretch it. If you are using something, how do you know what you are missing, without a model of sufficient scope to serve as a guideline? Further, it is difficult to simply combine two or more narrow theories in hopes of forming a theory of adequate scope, as there are obstacles in the form of terminological differences and differences in perspective, including differences in basic theoretical presuppositions.

It takes only a cursory glance at contemporary sociology to uncover the existence of many "versuses." These are generally dualities or dichotomies that are set up as contradictions or polar opposites. Some common "versuses" include, theory versus method, static versus dynamic, conceptual versus empirical, free will versus determinism, and structure versus process. To me, the existence of such "versus" dualities is symptomatic of the existence of exclusionary anomalies. These are anomalies that occur simply because the theory is not sufficiently broad to accommodate both conceptions but rather poses them as contradictions, or neglects one all together. While I agree with Giddens (1979) that the interaction between the synchronic and diachronic must be analyzed, or that the division between synchrony and diachrony must be transcended as Archer (1985, p. 60) puts it, I do not think that many of the theoretical approaches now extant in sociology are broad enough to do this. I think that systems theory is sufficiently broad and is up to this task, and this is one of its chief merits (among the many others that have been mentioned).

A final word on the relationship of the new systems theory to mainstream sociological theory is in order. To me, the most ironic and unfortunate thing about macrosociological functionalism was that it depended so heavily and unnecessarily on the invalid concept of equilibrium. This misplaced physical concept almost alone was responsible for excluding the study of conflict and social change from functionalism, and for the criticisms of functionalism such as teleology, tautology, determinism, and conservatism. What is so unfortunate is that as a macro theory of broad scope, functionalism had the opportunity to fashion an approach which, more than perhaps any other extant approach, did not exclude, preclude, prohibit, or contradict other theoretical approaches. Instead, what did it do? With all this potential at hand, it depended heavily upon equilibrium, one of the most preclusive concepts in sociology. As long as functionalism used equilibrium, social change was always an afterthought. The reason this is so ironic is because there are many narrower approaches in sociology that necessarily exclude phenomena simply because they are not broad enough to accommodate them. Functionalism was probably broad enough, it was just theoretically flawed.

Although I obviously cannot speak for all systems theorists, what the new social systems theory desires is rapprochement and integration (inasmuch as is feasible) with mainstream social theory. We do not wish to remain known as an approach that is in conflict with conflict theory, as functionalism is currently portrayed in many introductory sociology texts (see Robertson 1987). I envision a model broad enough to accommodate the study of *both* consensus and conflict simultaneously, as both exist in complex society.

How realistic is this vision? Quite realistic in my view. I urge you to cast your nets wide. The review in this chapter has shown that not only does the new social systems theory exhibit a number of points in common with mainstream sociology as well as a number of parallel developments, but the systems concept is de facto represented in mainstream sociology. For example, Archer (1985) champions morphogenetic theory for directly tackling the relationship between structure and action. Giddens (1979) uses the concepts of system and system integration extensively. Alexander (1985; Alexander and Colomy 1990) also discusses systems and equilibrium, while Collins (1975) critiques functional systems theory, and more recently (1988) devotes a whole chapter to systems theory, as does Turner (1991). I urge you to work with me in exploiting the points of contiguity to build a sociology that is integrative rather than divisive—a goal shared by both mainstream sociology and the new social systems theory.

PLAN OF THE BOOK

We have seen some of what mainstream theory has to offer, and have hinted at the depth and richness of the systems approach. The remainder of the book is devoted to demonstrating the contribution that the new social systems theory can make to sociology. Thus, I now turn in earnest to systems theory. The next three chapters (chapters 2–4) deal largely with review, metatheory, and critique, and with foundational work for a theoretical reconstruction. The four chapters after those (chapters 5–8) deal with the new social systems theory—living systems theory, social entropy theory, and autopoiesis. Chapter 9 is the aforementioned dual synthesis, and the conclusion.

More specifically, the book is organized chronologically. I will first consider the nineteenth century—"the age of equilibrium" (chapter 3). Chapter 4 deals with the genesis of modern approaches, including general systems theory, sociocybernetics, non-equilibrium thermodynamics, information theory, and artificial intelligence. Although introduced by Bertalanffy (1967) and Buckley (1967) and discussed in an article by Ball (1978), these important developments are largely unknown to sociologists, at least for the last twenty years.

Chapter 5 discusses living systems theory, while chapter 6 discusses social entropy theory and chapter 7 presents a congruence of the two. Chapter 8 discusses the current debate over autopoiesis. Chapter 9, as previously noted, concludes the volume with an attempt at overall synthesis, and a view of the "the road ahead."

CHAPTER 2

Social Systems Theory

This volume is based on the premise that systems theory has a distinct and perhaps even vital contribution to make to sociology. There are a number of contributions offered by the systems approach that are not forthcoming from any other sociological perspective today. One of these is sheer breadth, and the advantages that it entails. Only systems theory, among all the specialities in sociology, seems to hold realistic promise of an integrative framework necessary to combat what many see as excess fragmentation or overspecialization (hyperspecialization if you will) in American sociology (see Stryker 1979; Rossi 1980; Moore 1981).

Another value of a broad formulation is that it allows comparative analysis, both of the same society at different points in time and of different societies at the same point in time. Still another benefit is that a broad formulation allows one to focus on relationships as well as components. While the study of relationships is represented in sociology through network theory, there is a big difference between network theory and systems theory. The former abstracts relationships from the whole, while the latter studies not only the relationships themselves, but also their systemic context, including the environment.

Perhaps as important as systems theory's promise for helping to unify sociology is its promise for linking sociology to other disciplines. One of the chief goals of systems theory is to expose and avoid duplication of effort, as when researchers in different fields (or perhaps in the same field) are doing essentially the same research, perhaps using different words or labels, without knowledge of the other's work.

It is also important for the discipline to use the whole society as the basic unit of analysis. After all, this is the science of society. While sociology should share specialities with its sibling disciplines, it should also have a distinct core. Other social sciences generally do not study the entire society, leaving this to sociology.

Too often sociology also neglects this task, sometimes because of sheer complexity. Systems theory offers a framework for this crucial task.

A broad approach also has distinct theoretical advantages. It seems probable that anomalies found in narrow theories may be a function of narrowness, and that a broader theory would be less restrictive. Approaching this issue from the other direction, it is clear that broad theories such as social entropy theory (Bailey 1990) allow whole areas of investigation (such as the relationship between Q- and R-analysis) which are simply precluded by the narrower formulations extant in American sociology. Such a broad formulation allows not only the exploration of broader theoretical questions, but also holds more promise for the realistic integration of theory and method, and for true multilevel research.

These are but some of the advantages of the systems approach for sociology, and surely not an exhaustive list. Further reasons for studying the systems approach are discussed later in this chapter. The idea here is to merely show that some of the contributions of systems theory are not being fulfilled, and probably cannot be fulfilled, by any other extant approach in American sociology. These include simultaneous analysis of relationships and components, and of parts and wholes, the study of relationships between the system and its environment, the study of boundary problems, and attempts at unifying and integrating sociological theory.

Whatever Happened to Systems Theory?

Some sociologists may reject the unifying and integrative potential of a broad approach, preferring the pluralism of specialization. Others, however, surely worry about excess fragmentation, and perils such as lack of cumulation which may accompany it. They welcome an approach such as systems theory which seeks not to supplant more specialized views, but merely to link them. Thus, the broad view becomes itself part of the specialization, specifically, a broad view to complement the existing narrow views.

Systems theory offers a sorely needed broad view, but does not preclude the development of other, nonsystems, alternative broad perspectives. Why, then, is so little written about systems in American sociology of the 1990s? When those of us interested in systems theory mention our work, a common response from senior sociologists is, "Whatever happened to systems theory? I thought it had some potential about twenty years ago, but I have not heard much about it lately." Even more disquieting is the query sometimes heard from more junior sociologists, to wit, "What is systems theory?" Why is an approach with some potential to combat, or at least complement, the problem of overspecialization in sociology so neglected?

One answer is that many sociologists think that systems theory is dead. Is it? The answer is no, and this volume is dedicated to documenting this in some detail. Many exciting new systems developments, including nonequilibrium approaches, living systems theory, sociocybernetics, social entropy theory, and autopoiesis, deserve the attention of sociologists. But if exciting developments exist, why do sociologists think the field is dead, and what can be done to overcome this false perception? The answer to the first part is that sociological systems theorists have simply lost contact to a large degree with other sociologists.

The goal of this volume is to bring systems theory back into sociology where sociologists can evaluate it. The goal is not to sell the approach or force it upon anyone, but simply to make it accessible so that sociologists now unfamiliar with recent developments can evaluate and judge for themselves.

Why is the average sociologist seemingly unaware of the latest developments in systems theory? There are many possible reasons. Clearly, Merton's (1949) rejection of "grand" theory in favor of middle-range theory helped start a swing away from broad, global formulations. This trend has continued, with the predominant new American theoretical formulations of the last quarter-century being microformulations such as interactionism and ethnomethodology (see Alexander et al. 1987). Sociological systems theory has also been stung by critiques, beginning with the many criticisms of the equilibrium model of Parsonian functionalism (see, for example, Gouldner 1959; Lockwood 1956) and continuing with more general critiques such as that of Lilienfeld (1978). While these critiques had validity at the time they were written, they are all obsolete, and do not address the predominantly nonequilibrium thrust of contemporary systems theory. Further, they do not deal with the most exciting recent developments such as sociocybernetics, living systems theory, social entropy theory, or autopoiesis.

One can speculate as to many other reasons why systems theory has been neglected in sociology. Some sociologists may see it as too positivistic, too global, too confusing, too difficult, too ambiguous, too conservative, too this or too that. Whatever the case, it is clear that most of the objections seem to be based primarily on older, now obsolete versions of systems theory. Perhaps the main problem facing sociological systems theorists is that (at least until the publication of *Social Entropy Theory* by Bailey in 1990) there has been no comprehensive systems publication by an American sociologist since Buckley (1967). This volume seeks to rectify this situation by critiquing the latest developments in systems theory, so that sociologists may assess their efficacy. Some will surely be amazed to find how radically these latest developments depart from the Parsonian functionalism (e.g., Parsons and Shils 1951) that comes to mind when they think of sociological systems theory. Why did sociological systems theory develop and then (apparently) die? At least one reason was its overreliance on equilibrium, and the consequent pervasive criticism that it was conservative and not responsive to social change.

Contemporary sociological systems theory is alive and well. It deserves a voice, and a position along with the many other formulations in the theoretical armory of sociology. Ritzer (1988, pp. 193–94) says, "One of the more interesting developments in sociology was the meteoric rise and equally meteoric fall of systems theory. . . . Systems theory seemed quite attractive to sociologists in the 1960s." The inception of systems theory in the 1960s seems clearly premature in retrospect. Most of the interesting developments reported here are products of the 1970s or 1980s. If one were seeking a recipe for constructing theory, it might be to stir all the ingredients and wait twenty years. The twenty years have passed, and the results cry for evaluation by sociologists.

Those who criticized equilibrium will be interested in the latest trends in nonequilibrium analysis. Those who dismissed systems theory as ideologically conservative (see Lilienfeld 1978) may contrast the manner in which social entropy theory (Bailey 1990) includes analysis of minorities and women with the manner in which they are necessarily excluded by narrower specialties. Those who found systems theory incapable of dealing with change, and thus favoring the status quo, may be surprised to see that *Living Systems* (Miller 1978) discusses process as well as structure in most chapters. Those who found earlier systems formulations too "macro" may be surprised by the analysis of individuals in *Living Systems* (Miller 1978) and *Social Entropy Theory* (Bailey 1990). Those who faulted systems theory for promising multileveled analysis but not delivering it will acknowledge that living systems theory (Miller 1978) comprises the most comprehensive multilevel analysis constructed by any perspective to date. Those who found systems too deterministic may marvel at the discussions of control and self-steering in the new sociocybernetics (Geyer and van der Zouwen 1986). Those who complain that there are no developments in theory may find themselves awed by the evocative seminality of autopoiesis, and caught up in the debate over its applicability to sociological phenomena.

The considerable number who always liked systems theory but had not seen anything on it in some time will find this an opportunity to welcome an old friend. Those who wanted to study systems theory but feared it to be an unpopular path may find solace in the strength of the various systems perspectives presented here. Lastly, those who were convinced by Merton's (1949) original argument that broad theorizing was premature, and that sociology lacked the prerequisite, careful empirical work of other disciplines, should note that in the past forty years much research of this type has been conducted, thanks in part to Merton's admonitions. Those who argue that systems theory is overly abstract and complicated may be harder to answer. Since sociological systems theory is admittedly science based (see Ritzer 1988, p. 194) it is sometimes complex. While I cannot eliminate legitimate complexity, I can endeavor to reduce ambiguity and overlap as much as possible.

This volume makes it clear that the earlier criticisms of systems are as obsolete as formulations which engendered them. The emphasis here is on nonequilibrium approaches, which easily facilitate the study of change. The old systems theory was simply not asking questions that sociologists were interested in—the new systems theory does. This volume presents a theoretical monograph which critically evaluates the latest approaches. The volume seeks to synthesize and compare. I will show which elements of the new systems theory are compatible and which are not. The complementary ones will be combined, and all will be synthesized and condensed as far as possible.

What Is a Social System?

What is a social system? That question has frustrated many applications of systems theory in social science. Since one of the goals of systems theory is to synthesize different definitions of phenomena into one coherent concept, it behooves us to do the same thing with the concept of system. The task is simple enough if one is willing to master certain component concepts. Simply put:

> A system is a bounded set of interrelated components that has an entropy value below the maximum.

Although abstract, this definition is both concise and generic.

Entropy can temporarily be defined as the *degree of disorder* in a system. Operationalized in terms of probability, maximum entropy is the random or uniform distribution, where every possible case is equally probable. Minimum entropy is generally zero when measured statistically, and is the probabilistic *opposite of randomness*. This is where only one case occurs out of all possible cases.

While maximum entropy is the statistically *most* probable state, it is the theoretically *least* valuable state, as it represents complete disorder or lack of order, and thus no predictive or explanatory power. While minimum entropy is the statistically *least* probable state, it is the theoretically *most* valuable state, as it represents complete order or lack of disorder, and thus *perfect* predictability and explanation (100 percent of the variance is explained).

What is the empirical condition of the typical social system? Obviously it is not maximum entropy, as this is the null state. While technically one could still maintain that a system exists at this level, in reality it is the death state. The parts show no coordination, interrelationships among parts are zero, and one is unable to predict anything about systems' properties or actions. In a real sense the system does not exist, although if maximum entropy is not permanent, it is not impossible that under certain conditions the moribund system can be revitalized.

Do most empirical systems display minimum entropy then? No, because such a state of perfection is rarely found empirically, The succinct generic definition of a system presented above may especially for systems as complex as are most social systems. In order to attain minimum entropy the system would have to func-

FIGURE 2.1.
Range of Entropy Values in Social Systems

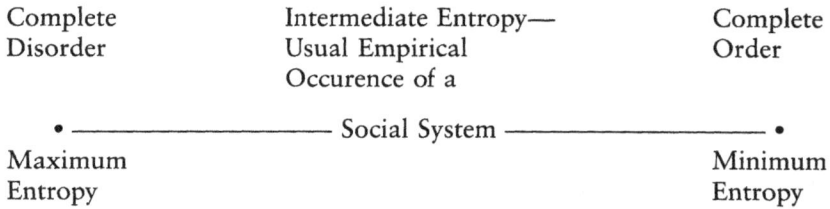

Complete Disorder	Intermediate Entropy— Usual Empirical Occurence of a	Complete Order
• ———————	Social System ———————	•
Maximum Entropy		Minimum Entropy

tion perfectly. All its parts would be perfectly interrelated, and we could potentially predict and explain everything we ever wanted to know about the system.

Obviously, most actual social systems are intermediate to these two extremes. By definition they are below maximum entropy, but unless perfect, they are above minimum entropy. This means that the correlations among systems variables are intermediate (for example, $R^2 = 0.37$) and predictability is fair. It is clear not only that the system is functioning, but that it is functioning quite imperfectly.

Definitions of Systems

The succinct generic definition of a system presented above may seem somewhat unorthodox to those familiar with more traditional definitions. Examining some traditional definitions, we see that they are easily subsumed by the generic one. For example,

> A *system* is a set of interacting units with interrelationships among them. The word "set" implies that the units have some common properties. These common properties are essential if the units are going to interact or have relationships. The state of each unit is constrained by, conditioned by, or dependent on the state of other units. (Miller 1978, p. 16, italics in the original)

According to Bertalanffy (1956, p. 3) a system is "a set of elements standing in interaction." While Parsons and Shils say:

> The most general and fundamental property of a system is the interdependence of parts or variables. Interdependence consists in the existence of determinate relationships among the parts or variables as contrasted with randomness of variability. In other words, interdependence is *order* in the relationship among the

components which enter into a system. (1951, p. 107, italics in the original)

Hall and Fagen say that:

A system is a set of objects together with relationships between the objects and between their attributes. (1956, p. 18, italics in the original)

One last definition is by Berrien, who says that:

A system is defined as a set of *components* interacting with each other and a boundary which possesses the property of filtering both the kind and rate of flow of *inputs* and *outputs* to and from the system. (1968, pp. 14–15, italics in the original)

The commonalities in these definitions are obvious.

1. All specify some basic units of the system, although names vary. These units may be called components, parts, units, characteristics, variables, attributes, and so forth, with some differences in the nuances of meaning.

2. All specify connections of some sort among the units, although again the names may vary. These connections may be called relationships, interrelationships, connections, correlations, and so forth.

3. All specify or imply that the relationships are nonrandom (that is, that entropy is below the maximum). For example, Miller (1978, p. 17) states that a veridical system "is a nonrandom accumulation of matter-energy," while Parsons and Shils (1951) also write of departure from "randomness of variability." Elsewhere, Miller (1978, p. 18) says that living systems "maintain a steady state of 'negentropy' even though entropic changes occur in them as they do everywhere else."

4. All allow the existence of boundary, although it may or may not be specifically stated or implied.

5. All allow or presume the existence of an environment outside the boundary, although it may or may not be specifically stated or implied.

TYPES OF SYSTEMS

Like many other phenomena, those systems meeting the generic definition display variability on a number of dimensions, thus forming types. The plethora of systems types can lead to confusion. In order to understand these varieties we must have at least a cursory look at some underlying dimensions. The most systematic way to do this is to apply Lazarsfeld's (1937) method of *substruction*. Substruction proceeds by identifying the basic dimensions underlying salient types. One can begin this process by examining several prominent types from among the many that exist.

1. *Conceptual System:* A system whose basic units are words or symbols (Miller 1978). This is also called a *pattern system* by Kuhn (1974). Since the units are symbols rather than concrete entities, Miller distinguishes this type from an empirical system. He says that his book *Living Systems* is an example of a conceptual system. A mathematical model such as a set of simultaneous differential equations would be another example.
2. *Concrete System:* Miller (1978, p. 17) says that "a *concrete, real* or *veridical* system is a nonrandom accumulation of objects in physical space-time." This type is called an *acting system* by Kuhn (1974).
3. *Abstracted System:* Miller (1978, p. 19) says that an abstracted system has relationships rather than objects as the basic units of analysis. Used mostly in social science and biology, this form reverses nouns and verbs, with the role or position being the system unit rather than the individual object. For example, a concrete system would assert that Washington (object or person) was president (role or position). Here the person is the unit and the presidency indicates what role he or she plays in the system. In the reversed abstracted system, one would say that the presidency (role) was occupied by Washington (object). Here the emphasis is on the role of presidency as the basic system unit, and the person is relegated to secondary status as the incumbent.

Whether the person (concrete system) or the role (abstracted system) is the most efficacious for the study of society has been the subject of intense debate (see Miller 1978; Kuhn 1979; Parsons 1979; Bailey 1981, 1983). There are a number of arguments on each side. Among the arguments in favor of using concrete systems are that they are easier to measure, are used in physical science and thus facilitate interdisciplinary study, facilitate boundary determination, and are more familiar.

Among the arguments in favor of abstracted systems are that the role, not the individual, is the proper unit of society, that roles are more permanent than individuals, that each individual has multiple roles, and that emphasis on the individual detracts attention from social and cultural phenomena, and can lead to a reductionist emphasis on biological factors and on physical time/space considerations (see Parsons 1979).

While Miller (1978) argues strongly for the use of concrete systems, Parsons (1979) is equally adamant in insisting that abstracted systems be used. The integrative view espoused by Bailey (1990) says that a more general approach will accommodate *both* perspectives. Bailey begins with a concrete system and then shows how abstracted systems are generated.

4. *Isolated System:* A thermodynamic system in which neither matter nor energy can penetrate system boundaries. This is an extreme form of closed system (see Hall and Fagen 1956).
5. *Closed System:* A thermodynamic system in which energy but not matter can cross system boundaries. This is "less closed" than an isolated system (Hall and Fagen 1956).
6. *Open System:* A system in which both matter and energy can be transplanted across system boundaries (Hall and Fagen 1956).

The definitions of isolated, closed, and open systems are absolutely crucial for applying systems theory to sociology. Unfortunately, they have been rendered obsolete (as originally stated in thermodynamics) by two developments: Einstein's equation of matter and energy, and the emphasis (since the mid-1940s) on information in addition to matter and energy. The equation of matter and energy led Miller (1978) to write both quantities as one hyphenated quantity (matter-energy). See Miller (1978) for further discussion. This conflation effectively erases the distinction between isolated and closed systems, so the term "closed" is used almost exclusively in general systems theory. It is also understood that a closed system is closed not only against matter-energy transfers, but also against information transfers. An open system in the modern sense possesses boundaries which allow not only matter-energy but also information to cross. In addition to transfers of information or "negentropy," open systems allow transfers of entropy, according to Klapp (1975, 1978). All living systems, and thus all social systems, are open systems.

7. *Regulated System:* A system in which matter-energy and/or information flows can be regulated. All social systems are regulated systems. All social systems must be open some of the time. If not, the matter-energy (for example, food) and information necessary to sustain life cannot be obtained. On the other hand, it is doubtful if any living system or social system could be permanently open. There are times when matter-energy and information flows into or out of the system must either be reduced or eliminated. For example, an organization must open its boundaries to matter-energy input in the form of personnel until the proper level of staffing is attained. It must then close or regulate these boundaries so that excess staff members do not enter, but also so that necessary staff members do not leave without replacement.

Note that there are various forms of regulation. A regulated system at a given time may regulate only matter-energy (for example, personnel security) or information (for example, computer security) or both. In addition it may regulate only certain types of matter-energy or certain types of information (for example, there may be security precautions on computerized information, but not on books, letters, or telephones). Further, one can distinguish between symmetrical and asymmetrical regulation. Symmetrical regulation consists of regulating matter-energy and/or information flows in both directions (both input and output). Asymmetrical regulation is either in-regulation (input) or out-regulation (output). For further discussion of boundary openings and closing and their effect on entropy levels see Klapp (1975, 1978), Comeau and Driedger (1978), and Bailey (1990). It should be mentioned that "regulated system" is a new term which is badly needed and has been coined here. There are similar concepts in the literature, such as the notion of controlled system in cybernetics, and the steering and self-steering systems of the "new" cybernetics (see Geyer and van der Zouwen 1986).

8. *Totipotential System:* A self-sufficient system. Few large social systems are truly totipotential. A true totipotential system would probably close its boundaries to all imports. It would thus be an asymmetrically regulated in-system. While there is probably no society which prohibits all imports, some which possess rich natural resources might be able to do so if politics allowed. Also, some individuals or tribes controlling rich natural resources might be totipotential, at least in specific time periods. For further discussion of totipotential systems see Miller (1978).

9. *Autopoietic System:* A self-reproducing system. In the least restrictive sense of the term, this would be a system such as a city which contains mechanisms for being self-organizing, if not fully self-reproducing. It contains mechanisms for organizing internally, that is, for assuring adequate supplies of food, information, and so forth. Most living systems, including social systems, can be seen as autopoietic in this least restrictive sense. In the most restrictive sense, autopoietic systems are systems such as cells which actually produce the components to reproduce the system (Maturana 1981; Varela 1979). In this restrictive sense the system not only organizes internally, but literally self-reproduces. There is less agreement, and in fact intense debate, over whether social systems are autopoietic in this stricter sense, although there is general agreement that they serve as hosts for autopoietic systems such as cells (see Mingers 1989; Robb 1989a).

In general, autopoiesis involves the regulation of supplies so that internal entropy does not proceed toward the maximum as dictated by the second law of thermodynamics (Bailey 1990, p. 52). An autopoietic system takes in proper levels of matter-energy and information to insure levels of negentropy rather than entropy. The study of autopoiesis is a relatively new but important development in social systems theory to which the whole of chapter 8 of this volume is dedicated.

10. *Hierarchical System:* A system which has two or more internal echelons of control or command is said to be an hierarchical system. For example, an organization which has a president, vice-president, manager, and supervisor, all have four echelons of control. A nonhierarchical system will have only one level, as in a friendship dyad or other small group. Miller (see Miller and Miller 1992) studies living systems at eight evolutionary levels—the cell, organ, organism, group, organization, community, society, and supranational.

In Miller's formulation, the eight levels form a hierarchy in the sense that each level includes the contiguous level below it as a subsystem. Another distinction in Miller's scheme is that *within* a given level, the first four levels—cell, organ, organism, and group—do not possess internal echelons (hierarchies). The next four levels do—the organization, community, society, and supranational.

Thus, one can determine whether a particular social system qualifies as a "group" or an "organization" *not* by counting the number of members, but by determining whether an internal hierarchy exists. If such an internal hierarchy exists, the system is an "organization." If it is lacking, the system is a "group." While system size is not the definitional factor, it, of course, is correlated with the existence of a hierarchy. Miller (1978) uses the term "echelon" to distinguish the internal hierarchy of an organization from the external hierarchy formed by the eight levels from cell to supranational.

To clarify, all organs contain cells as subsystems. All organisms contain organs as subsystems, which in turn contain cells as second-order subsystems. All groups contain organisms as subsystems, which in turn contain organs as second-order subsystems and cells as third-order subsystems. All organizations contain internal hierarchies (chain of command) called echelons (president, vice-president, secretary, and so forth). These echelons are *not* subsystems. Rather, the organization still contains groups as first-order subsystems, organisms as second-order subsystems, organs as third-order subsystems, and cells as fourth-order subsystems. The same logic can be used to trace subsystems for communities, societies, and supranational systems. There is considerable interest in hierarchy theory as a specialty within systems theory, and it has generated a sizeable literature (see for example, Pattee 1973).

TYPOLOGY

Having examined ten salient types of systems, it is now time to identify the formal dimensions underlying each, and to construct a formal typology. Bear in mind that not all types of systems have been examined, but only the most prominent. Other types could be added to the typology (with some modification) if desired.

The substruction technique recognizes that each type concept is formed by combining several underlying dimensions. What dimensions underlie the trichotomy of conceptual-concrete-abstracted systems? Upon examination one sees two dimensions: (1) the empirical-conceptual-operational (indicator) dimensions; and (2) the Q-R dimension. The first dimension separates the empirical level from the conceptual level. This effectively separates the concrete system from the conceptual system (although the abstracted system remains problematic). The first dimension is a classical distinction which generally recognizes only two distinct levels of analysis, although each may be called by a plethora of names. For example, the conceptual level may be called theoretical, heuristic, hypothetical, and so forth. The empirical level may be called concrete, empirical, ideal, "real world," and so forth (see Blalock 1968; Costner 1969; Bailey 1984c).

Dimension #1: Perceptual (X)-Empirical (X')-Indicator (X")

I have shown elsewhere (Bailey 1984c) that there are really three levels of analysis instead of two. The first is the truly conceptual (X) and it is merely a mental perception. The second is the empirical (X') level. The third is the measured indicator or operational level (X'') which is a combination or mathematical mapping of the first two. Theorizing can take place on X and data gathering on X', but true measurement or data analysis can only occur on X''. For example, your personal perception of a small group as a system is X, the actual empirically existing group of persons is X', and the verbal or mathematical model that you are able to construct is X''. The three-level model is shown in figure 2.2.

Note that if our perception X) of the empirical system (X') were adequate (path a) then the model (X'') would be unnecessary. After all, the model (X'') is only a tool for understanding, and true

FIGURE 2.2.

The Three-Level Measurement Model as Applied to Systems

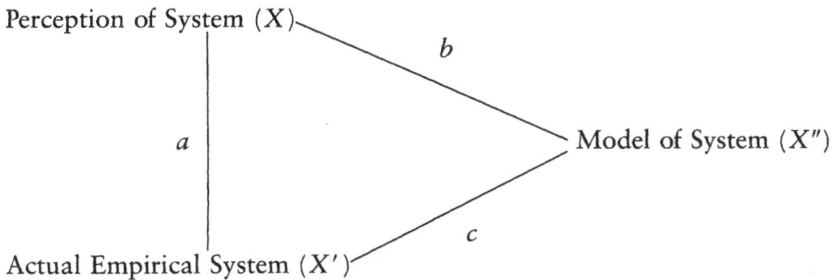

Perception of System (X)

Actual Empirical System (X')

a

b

c

Model of System (X'')

understanding must ultimately occur for each individual re-searcher on the conceptual level (X). However, modern social systems are so empirically complex that direct observation and full understanding along path *a* is rarely possible (perhaps it is possible for dyads or other small systems). Thus, while physiologists might directly study path *a* of figure 2.2 for the first two of Miller's (1978) eight levels (the cell and organ), it is doubtful that the next six levels can be studied in this manner. Rather, one must first observe some facets of X' to inform the thought processes (X) as much as possible. Then one constructs an *untested* model (X''_u) which can take various forms. It may be a verbal thesis such as a book, a mathematical model, or a theory or set of hypotheses suitable for testing. Then, data can be gathered from the empirical case (X') to test or further inform the verbal thesis, mathematical model, or theory (X''_t). The process can then be replicated through various sequences of the three paths to further refine one's knowledge of systems.

Two points of clarification are necessary. It is clear that Miller's (1978) concrete system is an empirical system (X') on this dimension (dimension #1), but his conceptual system is on the indicator or model (operational) level (X''), while he does not specify a parallel system on the conceptual level (X). This is easy to specify, and can be labeled a perceptual system. As a combination of the other two, the indicator level (X'') is exceedingly easy to conflate with either of the other two (either X or X'), and this is often done in practice (see Bailey 1990). For example, both X and X'' might be labeled conceptual systems, or both X' and X'' might be labeled empirical systems. This conflation of the three levels, or omission of one level, should be strictly avoided for clarity and precision of analysis. For further discussion of this complex subject see Bailey (1990). Note also that the three "levels" of the three-level model are analytical dimensions, and are not to be confused with Miller's eight levels which constitute a true hierarchy (each higher level incorporates all of the levels below it).

Dimension #2: Q-R Analysis

While the first dimension of the typology effectively separates the concrete and conceptual types, the position of the abstracted type remains ambiguous. As an analogy to statistical models, this means that "not all variance is explained by one dimension" and another dimension is needed. For example, Miller (1978) says that the unit of the abstracted system "may or may not" be empirical (X'). This revives the age-old controversy in social science regarding the question of whether certain social phenomena are to be regarded as empirical entities or alternatively as "social constructs." For example, what is the status of the social role? Parsons (1979) contends that the role, not the individual, is the proper unit of analysis of the social system (see also Luhmann 1986).

The answer is that whether one regards the role or other phenomena as constructs or empirical entities is somewhat irrelevant from the standpoint of the first dimension, as this dimension will accommodate it in any case. The problem in classifying the abstracted system is that if the role (for example) is not an "object," and thus is not the unit of a concrete system, what is its status, particularly vis-à-vis an object?

Parsons (1967) says that the role should be the unit of analysis for a number of reasons—multiple roles for one. Mr. Smith is Mrs. Smith's husband (role 1) and also is the mail carrier (role 2). According to Parsons, emphasis on the person as the unit may mean losing sight of some of these roles. The roles, according to Parsons, are the basic building blocks of the social system, and not the individuals.

The question still remains: what is the epistemological or ontological status of the role? This is a crucial issue in resolving the debate over the proper unit of the social system. Miller (1978) refers to the role as a "relationship." Yet upon examination it is clear that entities (roles) such as the marital status and occupational status discussed by Parsons (1967) are simply regarded routinely by sociologists as "variables," and are routinely gathered on almost every contemporary social survey. More specifically, both are classified as nominal or categorical variables.

At this point, the relationship of the abstracted system to the concrete system becomes clear. The "roles" of marital and occupational status are empirical, and thus empirically verifiable. Thus, abstracted systems, like concrete systems, exist on the empirical level (X' on dimension #1). However, this does not preclude someone constructing abstracted systems on either the conceptual (X) or operational (X'') levels as well, and in fact each empirical system will generally have parallel counterparts on these two levels. But while both systems are empirical, *the concrete system is comprised of interrelated objects,* while in contrast *the abstracted system is composed of interrelated variables.*

There is a crucial methodological and interpretative difference here. While Hall and Fagen (1956) define a system as having interrelationships between both the objects and their attributes simultaneously, in reality this specification is rare or nonexistent, particularly in social science. This is clearly shown by reference to table 2.1. The majority of statistical sociological research utilizes correlations among variables. A common study computes r for columns of the matrix in table 2.1 (for example, correlating variable 1 with variable 2, or variable 1 with variable 3). This is essentially the approach used in Parsons's abstracted system, where the units are standard sociological research variables such as occupation

TABLE 2.1.
The Score (S) Matrix
Variables

Objects	1	2	3	·	·	·	N
Jim	$Score_{11}$	$Score_{12}$	$Score_{13}$	·	·	·	$Score_{1N}$
Bill	$Score_{21}$	$Score_{22}$	$Score_{23}$	·	·	·	$Score_{2N}$
Jane	$Score_{31}$	$Score_{32}$	$Score_{33}$	·	·	·	$Score_{3N}$
·	·	·	·	·	·	·	·
·	·	·	·	·	·	·	·
·	·	·	·	·	·	·	·
O	$Score_{o1}$	$Score_{o2}$	$Score_{o3}$	·	·	·	$Score_{oN}$

and marital status. *The abstracted system is a variable system—the units are variables and the relationships are relationships among variables.*

This is definitely *not* the case for concrete systems. Concrete systems are based upon correlations between objects (*rows,* not columns, of the matrix in table 2.1). An example would be a correlation between object 1 and object 2. Generally the objects are persons (organisms in Miller's approach).

The correlation of objects is called *Q-analysis,* while the correlation of variables is called *R-analysis.* Notice that there is *only one* internal set of scores in table 2.1. Thus, both *R-* and *Q-*analysis deal with exactly the same data set, they just start from different points, somewhat like two farmers plowing the same ground, one from North to South and the other from East to West. Referring back to figure 2.1, it is apparent that in the case of maximum entropy, there are no correlations *regardless* of whether one correlates columns (*R*) or rows (*Q*). Thus, the question of whether to use concrete (*Q*) or abstracted (*R*) systems is moot for maximum entropy, as no living or social system exists empirically. In contrast if one had minimum multivariate entropy for all *N* objects and *M* variables, then all *Q*-correlations would be 1.0 and all *R*-correlations would be 1.0. Here again the distinction between concrete and abstracted systems would be relatively unimportant, as perfect predictions could be made for either (although the interpretations would be different). In the usual intermediate case, both *Q-* and *R*-correlations would be of intermediate utility in enabling explanation and prediction. Here the interpretations become more problematic in either case, and it is more difficult for adherents of one approach to comprehend the other.

In any case, the relation of *Q* to *R* is mathematically unambiguous—one is simply the matrix transpose of the other (see Bailey 1972). That is, to go from *Q* to *R* or from *R* to *Q,* simply turn the matrix in table 2.1 on its side (transpose it). Interpretatively, it is more difficult to go from *R* to *Q* or from *Q* to *R.* Most social scientists deal with objects or groups of objects in their verbal theories, as when they speak of individuals, dyads, families, cities, and so forth. This is essentially a verbal form of *Q*-analysis. They also have little trouble dealing with variables (*R*-analysis). For example, they will speak of an individual (*Q*) and his or her characteristics—sex, race, age, occupation, and so forth (*R*).

However, moving from verbal theory to statistical analysis in social science, one sees *almost total reliance* on the interpretation of R-analysis. Social statisticians report their sample size (N). This is where Q-analysis generally ends. They then almost invariably utilize R-statistics which seek correlations among variables, for example the correlation between education and income, or between race and education. There are only a few statistical analyses which focus on objects, for example, some social-mobility and migration models such as the mover-stayer model (Spilerman 1974); some sociometric models such as blockmodelling (e.g., White et al. 1976), and Q-factor analysis or cluster analysis (Butler and Adams 1966). It is no wonder then that sociologists are prone to R-analysis (abstracted systems) because that is virtually all they have experience with. In biology the situation is quite different, as biologists use Q-analysis routinely, and so systems analysts from biology would be quite comfortable with concrete systems (Q-analysis). For further discussion see Bailey 1983.

The point here is that the link between Q-analysis (concrete) and R-analysis (abstracted) is unambiguously clear, as table 2.1 shows. This means that no one need be afraid of beginning with a concrete system for fear that this will preclude abstracted analysis. One can always proceed from concrete to abstracted analysis, and in a sense that is what we do in statistics. However, it is somewhat more difficult to proceed from abstracted to concrete for the simple reason that the *variables* or concepts of abstracted analysis are almost universally conceptualized as properties or characteristics of some *objects* (Q). If one begins with the concrete or Q-system, it is generally relatively easy to identify the respective variables for each object, and thus to proceed to abstracted or R-analysis. However, if one begins with analysis of a set of variable values (abstracted or R-analysis), it may be more difficult to identify the objects for each variable. There are also some technical statistical problems in switching from R- to Q-analysis, or vice versa (see Bailey 1983).

Dimension #3: Closed-Regulated

While the first two dimensions involved complex epistemological issues, the remaining ones are fortunately somewhat easier to understand. In the majority of cases the prior discussion suffices to illuminate the crux of the matter, and we need only to elaborate here. The bulk of this elaboration consists of clarifying each dimension of the typology in the context of the other dimensions.

It is clear that the closed-open (regulated) distinction applies primarily to the empirical level (X') of dimension #1, and to the concrete (Q) case of dimension #2. Take Miller's 1978 empirical example of the bakery as a concrete system. Here the boundaries need to be regulated so that the proper personnel are admitted, and the proper information is received (recipes, costs of supplies, and so forth). Note that these all are matter-energy or information inputs. However, the boundaries cannot be held permanently open, as the concrete system does not need extra supplies, extra personnel, or a surfeit of information (or extraneous information).

Why is not the closed-regulated distinction just as applicable to perceptual systems (X) or models (X'')? In a sense it is, as we obviously cannot allow extraneous thoughts to invade our perceptions (X) of the systems we seek to comprehend, nor allow extraneous symbols into our mathematical models (X''). However, in a real sense the X and X'' levels are formulated by us as tools to understand the empirical world (X'). It is at the empirical level where the terms "closed" and "regulated" have real meaning and important ramifications, as they are crucial for the understanding of the regulation of matter-energy and information, and subsequently the regulation of entropy.

Dimension #4: Totipotential–Not Totipotential

Technically speaking, few social systems are truly totipotential or self-sufficient. The largest nations perhaps could be (with a lowered standard of living), but they take in so many imports that they do not appear totipotential. Smaller systems which probably could not function except as subsystems within larger systems are clearly not totipotential. This includes body organs and departments (for example, accounting or personnel) within larger systems. Note that as in the case of the closed/regulated distinction (dimension #3), this distinction applies primarily to empirical systems (X').

Dimension #5: Autopoietic–Not Autopoietic

An autopoietic system has internal control mechanisms which enable it to insure entropy levels below the maximum. In large systems this entails not only boundary regulation so that inputs and outputs of matter-energy and information are adequate, but also regulation of complexity, so that the system is not overwhelmed by the complexity of the logistic tasks facing it.

A system need not be totipotential to be self-organizing. In fact, even a small system or subsystem can be self-organizing in the context of a benign environment or host system. Further, a small system has the distinct advantage of not having to deplete its precious entropy-combating energy by dealing with internal complexity. However, a complex system which nears totipotentiality obviously has more potential for survival than a small system which suddenly finds itself in the midst of a hostile context. Note that once again, as in the cases of dimensions #3 and #4, this dimension refers primarily to the empirical case (X') and dimension #1.

Dimension #6: Hierarchical–Not Hierarchical

This dimension has already been discussed in some detail. Social systems are hierarchical almost by definition, as they tend to include as subsystems those smaller units which also meet the definition of a system. Thus, a supranational system such as the United Nations is hierarchical in the sense that it is inclusive of several telescoping levels, with each lower level including others still lower, until the level of the cell. Notice that the analysis is now limited to living systems. The notion of hierarchy is somewhat more complicated for nonliving systems such as machines.

The United Nations is a system. It includes societies (systems) which include cities (systems) which include organizations (systems) which include groups (systems) which include individuals, and so forth. Above the level of the group, each of these systems has its own internal hierarchy (echelon) of power and control (president, vice-president, and so forth) that is separate from the other levels mentioned.

While dimensions #3, #4, and #5 (closed-regulated, totipotential–not totipotential; autopoietic–not autopoietic) are most useful when applied to empirical social systems (more specifically, to concrete systems), the hierarchical-nonhierarchical distinction has application to the model or operational level (X'') of dimension #1 as well (but less application to perceptual systems [X]). Not only do empirical systems have hierarchies, but statistical models used to study them can have hierarchical properties of their own as well. It was noted above that most statistical models in social science are R-models. This is not true in biology, particularly in numerical taxonomy where Q-cluster analysis is widely used (see Sokal and Sneath 1963; Sneath and Sokal 1973; Bailey 1975, 1982, 1983, 1985). These techniques essentially model concrete systems by replicating the states through which the system developed. If the empirical system (X') developed through hierarchies, then the model (X'') mirrors this through its own hierarchies (see Bailey 1985, 1987).

Analysis

Development of the systems typology is now complete, and the full typology is shown in figure 2.3. It must be reiterated that this typology is not exhaustive, and that other system types can surely be found. Nevertheless, this schema does seem to incorporate the major types most useful in the application of systems theory to social phenomena. Further, since the typology has just been derived, identification of empirical cases has not yet been attempted, and so concrete examples of the ninety-six cells are not presented in figure 2.3. Perusal of the six dimensions reveals that the first two, while undoubtedly the most complex and difficult to understand, are also of most value in clarifying some of the confusion surrounding the utilization of extant systems concepts such as conceptual systems, concrete systems, and abstracted systems. Without employing the logic of substruction (Lazarsfeld 1937) and searching for *underlying* dimensions of the principal extant types as was done here, it is doubtful that the two basic dimensions underlying these three important types could have been discovered. These dimensions are the perceptual (conceptual-indicator-empirical; dimension #1) and the Q-R (dimension #2).

The use of these two dimensions immediately clarifies and demystifies the two major concepts of use to social systems theorists—the concrete system and the abstracted system. The concrete system is immediately identifiable as an empirical (dimension #1) Q-system (dimension #2). The abstracted system is clearly an empirical (dimension #1) R-system (dimension #2) *in most cases*. Miller (1978) notes that abstracted systems can also include symbols which do not have direct empirical representations. In

FIGURE 2.3.
A Typology of Systems

			Q						R					
			X		X'		X''		X		X'		X''	
			H	NH	H	NH	H	NH	H	NH	H	NH	H	NH
Closed	Totipotential	Autopoietic												
		Not Autopoietic												
	Not Totipotential	Autopoietic												
		Not Autopoietic												
Regulated	Totipotential	Autopoietic												
		Not Autopoietic												
	Not Totipotential	Autopoietic												
		Not Autopoietic												

H = Hierarchical. NH = Nonhierarchical

such cases an abstracted system remains an R-system (dimension #2), but could be an operational system or model (X'' on dimension #1), or even possibly a mixture of model (X'') and empirical (X'). Bear in mind, though, that *each* empirical system (X'), whether R or Q, can have counterparts at the conceptual (X) and operational (X'') levels insofar as we can perceive it and model it. It appears also that some abstracted concepts such as the role, inasmuch as they can be written either as characteristics of individuals (for example, occupations), or as units of society, serve as micro-macro links. This fact is valuable, but complicates the analysis. Due to its complexity it will be discussed in further detail in chapter 6 (see also Bailey 1990).

The raging debate over concrete systems (Miller 1978) versus abstracted systems (Parsons 1979) has now been defused by this typology, which shows the former to be the Q-analysis favored by biology and the latter to be the R-analysis favored by sociology. Further, the links between the two are clear, and one is no "better" than the other. There are some pragmatic reasons for starting with concrete and going to abstracted, but both are used in social entropy theory, and will be discussed in chapter 6 (see also Bailey 1990).

The debate over autopoiesis (dimension #5) is not similarly solved by this typology, but is continued in chapter 8. The typology here is original in the combination of the six dimensions, and is still exploratory. There is undoubtedly some overlap or correlation among dimensions #3 (closed-regulated), #4 (totipotential–not totipotenital), and #5 (autopoietic–not autopoietic). It is entirely possible that with refinement and the analysis of empirical cases, one or more of these dimensions could be eliminated from the typology as redundant (while still being worthy of analysis in its own right). However, for now the heuristic value of the typology consists of the combination of six major dimensions underlying the most important types of systems studied in social systems theory.

WHY STUDY SYSTEMS?

It is now profitable to return to the question of the value of systems theory in sociology. A number of reasons for studying systems have already been given. Inasmuch as social groups meet the definition of a system provided above, all sociologists are systems theorists, and from a strictly chauvinistic perspective, anyone who does not use a systems model is perhaps using a model too narrow to be sufficiently isomorphic with the reality being modeled. Social entropy theory (Bailey 1990) begins with the premise that the model must be as complex as the phenomena being modeled, and one cannot study complex modern society with a simple model.

From the perspective of this volume, the study of social systems is inevitable. Systems as defined here in their variety are ubiquitous in society. As has been seen, from a supranational entity such as the North Atlantic Treaty Organization (NATO), the United Nations, or the European Common Market, through the society, the city, the organization, the group, and the individual, all of these meet the definition of a system. More specifically, they are easily seen to be concrete, regulated systems. Most are hierarchical, and perhaps demonstrate varying degrees of autopoiesis. Since this is the case, the proper question is not *why* one studies social systems, but how one could *avoid* studying them and still have social science. Since systems are ubiquitous among social phenomena, it is clear that they must be dealt with in some form by any social science, for a system by any other name is still a system.

While one cannot avoid studying social systems to some degree if he or she wishes to understand social phenomena, there are of course many approaches to the study of these phenomena. Many scholars, and even disciplines, do not perhaps recognize the systemic qualities of the phenomena they are studying, because they are not studying the whole phenomena but only certain aspects of it. This is clear in social science.

Often social scientists make no attempt to understand a concrete system (for example, an entire society) by simultaneously studying all of its properties (characteristics or variables) and their interrelationships. Rather, they study only one of a few variables at a time in a piecemeal or unintegrated fashion. For example, economists study monetary phenomena, political scientists study political phenomena, psychologists study psychological processes, and so forth. By abstracting these phenomena away from the objects that they are properties of (the individuals or groups) and away from other variables, the various disciplines often effectively obfuscate the holistic and systemic nature of the phenomena they are studying.

For example, the economist who is studying supply/demand equilibrium or the level of the money supply in various labeled categories ($M1$, $M2$, and so forth), is diverting attention away from the systemic nature of the social system in at least two ways: (a) by separating a variable (money) from the Q-system generating and utilizing it (the society), and (b) by separating this particular R-characteristic or variable (money) from other R-variables. When other variables are brought into the analysis in social science they are often introduced in an additive fashion (rather than first considering a whole set and then excluding some).

For example, it is typical in common social science to talk briefly about Q-analysis in the form of a *sample* (for example, 460 males, ages forty to sixty), and then move directly to R-analysis by first emphasizing a single "dependent" variable (such as annual income). Then, "independent" variables are added in a sequential fashion. For example, education, then race, then gender, and so forth. This is an efficient and parsimonious strategy in the sense that one surely does not study anything that one does not have to. However, a major problem with this analytical strategy is that focus is displaced from the systems viewpoint. Rather than focusing upon understanding human social groups in all their ramifications, one is focusing on narrow technical concerns. Very often attention becomes unduly focused on technical characteristics of the model, such as whether the sample is of sufficient size and was gathered correctly, whether the level of measurement meets the assumptions of the statistical model (for example, whether nominal variables such as race and occupation can be incorporated in a ratio model), whether there is multicolinearity or interaction, and so forth.

Not only do these narrow technical concerns *divert* attention from social phenomena, but they are essentially X'' (model) concerns rather than X' (empirical) or X (conceptual) concerns. That is, in terms of the three-level model of dimension #1, the basic purpose of social research is to use conceptual reasoning (X) to understand empirical phenomena (X'). The model (X'') is used basically just as a tool for accomplishing this major purpose. However, much sociological literature is sidetracked into focusing on these misplaced concerns (see Bailey 1990).

It would be distinctly preferable from the standpoint of systems theory to first describe the whole system (group, city, organization) instead of only a sample of persons abstracted from some system. Statisticians often speak of a population or universe from which their sample is drawn. The problem from a systems perspective is that this universe *may or may not coincide with some concrete system*. After the whole system is described, then the important variables affecting its operation should be described, with an emphasis on their relationships and interrelationships, including multiple interactions and nonlinear relationships.

At this point, if the scholar feels unable to proceed holistically, certain variables can be divided out from the whole, while the researcher retains some perception (X), however vague it may be, of the empirical concrete system (X') being studied. The scholar should do this, rather than concentrate heavily on secondary model characteristics (X'') such as multicolinearity, homoscedasticity, or other properties which are chiefly properties of the model rather than of the phenomena of interest, and thus are of secondary relevance to the real purpose of the study.

At this point the reader might well say that there is a flaw in this strategy, as no sound strategy for dividing variables from the whole exists. Thus, the systems approach is lacking. This illustrates the sort of non sequitur often applied to systems theory and other new approaches. It is certainly true that a comprehensive strategy for variable selection is lacking in the systems approach, but this is similarly true for all social research, both divisive and additive, holistic or piecemeal, theoretical or statistical. Thus, the non sequitur lies in attributing to the systems approach that which is a general failing of social inquiry.

Reconsider what may be the typical sociological approach to research. An investigator discusses a "population" and then discusses a subset of this population (a "sample") which is said to adequately represent the whole population, with "adequacy" generally defined in a statistical sense. The investigator generally presents an R-statistical analysis based on a relatively few variables. These are generally labeled as "dependent" (usually only one or a very few) and "independent" (the remainder).

There are two basic deficiencies here, not only from the systems perspective, but from the standpoint of a general comprehensive understanding of social phenomena as well. These can be labeled the Q-problem and the R-problem, indicating whether the problem lies respectively in the realm of objects (Q) or variables (R). The two problems are highly interrelated, as might be expected.

1. *The Q-Problem.* The genesis of the Q-problem lies in the conceptualization of a "problem" rather than a system. One might ask, from the standpoint of comprehensive social inquiry (which we can call "positive positivism"), just what epistemological or ontological position does the concept of "population" or "universe" have in our inquiry? It is positivistic to be sure. Yet, just what does it represent either theoretically or methodologically? While the theoretical relevance of statistical populations undoubtedly varies widely, in general the population is selected on a rather ad hoc basis. Its selection may be somewhat "theory driven." For example, a student of race relations may select the black population of the United States as the population. However, a population such as the black population of the United States or female population, or aged population, is not a theoretically or empirically coherent body or organization, but just a statistical category. This is because it lacks internal relationships or networks.

Contrast such a statistical category with an organized unit such as a concrete system. If the population is not a statistical category but in fact a concrete system (such as the American Association of Retired Persons or AARP), then it is more theoretically and empirically meaningful. Why is not a statistical category more meaningful (or just as meaningful as a concrete system) for social inquiry? Simply because the relations of interest (R-relations) are not produced by the statistical category, but by some concrete system. Thus, the process of wresting the statistical category from its material basis clouds or destroys the very relationships we wish to understand.

2. *The R-Problem.* To make this point clearer, turn to the R-problem. Suppose a researcher in race relations selects the black population of the United States as the universe, and selects an "adequate" sample from it. What is the next step? Typically, one specifies variables (R) such as income (dependent) and education (independent). Suppose that all educational and income levels are relatively low (attenuated), and that the correlation between variables is low. It is then concluded that education does not translate into income gains for blacks to the degree that it should. It is then further concluded that blacks are the victims of racism.

There is a major problem here. The problem is that the perpetrators of racism are not even the people in the statistical category that is being studied (unless the racism is self-inflicted). Rather, the perpetrators are outside the sample, and we have not studied them. At the end of the analysis, it is found that the basic gain of the study is mere description. We have inferred effects (of racism among blacks) but have not even studied the persons causing these effects (whites). We have documented that blacks have low education and income (which was probably never in doubt). However, we have not made significant theoretical gains, and we have probably encountered a host of methodological problems (sampling method, sample size, level of measurement, and so forth) that cloud the validity of the analysis.

How can the Q- and R-problems be solved? The first (Q) is rectified by choosing as the statistical universe the population of some concrete system such as a city, organization, or so forth. Why does this help? Because it is the Q-*system* which generates the relationships between variables (R-relationships) through its orderly action. *Orderly action by actors (Q) lead to statistical order among variables (R).* This is the principal axiom of the current approach to social systems theory (see Bailey 1990). To illustrate, if the employers in a society (Q) treat blacks randomly when blacks apply for jobs, then race and occupation will show no statistically significant correlation (R). On the other hand, if bosses (Q) give all blacks low-paying positions, then race and occupation will be significantly related (R).

Figure 2.4 shows six types of statistical universe or population. Only type 1 is adequate. In this type the boundaries of the population are synonymous with the boundaries of the concrete system that generated the order to be studied (for example, the population of London). All of the others are inadequate. Type 2 is underbounded, so that important actors (Q) and perhaps important variables (R) are omitted. Type 3 is overbounded, so that extraneous actors (Q) and variables (R) may be added to the concrete system. Type 4 is completely outside of the concrete system, so that the nature of the actors (Q) and variables (R) is unknown. The data are not from the system of interest, but it is not known what system they do represent. Type 5 is an overlapping population, so that the two concrete systems may be mixed. Type 6 is the statistical category or cross section discussed earlier, where some of the actors (Q) from within the concrete system's boundaries are selected on the basis of some factors, often ascribed variables (for example, race or gender) but sometimes achieved variables (for

FIGURE 2.4.
Congruence Between a Concrete System and a Sampling Universe or Population

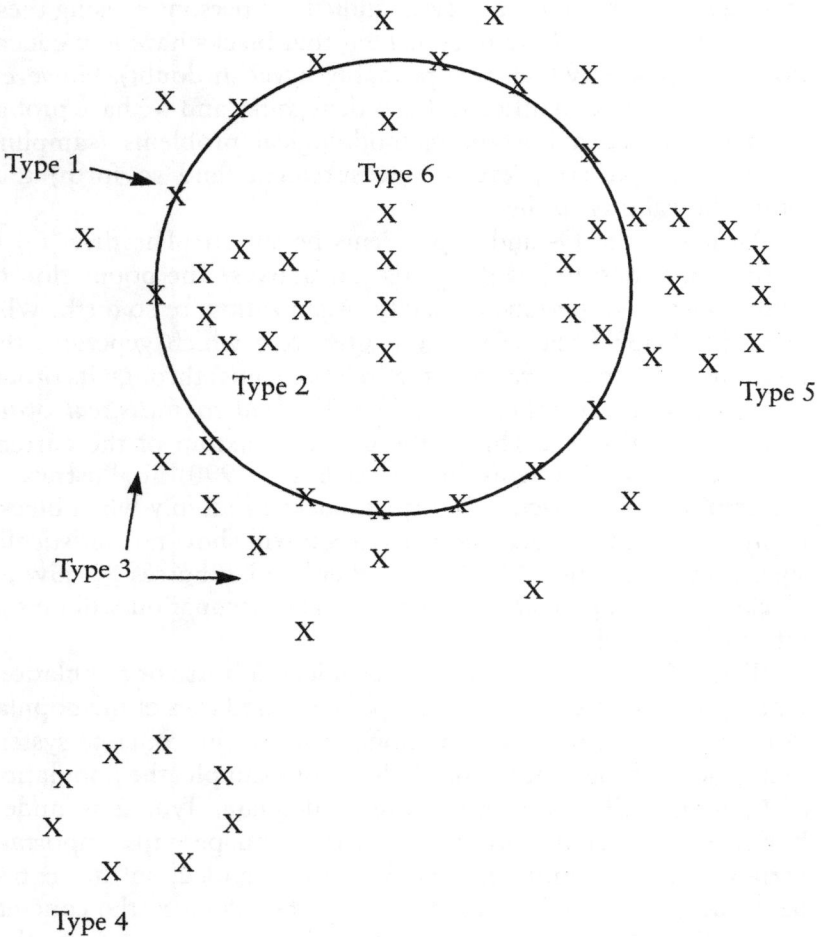

example, education or income). Thus, the solution to the Q-problem lies in coordinating the universe with the boundaries of the concrete system.

Fortuitously, many social researchers do this probably without even realizing the significance of what they are doing. Census data tends to be organized on the basis of concrete systems (sometimes) by choosing systems boundaries (such as national boundaries) to define the population or universe, and this helps immensely in guiding statistical researchers in the right direction. It is not suggested that all researchers must choose universes consistent with the boundaries of concrete systems, but only that they recognize the issues involved.

How can the R-problem be rectified? Perhaps not as easily as the Q. In Q-analysis one has the advantage of dealing with a finite set of objects (for example, persons). Although the exact number in the population or universe may not be known correctly, at least it is some finite number. The same is not true for variables (R-analysis). Unfortunately, researchers often have no clear idea of the universe of variables that they are dealing with. For example, what is the universe of variables in a study of race relations? How can one select a representative sample of these variables, the same way one selects a random sample of persons from the universe? This would entail knowing in advance the universe of every variable relevant to the study, and this is generally unknown. Since one has no clear representation of the universe, sampling is precluded. Rather than divide some known universe, or select from a universe of salient variables, the usual procedure is to begin with a few variables (supposedly "theory driven") and then perhaps to add more if the analysis proves inadequate.

Unhappily, systems theory has no answer to this general problem any more than do other approaches. It is plagued with the problem just as they are. However, systems theory does offer not only recognition of the crucial issues involved (which some of the other approaches seem to lack), but also offers a more comprehensive model, and thus a comprehensive array of variables (though not presented as a comprehensive universe). See *Living Systems* (Miller 1978) and *Social Entropy Theory* (Bailey 1990) for comprehensive sets of theoretically derived variables.

A problem in social science is that study selection is sometimes idiosyncratic or ad hoc. Researchers choose research projects for a variety of reasons, often highly personal (for example, they may study the adaptation of Eastern Europeans in the United States because their parents were migrants from Eastern Europe). Examples of this sort of research abound in social science, with some seemingly based on little or no theory (the researcher seems to search for theory after selecting the research topic, rather than letting the theory inform research formulation). Many other studies which are based on theory derive from very narrow perspectives. These perspectives often seem to have no effective built-in assessment procedures, so that if a wrong turn were taken years ago, the subsequent literature could follow it down this wayward path for years to come in a sort of inifinite regress. The plethora of narrow approaches extant in sociology today generally lacks a unifying framework, and raises the specter of intellectual paralysis (Rossi 1980).

The systems approach is neither a panacea nor a replacement for the extant sociological approaches. It is merely a complement to them. Further, there is certainly nothing wrong with studying narrow topics that one has a personal interest in. Even theorists with very broad perspectives have done this (for example, Weber, author of the *Protestant Ethic and the Spirit of Capitalism,* had a mother who was an ascetic Calvinist—see Ritzer 1983). The danger lies in drifting away from a comprehensive analysis of perspectives which deal with fundamental issues crucial to the understanding of society. There are myriad special topics that deserve to be studied by social scientists, as long as others complement this work by inquiring into the fundamental processes and structure of the society as a whole.

TOWARD A POSITIVE POSITIVISM

A positive positivism is definitely needed in sociology now. By that I mean an approach which encompasses the best and most powerful features of the canons of scientific method. I am taking positivism here "writ large" as the general scientific approach, and not as a narrower approach as in the Frankfurt School. Scientific method, with an emphasis on critical inquiry, rigor, specificity, and verification, is clearly a nonnegative or neutral force, just as is the generic definition of a system (see Klir 1969). It is only through their specific applications that perspectives such as positivism and systems become objectionable to some. These approaches have come in for their share of criticism in the past. They are not necessarily linked. One can of course have positivism without systems. Conversely, nonpositivistic systems theory is feasible, and currently exists in the form of "soft systems" (see Checkland 1981, 1985), although many critics have probably never seen it and so are skeptical.

I believe that the hallmark of *any* positive approach to sociology, whether it is labeled positivistic, subjective, qualitative, humanistic, and so forth, is that it helps us understand some facet of society without impinging negatively upon other approaches or their adherents. In other words, an approach should make a positive contribution to knowledge without criticizing others. Too often in the past, positivism (or the "scientific approach") in general and systems theory in particular has been a negative force by failing in one of these two key areas—either not contributing significantly to knowledge, or by criticizing others. One salient form of criticism (perhaps indirect criticism) is to formulate an approach which precludes alternative approaches or denies them their legitimacy. Positivists and systems theorists in sociology have no monopoly on this, but they have certainly been guilty of it on more than one occasion.

Negative Positivism

I will briefly critique some past social systems approaches, assessing the degree to which they were scientific or "positivistic," their failings in the two areas specified (failing to contribute or precluding valuable approaches) and some criticisms that have been made of them. This is only a cursory overview, as these matters are dealt with in more detail throughout the book.

Is Systems Theory Positivistic? As shall be seen in chapter 3 and elsewhere, some principal nineteenth-century systems approaches definitely were positivistic and based directly on physical science. Spencer was writing *First Principles* (1892 [1864]) during the 1850s when Clausius was developing the second law of thermodynamics (Clausius 1850). Spencer's extensive equilibrium analysis was written while he was still unaware of Clausius's work, and was based squarely on the first law of thermodynamics (the conservation of energy). Pareto (1935) was also a positivist (scientist). He wrote later than Spencer, and was a student of thermodynamics before turning to social science. He based his equilibrium analysis largely on principles of thermodynamics. Homeostatic theorists such as Henderson (1935) and Cannon (1929, 1932) were natural scientists who later in their careers developed an interest in social science. Both were influential in the work of Parsons (1951) and Miller (1978). Parsons denounced positivism and empiricism in general, along with the evolutionary approach (see his famous introduction to *The Structure of Social Action* in Parsons 1937).

But while Parsons eschewed quantification and decried empiricism and positivism, facets of his work, such as his definition of system (Parsons and Shils 1951, p. 107) have clear positivistic foundations, though somewhat masked by the strictly verbal presentation.

Contributions. The contributions of the social systems approach to this point are legion, despite the claims of some detractors. Perhaps the chief contribution of this school has been to keep alive the notion of a society *sui generis* as so strongly advocated by Durkheim. As Ritzer (1983) noted, and as is well known, many modern American sociologists have abandoned the broad view championed by classical sociologists in favor of narrow technical concerns and small-scale studies. There are many reasons for this. The systems approach has borne much of the burden for keeping alive some semblance of emphasis on a large-scale unitary society as the proper unit and subject of analysis. Other contributions of the systems approach include part-whole analysis, analysis of part-part relationships, study of feedback, study of micro-macro relationships, study of boundaries, and of course the study of function, and the relation of structure and process. Many of these topics have been studied also by nonsystems theorists, but often without the benefit of a comprehensive analytical framework.

The contributions of a systems framework toward integration and comparative analysis have previously been mentioned. Ritzer (1988, p. 451), in discussing Buckley (1967), adds four more: systems theory promises a common vocabulary to unify the behavioral and social sciences; systems theory is multileveled, and thus can be applied to large and small, to the objective and subjective; systems theory emphasizes the whole and the study of relationships, and thus provides an alternative to the piecemeal analysis commonly found; and the systems approach sees all aspects of the sociocultural system in terms of process.

It should be noted that the integrative thrust of systems theory is in line with recent integrative developments in sociological theory. Ritzer (1981; 1988, pp. 503–18; Ritzer and Bell 1981) has developed an integrative approach based on the notion of an exemplar. Ritzer (1988, pp. 484–502) also acknowledges the fragmentation that some have seen as a crisis in sociology, but states that there are currently many signs of growing coherence in sociological theory. As Ritzer (1988, p. 484) says, "That coherence centers around the emergence of considerable consensus that the central problem in sociological theory is the study of the relationship between macroscopic and microscopic theories or levels of social reality." The clear multileveled and broad approach of systems theory makes it very compatible with this trend and perhaps breathes new life into it. Micro-macro links are especially clear in both living systems theory and social entropy theory (see chapters 5, 6, and 7). While many approaches to the micro-macro problem approach it from the micro perspective, and some are quite reductionist (see Ritzer 1988, pp. 484–502), systems theory is somewhat unique in being broad enough to approach the problem from the "top" (macro) or from both the micro and macro perspectives simultaneously, and without the necessity of reductionism.

Failure to Contribute. Social systems theory can be faulted for its failure to contribute to social inquiry on some points, though the diversity of systems approaches makes it difficult to generalize. Some of these failures are well known, and I will have occasion to comment on them throughout the book. To summarize briefly, here are a few failures.

1. *The failure to operationalize.* This is most prominent in Parsonian functionalism (Parsons and Shils 1951). This is due both perhaps to the scope and complexity of the theory, but also to its manner of formulation, particularly the emphasis on abstracted systems. The analysis of abstracted systems clouds the isomorphism between the model (X'') and the empirical system (X') (path c of figure 2.2) and makes operationalization difficult or impossible. Failure to operationalize may not be viewed as a problem by Parsonians or other schools (for example, by German idealists), but it is a definite problem not only from the standpoint of positivism, but also of empirical inquiry in general.

2. *Failure to include subjective phenomena in the systems purview.* This is not so much a limitation of classical functionalism as of more adamantly positivistic approaches such as living systems theory (Miller 1978).

3. *Obsolescence.* The failure to keep up-to-date is clear in the

social systems approach. This is probably partially due to the criticisms the approach has received. Dealing with these has probably diverted the attention and sapped the strength of systems theorists (note the angry retort of Parsons [1961b] to yet another attack on equilibrium). Who knows, obsolescence could even be due to the alleged "conservatism" (see Lilienfeld 1978) of the systems approach, although the value of this allegation is in doubt, as shall be seen shortly. Whatever the reason, the systems approach is now both stagnant and in a state of obsolescence. Which came first is an example of the classic chicken/egg question (did obsolescence cause stagnation, or did stagnation lead to obsolescence?). Whatever the reason, the potential of this approach and its value to general sociology is too great for either stagnation or obsolescence to long continue. Recent contributions such as living systems theory (Miller 1978), social entropy theory (Bailey 1990), sociocybernetics (Geyer and van der Zouwen 1986), autopoiesis (Luhmann 1986) and the present volume, should help revitalize the field. The field as many sociologists remember it remains rooted in the nineteenth century, chiefly because of concepts such as equilibrium. It must be modernized, and this is the task of this volume.

Examples of obsolescence are myriad. One of the chief ones is overreliance on the concept of equilibrium. While some social phenomena are characterized by equilibrium, many are not. Modern systems theory will be nonequilibrium analysis (principally entropy theory). A related example of obsolescence is the use of closed systems models, which equilibrium analysis sometimes requires. Modern systems theory such as autopoietic theory transcends the classic open-closed dichotomy, as an autopoietic system is a living system (and thus open), but is organizationally closed (see chapter 8 of this volume and Luhmann 1986). Modern systems theory will combat these examples of obsolescence through nonequilibrium analysis, through emphasis on feedback, through emphasis on individual action and control, through emphasis on subjectivity (not all systems theorists will agree with this), through emphasis on operationalization, as for example through the methods of contemporary information theory—not to mention utilization of computerized models—and through the use of concepts from fields such as artificial intelligence. The new approaches presented in this volume are so different and fresh compared to the old approaches that the issue of obsolescence will be irrelevant.

BLOCKING

It is sad to see any perspective fail to realize its potential. It is sadder still to see it formulated in such a manner as to *block* certain lines of inquiry, and thus to hinder adherents of those perspectives from achieving their individual intellectual goals. This is particularly true if the blocking paradigm has a dominant position, as functionalism did during the 1950s. There were four principal blocks inherent in obsolete social systems theory, particularly Parsonian functionalism.

1. *Block #1*. Nonequilibrium analysis was precluded by Parsons. Examination of Parsons's definition of a system (Parsons and Shils 1951, p. 107) reveals conclusively that equilibrium (and a return to equilibrium if it were to be disturbed) *is actually incorporated in the very definition of system*. In fact, as used by Parsons, equilibrium is synonymous with order (this is unusual among systems theorists). Parsons took the concept of equilibrium for granted (as taught to him at Harvard by Henderson) and never questioned it, believing that without equilibrium, society would display no order, and thus social science could not exist (see Parsons, 1961a, 1961b). Needless to say, it bothered him mightily to see this revered and sancrosanct cornerstone of his approach questioned. However, this book will show conclusively that many examples of nonequilibrium functioning exist in social science phenomena. In fact, the modern systems approaches presented here seldom mention the word, and certainly do not include it as part of their definition of a system. There is no need to abandon the concept of equilibrium. It can be used in the limited instances where it seems to apply, but it cannot be allowed to block all approaches that contradict it.

2. *Block #2*. As a corollary of block #1, the study of process and social change (including revolutionary change) was impeded (but not entirely precluded) by the equilibrium notion. Block #1, equilibrium, has a host of blocking phenomena associated with it, including methodological and theoretical (see Bailey 1984a). This will be discussed further in chapter 3. The second block greatly hindered the analysis of social change. Although this charge has been made many times and defended many times (see Ritzer 1988), nevertheless it seems to be a very real block, since it implies

that change which does occur in one place in the system will be undone by other forces in the system which react so as to restore equilibrium. This has been thoroughly ingrained in the equilibrium model since Spencer (1892 [1864]) and Pareto (1935), and is based at least partially on a natural science equilibrium model as embodied by Le Chatelier's principle (Miller 1978).

3. *Block #3.* Another common criticism, also a corollary of block #1 (equilibrium) is that the role of individual action is minimal or invisible, so that the systems model seems to be deterministic. Again, if equilibrium will always be reestablished by institutional reactions from within, whenever it is disturbed from outside (clear in Pareto), then it would seem to matter but little what action any particular individual or even group takes. But the principal objection here would seem to be not so much that individuals do not have a role in shaping society, but simply that their role is never made clear in the equilibrium structural-functional model. If equilibrium always prevails, and is always reestablished if disturbed, then strong forces such as economic forces, biological forces, or physical or environmental forces would seem to prevail over individual action. This is seen by critics as not only antipersonal, but nonreasoning as well. It can be characterized as economic determinism, environmental determinism, and so forth. This is one of the most egregious of the obsolete features of old systems theory, and goes all the way back to the nineteenth-century writings of Pareto (1935).

According to Ritzer (1988) this aspect of functionalism is partly responsible for its reputation as a conservative model. It would seem that this criticism would be largely dispelled by spelling out in some detail (which both Spencer and Pareto did) just exactly *how* equilibrium is reestablished if disturbed. Later functionalism used the homeostatic model which showed in some detail how balance was reestablished. This did not satisfy critics either, as they charged that the functional model is more "simplistic" than homeostasis, and does not illustrate feedback mechanisms, for example (Giddens 1979). This is yet another example of obsolescence. Early theorists such as Spencer and Pareto wrote before the model of homeostasis was developed (Cannon 1929). Others generally wrote before cybernetics and feedback theory was fully developed, although some of Parsons's writing does include the "cybernetic hierarchy of control" (see Turner and Maryanski 1979). For further discussion see Bailey (1984a).

   4. *Block #4.* The fourth block is failure to operationalize, and has already been discussed. Some abstracted models such as *The Social System* (Parsons and Shils 1951) seem to preclude operationalization, while concrete models such as living systems theory (Miller 1978) allow it. Although living systems theory is not fully operationalized, it was formulated with this goal in mind, and does include many hypotheses suitable for testing, as does social entropy theory (Bailey 1990).

CRITICISM

There have been many criticisms of functionalism, and they need not be repeated here. Some of the principal ones are by Lockwood (1956), Gouldner (1959, 1970), and Horowitz (1962). For others see Turner and Maryanski (1979), Ritzer (1988), Lilienfeld (1978), Giddens (1979), and Bailey (1984b, 1990). Most of these can be traced in some way to the equilibrium concept, which will be discussed in detail in chapter 3. I will list the principal criticisms here. They can be divided into four basic categories: Substantive, methodological, theoretical-logical, and ideological.

Substantive.   Many substantive criticisms could be delineated, depending upon one's perspective, as is the case for most broad models. The major ones have already been discussed. They are:

1. Neglect of social change (emphasis on equilibrium, structure, and the status quo).
2. Neglect of individual action (determinism).

Methodological.   There are also a number of methodological criticisms. The principal ones are:

3. Failure to operationalize (already discussed).
4. Inadequate measurement of system state. Some measure of the system ranging from lack of integration to full integration is needed. Equilibrium implied that the system is always in a single state of integration (or soon will return to that state).

Methodologically, this precludes the need for developing a full-range measuring device, and thus severely limits and cripples the systems approach, which must be applicable to nonequilibrium cases as well. Again, this is an example of obsolescence. At the time the model was originally formulated, equilibrium reigned and was seldom if ever challenged, while nonequilibrium analysis was virtually unheard of. Now there is no excuse for obsolescence, and the model must be updated.

Theoretical-logical.   There are four chief criticisms here. They are:

5. Determinism, and the use of physical analogies, including generally inappropriate borrowing from natural science. This has already been discussed. Again, it is an example of obsolescence that can be largely overcome by broadening the model to include individual action and subjectivity, and by including modern nonequilibrium and cybernetic feedback techniques.

6. The use of tautologies. Why does this system survive? Because it is functional. How do you know it is functional? Because it survives. Perhaps some simplistic analyses were vulnerable to this sort of tautology, but only certain functional-equilibrium models. The general model of systems is not particularly vulnerable to this flaw (see Turner and Maryanski 1979).

7. The use of inappropriate teleology.

8. Problems of causality. These two can be discussed together. Systems theory in the form of functionalism has been criticized for being teleological, or discussing "purpose" or even "divine purpose" (see Turner and Maryanski 1979). For these reasons, many functionalists took pains to avoid the term "purpose" (see Merton 1949; Davis 1959). Miller (1984) reported that extended intellectual discussions of systems theory inevitably turn to the notion of teleology, which he regards as an unsolvable philosophical or metaphysical issue that is out of place in a scientific ("positivistic") systems theory. The truth is that nothing in the systems definitions presented above requires or even hints at teleology (see also Bertalanffy 1967). This problem is introduced with functionalism, primarily with functions designed to produce equilibrium. I see no reason for a modern systems theory to be vulnerable to charges of teleology.

The problem arises when the existence of an institution (part) is explained by its *function* for the society (whole). Such whole-part analysis is not typical of systems analysis, but is somewhat unique to functionalism (see Turner and Maryanski 1979). Modern systems analysis portrays the system as a set of simultaneous differential equations, as

$$X_1 = f(X_2, X_3, X_4, \ldots, X_n) \tag{2.1}$$

$$X_2 = f(X_1, X_3, X_4, \ldots, X_n) \tag{2.2}$$

$$X_3 = f(X_1, X_2, X_4, \ldots, X_n) \tag{2.3}$$

$$X_4 = f(X_1, X_2, X_3, \ldots, X_n) \tag{2.4}$$

$$X_n = f(X_1, X_2, X_3, \ldots, X_{n-1}). \tag{2.5}$$

Here causation is mutual—all variables *cause* all others. Each is in turn a dependent and independent variable. But simple functional models portrayed the variable to be explained (E) as serving a function (F) which obviously could not be served until E existed, so that

$$E \rightarrow F \tag{2.6}$$

with F the reason for E. This seemed backward to traditional positivists and causal theorists within sociology who say that E should be explained in terms of a prior *cause* (P), or

$$P \rightarrow E \tag{2.7}$$

(see Dore 1967).

Dissatisfaction is further exacerbated by the lack of understanding of how or why the process occurs. This is ameliorated by modern concepts of feedback in sociocybernetics, so again the problem is obsolescence. Further, Stinchcombe (1968) showed how to diagram the functional process in a feedback loop which can be interpreted in terms of causation. All of these developments (feedback, causal interpretation, simultaneous differential equations) all defuse the charges of teleology and improper use of causality.

Ideological. The major ideological charge has been conservatism (see Lilienfeld 1978). As corollary problems, Lilienfeld sees scientific pretension, abstraction, and love of analogies; he also finds systems theory highly speculative. The criticism of conservatism and systems theory as an ideological consciousness stem primarily from two main factors in functionalism: dominance of the status quo, and determinism or impersonality stemming from the relative neglect of individual action. Both of these are due rather directly to equilibrium, and these charges are groundless in the contemporary approaches discussed here. Thus, the criticisms are now as obsolete regarding the new systems theory as is the work which engendered them. The charges of conservatism, authoritarianism, and so forth, are unfounded in nonequilibrium models such as social entropy theory which give prominence to process and to individual action, and which in a real sense challenge the status quo (including the status quo in sociological theory) rather than supporting it.

A Positive Positivism

Nineteenth-century sociology can be characterized by Durkheim's term "mechanical solidarity." The discipline was much more holistic and unitary than today, and the major journals could accommodate most perspectives satisfactorily. Indeed, as late as the late 1930s, all delegates to the American Psychological Association annual meeting in Ann Arbor, Michigan, could be included in a singe group photograph (Miller 1984). Psychology graduate students could then realistically be expected to know all of the English literature in their field, and most of the German and French (Miller 1984). This is about the time that the systems theory that most sociologists are familiar with (structural functionalism) achieved its primary form.

Now Durkheim's organic-solidarity stage is fully upon us. Sociology and related fields are widely specialized and divided (if not fragmented). Many of these specialities seemed to have exhausted their fertility and have peaked, if not stultified. We may well be moving rapidly from the age of specialization into a third new stage—the age of integration (see figure 2.5).

The age of integration is the systems age. It is characterized by badly needed integrative frameworks such as systems theory or the micro-macro problem (see Ritzer 1988, pp. 484–518). Today we

FIGURE 2.5.

Three Stages of Disciplinary Development in Social Science

Mechanical	Organic	Integrative
(Relative Homogeneity) →	(Relative Fragmentation) →	(Synthesis)
1850–1950	1950–1990	1990–?

see a situation where the same symbol or term is used for a number of different purposes, all with different meanings. For example, the *H* statistic in information science has a number of different interpretations, as does *Q* (see Bailey 1975, 1984). The *H* statistic is variously interpreted as information, entropy, uncertainty, and surprisal (see Bailey 1983). The modern systems movement is dedicated to increasing scientific efficiency and the cumulation of knowledge—both between and within disciplines—through the amelioration of such inconsistencies or anomalies. Systems theory will make scholars aware of these problems.

But while the development of the division of labor did not eradicate mechanical groups, the development of integrative solidarity will not erase the two prior forms (mechanical and organic). Indeed, Miller (1984) estimates that only 5 percent of scholars in a university should be in integrative (for example, systems) research, while the other 95 percent should remain in various specialties. The issue can be put succinctly: social phenomena display both similarities and differences. Specialization often emphasizes the *differences* in subject matter, and pits one specialty against another to some extent, or at least separates the specialties to the point where it is difficult for one scholar to master very many, as each has its own literature, jargon, methods, and so forth.

Virtually *all* scholarship at the present time (probably 99 percent) is devoted either to the study of differences (through division into disciplines and subdisciplines) or to the study of similarities within very narrow confines. Ironically, even these narrow disciplines and subdisciplines are now in many cases facing the specter of information-input overload (Miller 1978). Unless some of the plethora of research now being generated can be synthesized, it will likely be lost to all but a few, and cannot be used effectively. The modern systems movement seeks to rectify this. It has been developing since the 1950s. This is ironically just about when functionalism peaked. Most functional systems theory predates modern systems analysis and did not profit form it, thus its obsolescence.

The modern approach to social systems theory abhors labels, and for good reason—labels divide while systems theory seeks to integrate. Labels are often counterfeit type-concepts, with a name but no content. However, since any systems approach will inevitably be labeled (often as structural-functionalism, see Ritzer 1975, p. 67), I will label it here before others get a chance to.

Social systems theory defies extant labels because it is such a broad approach. It is positivistic in some respects. It embraces quantification, adopts the scientific canon of verification, and seeks links with other disciplines, including natural science. It is *not* positivistic in the traditional sense in several ways. It embraces metatheorizing, and is often very abstract. It also explicitly includes the study of subjective phenomena such as values and perceptions (see level X of figure 2.2) and embraces qualitative as well as quantitative approaches (although some parts of systems theory such as living systems theory explicitly exclude the subjective as nonscientific). It is macro, but also emphasizes individual action within the macro context.

Such a positive positivism is much less judgmental and exclusionary than most specialities within sociology. Systems scholars are often free to pursue rather idiosyncratic research. Positive positivism seeks the best of all worlds. This is difficult, but potentially rewarding. By casting our net wide, we avoid exclusionary anomalies stemming from narrow models which simply lack sufficient breadth to prove efficacious (see Bailey 1987). Anomalies can often be identified by looking for antagonistic words such as "versus." They abound in social research—"micro versus macro," "structure versus process," "qualitative versus quantitative," "theory versus method," and so forth. Such dualities are the epiphenomena of divisive "specialized" research which cuts boundaries that often avoid or cut across important intellectual issues. These anomalies are virtually unknown in an adequately broad model. Such a model seeks to avoid the blocks found in earlier social systems models, principally equilibrium-functionalism. Modern analysis is exemplified by SET (social entropy theory), which seeks to *include* rather than *preclude*.

Of course there is a cost to avoiding narrow hyperspecialization. The cost of avoiding blocks is to provide a skeletal framework which is of such generality that it neglects detail, and this is perhaps objectionable to specialists who may spend virtually their whole lifetime on the fine points which systems theorists quickly gloss over in passing. However, if truly nonblocking, the general model allows for a wealth of detail to be filled in later by those who desire to do so. There is nothing wrong with division and detail unless they lead to intellectual paralyzation through fragmentation (Rossi 1980) and de-emphasis of the "big picture."

A properly constructed (semi-)positivistic panorama will provide the overview needed for proper problem selection in sociology. This positive positivism is based on a nonequilibrium analysis of concrete systems. Parsons (1979) seemed to fear concrete analysis because he felt it led to dangers of biological and physical reductionism, with subsequent de-emphasis of the salient social and cultural factors central to sociology. He simply desired an effective sociology. The problem is that his analysis is so free-floating as to be of limited utility as an integrative tool for the 1990s.

Consider an analogy. Suppose you go to a hot-air balloon festival and observe two balloons. One is tethered securely to the ground, and the other is untied and beginning to float without occupants, but its tether is within reach. Your reaction will most likely be to tether it so that it is accessible. Parsons's free-floating systems (biological, psychological, societal, cultural) are difficult to operationalize, or to apply simple systems concepts such as "boundary" to. This again is an unnecessary "block," or preclusion of valuable lines of inquiry. By studying the holistic concrete system in its physical environment, one has access not *only* to the abstracted variables Parsons favored (which is important), but to other valuable intellectual pursuits as well. Referring to figure 2.2, positive positivism emphasizes study of the isomorphism between the empirical concrete society (X'), our perception of it (X), and our models (verbal, mathematical, computer simulations and so forth) of it (X'').

Such a positive positivism opens up opportunities rather than closing doors. It allows us to select from an array of intellectual problems involved in truly understanding how society operates on a day-to-day basis. This is the mission of sociology. A positive positivism allows us a wide array of tools. Abstracted analysis is easy to get to from a concrete grounding (see Bailey 1990). Further, the concrete systems model facilitates access to the natural sciences, allowing us to use valuable tools from biology, physics, and chemistry when desired. This moves us one step closer to fulfilling Comte's [1830–1842] prophecy of sociology's role in the community of science.

COUNTERPOINT

We have seen in chapter 1 some of what mainstream theory has to offer, and a hint at what the systems approach can offer. These issues are covered in more depth in this chapter. This chapter demonstrates in more detail the depth and richness of the systems approach. Among the tools not available in extant mainstream theory are the Q-R and three-level distinctions, as well as the typology of system types. These tools will be used throughout the volume in developing the contribution of systems theory.

One of the chief contributions of this chapter is the discussion of blocking. This discussion begins the reconstruction of the systems perspective. I noted in chapter 1 that, like neofunctionalism, the new systems theory had experienced elaboration, revision, and reconstruction, as discussed by Alexander and Colomy (1990). The successful reconstruction of the systems perspective hinges on a critique of equilibrium, and a reconstruction in terms of nonequilibrium dynamics. This is continued throughout the volume, and enables the new systems theory to share a place in the third postwar phase of theory development (the first phase was functionalism, the second was the rise of micro theory, and the third is the new synthesis—see Alexander and Colomy 1990).

Another contribution of the chapter is the distinction between concrete systems and abstracted systems. It is not size nor scope per se which makes the system concept subject to hypostatization, but rather how the concept is defined. The new systems theory, unlike Parsonian theory, uses a concrete system to begin with. This largely frees us from the problem of hypostatization, and allows us to study "real people" as Collins (1975) advocates. The emphasis on boundary is also important in avoiding hypostatization. The new systems approach also explicitly introduces time and space into the analysis, and so is in line with Giddens's suggestions for theory. Further, the typology shows a breadth and richness not heretofore available in extant mainstream theory.

In contrasting my definition of systems with that presented by Giddens (1979, p. 66), notice that mine is much more general and subsumes his. Mine emphasizes entropy, stresses relationships, and explicitly recognizes boundaries. Giddens's definition neglects boundaries and has no mention of entropy or other measures of system state.

To compare definitions:

Bailey: A system is a bounded set of interrelated components that has an entropy value below the maximum.

Giddens: "[A system is] reproduced relations between actors or collectivities, organised as regular social practices." (Giddens 1979, p. 66).

Notice that Giddens, by stressing relations between actors, is defining a concrete system rather than an abstracted system in Miller's (1978) terms. Thus, he is departing from Parsons's (1979) tradition of using the role as the basic systems unit, and is seconding the tack taken by Miller and myself.

Notice also that the two definitions (Giddens and Bailey) are quite compatible. Mine is much more general, and both apply only to living systems. Further, the reproduction that Giddens includes in the system definition is stressed by Bailey (1990) in the application to social systems, but is not specifically included as part of the basic, and more general, definition. In other words, I accept Giddens's definition of system and consider the two definitions to be compatible and complementary.

As for metatheorizing, a number of Ritzer's forms are evident here. Inasmuch as this chapter (particularly the critique of equilibrium and the typology) is a prelude to the development and synthesis of the new systems theory, the analysis is M_P (prelude to the development of sociological theory) in Ritzer's (1990a, p. 18) terms. The first variant of M_U (the internal-intellectual approach) is also evident in the analysis of equilibrium and in the use of the Q-R and the three-level models. The typology might also be labeled M_O if it is considered to be sufficiently "overarching" (see Ritzer 1990a, p. 18).

CHAPTER 3

The Age of Equilibrium

The age of equilibrium was glorious. Systems not only existed, they flourished. A system was not only a collection of parts, but was whole. This holistic quality was irrefutably demonstrated in the existence of equilibrium. In separate proximity the parts were mere parts. When related they were a whole—but a mere whole as a sum of the parts. However, when the system was in equilibrium, not only was it in a glorious state of stability, it also possessed a quality that no part alone could ever possess. Only systems have equilibrium, as it is a group property not an individual one. Equilibrium is thus an emergent phenomenon, it exists above and beyond the mere sum of parts and their interrelationships, and thus should prove to skeptics once and for all that the whole truly is greater than the sum of its parts.

The age of equilibrium lasted roughly from 1850 to 1950. It was glorified as it was born, and was vilified as it waned (is this not a classic case of the paradigm of life—entropy decrease followed inevitably by entropy increase?). The age of equilibrium was born out of thermodynamics, particularly the study of heat flows through liquids (such as water) and gases. Before it reached its present state of existence it was destined to travel from the physics laboratory through many academic disciplines not only in the natural sciences but also in the social sciences.

Interesting enough, this hardy concept was not even mentioned in the basic canons of thermodynamics—the first and second laws.

First Law: Energy can neither be created nor destroyed.

Second Law: The entropy of the world increases to a maximum.

Not only is the concept of equilibrium missing, but the concept of entropy is prominent. Rather than being an interloper introduced in the twentieth century to compete with equilibrium for the attention of systems theorists, in fact entropy was there all along, displayed more prominently than the essentially derivative concept of equilibrium.

THE APPEAL OF EQUILIBRIUM

It has been speculated that one reason for the appeal of equilibrium in sociology was Durkheim's anxiety over the French Revolution. This may have been so traumatic for him that he turned to a theory stressing social integration (Johnson 1975) and others, notably Parsons, followed him in stressing this concept. The French Revolution may well have influenced Durkheim, and thus his followers through him, but if so it was only one of many currents of equilibrium thought flowing in social science as well as in physical science.

To fully understand the appeal of equilibrium and the breadth of its influence, we must deal with the confusion surrounding it. The chief problems with equilibrium in sociology are easily stated: it has never been properly operationalized; in its physical definition it is not necessarily synonymous with integration; it is respectively an heuristic, a construct, and an empirical notion; and it is subject to reification. This maze of complications is best understood by repeating figure 2.2 for equilibrium, as in figure 3.1.

Perceptual (X)

Equilibrium as a perception or construct (X) is glorious in the eyes of the beholder. It conjures up grand notions of stability, peace, order, harmony, smooth relationships, etc. In contrast, the perception of entropy has quite negative connotations for many persons (death, disorder). A colleague of mine suggested quite seriously that perhaps the reason that Boltzmann (originator of the basic equations for statistical entropy) committed suicide was that working with the concept of entropy made him so gloomy. Equilibrium, however, has the generally positive connotation of balance (in lay terms). The appeal of equilibrium on the perceptual level (X) seems

FIGURE 3.1.
The Three-Level Model as Applied to Equilibrium

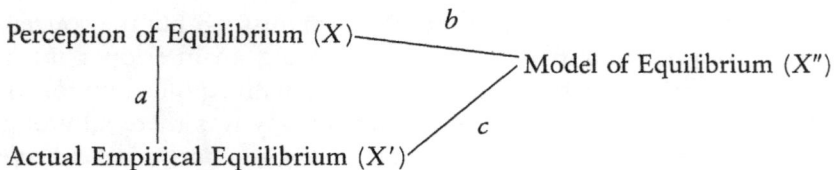

Perception of Equilibrium (X)———b———→ Model of Equilibrium (X'')

a

c

Actual Empirical Equilibrium (X')

unambiguous until one realizes that this popular connotation is in conflict with "true" equilibrium in thermodynamics, which is maximum entropy or systems death.

Empirical (X')

Equilibrium on the empirical level (X') is more difficult. Since equilibrium theorists never provided rigorous means for identifying the variables involved in equilibrium nor for measuring them, it is doubtful that the average person could ever "observe" the existence of equilibrium. What are the ramifications of this? Does this mean that equilibrium can never exist at the empirical level (X'), but only as a perception (X) or model (X'')? Or does it mean that empirical equilibrium is real, but is difficult to observe? A reading of either Pareto (1935) or Parsons (1951) shows that both considered equilibrium to exist empirically. This could probably be illustrated by a society lacking revolution or rapid social change. Even with such events, equilibrium would apply, as the society could be said to be in a transition to a new equilibrium (Parsons 1961b).

Social equilibrium was clearly an empirical phenomenon for Pareto (1935); see also Russett (1966). In fact, he chided statisticians such as Mill for not realizing that the return to equilibrium was the nature of the empirical phenomena they were observing. Specifically Pareto said that short wars waged by rich countries and mild earthquakes would soon subside, occasioning a return to equilibrium. The devastation of an earthquake and the relative calm after it were empirically observable.

However, a chronic source of confusion has always been the fact that many equilibrium theorists, including both Pareto and Parsons, always used the empirical level of equilibrium (X') in concert with the parallel heuristic model of equilibrium on the model level (X''). The vexing thing about this was that they often used the heuristic level as a sort of a safe haven into which they could retreat for safety when criticized on empirical grounds, only to run back to empirical pastures as soon as the critics had gone. This is clearly seen in Parsons's writing. When his use of equilibrium was criticized, he retorted that it was only a model or assumption, and did not need to hold exactly in the empirical case (Parsons 1951, p. 481). Yet, it is clear from Parsons's definition of system cited above that equilibrium is meant to hold empirically, and in fact does so by definition, as it is made synonymous with social order (Parsons and Shils 1951, p. 107). If one agrees with Parsons's definition of system, he or she is accepting the empirical notion of equilibrium. Pareto also vacillated between the X' and X'' levels, as discussed by Russett (1966).

Model (X")

The primary use of equilibrium in most disciplines has been as a model, generally a mathematical model. This is true in thermodynamics where equilibrium exists in an isolated system when entropy (S) is at a maximum, and there is no change in entropy $(dS = 0)$. This is the *basic definition of equilibrium,* from which all others have been derived in some sense. For a system with given levels of the extensive properties of energy and volume, equilibrium will only occur when entropy is maximized. Other equations can be written to describe equilibrium conditions for the respective variables for a given constant (nonmaximum) level of entropy. However, mathematically these equations are all equivalent.

It is crucial to note that the equilibrium state $(dS = 0)$ is maximum entropy or system death—*not* system "integration." It is maximum disorder, and so is the opposite of integration. While maximum entropy or equilibrium generally is not found empirically (it would be a disaster—literally death), it is a marvelous mathematical tool. It serves as the ultimate criterion point, and a perfect tool for deriving the exhaustive set of mathematical relationships among all variables in the system.

According to classical thermodynamics, isolated systems will tend towards equilibrium (the second law). Also, equilibrium can be conceived theoretically in terms of reversible processes, such as heat flowing from hotter to colder bodies (empirically valid or a natural process) *reversed* by an equal amount of heat returning, or flowing from colder to hotter bodies (not empirically possible, thus "unnatural"). The reversible system *only exists* on the X'' level (or X level) *but never empirically* (the X' level). However, as a theoretical and mathematical heuristic tool it is invaluable. Physicists study thermodynamic systems by imagining what would happen to entropy levels if a small amount of heat were added to the isolated system. The mathematical relationships can then be traced. *Notice that this is a logical contradiction. By definition isolated systems have impenetrable boundaries, so even a small amount of heat cannot be added empirically.*

However, such logical fallacies form the crux of the modeling procedure in the natural sciences and in some social sciences (for example, economics). The scientist says, assume this could be done (e.g., heat could be added to the isolated system), then what would happen? This is essentially the "if-then" form of the ideal-type strategy proposed by Weber (1947, pp. 83–84). Pareto consciously proceeded from this heuristic modeling procedure (X'') that he had studied in thermodynamics to the empirical case (X') in sociology. If the thermodynamic theorist is working with an isolated system, the system could not tolerate a massive violation by a large heat transfer across system boundaries, but a small "hypothetical" violation is tolerated.

What is the empirical analogue to this "hypothetical" disruption of equilibrium from outside forces? It is an empirical disturbance, but only a "small one," and of "short duration." This is how Pareto moved from "hypothetical" disturbances in physics that were purely heuristic (X'') to empirical (X') disturbances of equilibrium in sociology (such as short wars in rich countries). This is clear in the following quote from Pareto (1935, p. 1436):

> Somewhat similar to the artificial changes mentioned are those occasional changes which result from some element that suddenly appears, has its influence for a brief period upon a system, occasioning some slight disturbance in the state of equilibrium, and then passes away. *Short wars waged by rich countries, epidemics, floods, earthquakes and similar calamities would be examples. Statisticians long ago observed that such incidents interrupt the course of social or economic life but briefly;* yet many scientists, who have worked without the concept of equilibrium have kept meandering about in search of imaginary causes. Mill for one, wondered why a country afflicted for a short time by the curse of war soon returned to its normal state. (italics added)

Note both the short duration (as in physics) and the small impact (short war in a "rich" country). Thus, Pareto subtly made the verbal transition (not to mention translation to English) from thermodynamic hypothetical model (X'') to sociological empirical entity (X'). This is a long journey to be made in a few paragraphs, and its significance probably escaped most readers entirely. I leave it to the reader to judge whether Pareto's empirical equilibrium is real or merely a reification (see Russett 1966).

As an heuristic model (X''), the equilibrium model has no parallel. It is the criterion sine qua non. Consider the heuristic in economics. Price equilibrium is where the supply and demand curves meet. Such an heuristic device is probably most familiar to sociologists as an analogue to the ideal type (Weber 1947, 1949). Kuhn (1979) says equilibrium does not exist empirically, but empirical cases "approximate" it, and so it can be used to understand them. The degree of deviation of empirical cases from equilibrium can be studied. But how is this so? How can one compare an heuristic (hypothetical) case on level X'' with an empirical case on level X'? This is a question which has long stymied students of the ideal type. The answer is simple enough—simply identify the empirical counterpart (X'_e) of the equilibrium point (X''_e), and see how far the empirically observed point (X'_o) departs from the empirical point (X_e'), or

$$X''_e$$
$$|_c$$
$$X'_e \quad \overline{\quad d \quad} \quad X'_o$$

(measure distance d).

To summarize, equilibrium was initially conceived in thermodynamics as a property of closed (technically "isolated") systems. It was *never meant* to be applied to open systems such as social systems. Further, it is in most instances heuristic (level X''). Still further, "ultimate" thermodynamic equilibrium consists of realization of the second law, meaning that maximum entropy exists and the system is dead. This is the opposite of the "integration" that most sociological equilibrium theorists perceived (at the X level) for their concept. Thus, there is a real juxtaposition between the three levels—X is perceptually harmony to sociologists, X'' is the most valuable for most disciplines, and X' is generally a myth. One must be very careful not to mix the heuristic X'' and the empirical X', or to subtly slide from X'' to X' as Pareto did. A common form of this is reification. We will now turn to examination of equilibrium in more detail through focus on the work of major equilibrium theorists. Some were natural scientists and some were social scientists, but all had an impact on the notion of social equilibrium.

SOME MAJOR EQUILIBRIUM THEORISTS

> Nobody ever questioned it [equilibrium].
> —James Grier Miller 1986

Clausius

Clausius (1879) was instrumental in developing the thermodynamic theory leading to the second law. However, a cursory reading of Clausius finds virtually no mention of equilibrium (Bailey 1984a). Instead, his emphasis was on entropy (*S*). The second law was stated in terms of entropy, *not* in terms of equilibrium. Equilibrium came about if entropy ever *did* come to a maximum under the second law.

This would be the "ultimate," "true," or "maximum" equilibrium. It would subsume or render moot all of the other forms of physical equilibrium that have been conceived (e.g., "static," "stable," "rotational," etc.). All elements of the system would be "balanced," and at rest. There would be no internal pressure to disrupt the equilibrium, and technically no outside pressure as well (as the system is "isolated" in the thermodynamic sense, meaning that salient external forces cannot impinge upon it). This state of rest is the most "probable state" in statistical terms. From every viewpoint it is the ultimate equilibrium, and most likely nothing will disturb it.

The theoretical problem for social scientists is that this is not their treasured state of integration at all, but is the opposite, the maximum state of disintegration (maximum entropy). Under the condition of maximum entropy, the system has exhausted its internal energy (and information) and is dead. Unless some new infusion of energy from external sources could be found (impossible under the definition of an isolated system), there is no hope for revitalization. As stated above, this is the state where *S* is maximum and $dS = 0$. The "equilibrium" represented by the lack of change in entropy is null in the sense that the system is dead. While equilibrium equations can be written among the other properties of the thermodynamic systems (pressure, volume, temperature, etc.), they are *all* mathematically equivalent to the maximum entropy condition (that is, they only hold under the condition that S = maximum), and so they are just other mathematical ways of saying the same thing.

Spencer

Spencer was aware of the first law of thermodynamics but not the second. He worked with both of the reigning intellectual concepts of the nineteenth century—equilibrium and evolution. These constitute together a sort of dialectic, but it is difficult to tell in such a "chicken or egg" problem, just which is the thesis and which the antithesis. It would seem logical that evolution was considered the thesis, and this process of change spawned the antithetical structure of equilibrium. Equilibrium was generally referred to as "equilibration" by Spencer (1892 [1864]) thus giving it the connotation of process and thus lessening perhaps its degree of antithesis with the evolutionary process.

Spencer used a number of types of equilibration. These, along with the work of the major equilibrium theorists, are spelled out in detail elsewhere (Russett 1966; Bailey 1984a, 1990). Spencer's first form of equilibrium ("stable" equilibrium) is of little interest. Based on the first law (conservation of energy) it uses the analogy of a projectile (e.g., spear) thrown into the air, which as it loses energy, will fall to the ground and lie motionless (stable equilibrium). This is of little theoretical interest for sociology. His several concepts of moving equilibrium are much more theoretically sophisticated. The cessation of movement straight ahead, as in the example of the thrown spear, will result in "static" equilibrium when motion ceases. An object with its center of gravity at rest at its lowest point is said to be in "stable" equilibrium. Spencer's first type of equilibrium exemplifies both. His second form (moving equilibrium) incorporates both straight-ahead and rotational motion.

Spencer's third form (moving equilibrium) is perhaps his most interesting form, and is well described by this passage:

> Any system of bodies exhibiting, like those of the Solar System, a combination of balanced rhythms, has this peculiarity;—that though the constituents of the system have relative movements, the system as a whole has no movement. The centre of gravity of the entire group remains fixed. Whatever quantity of motion any

member of it has in any direction, is from moment to moment counter-balanced by an equivalent motion in some other part of the group in an opposite direction and so the aggregate matter of the group is in a state of rest. Whence it follows that the arrival at a state of moving equilibrium is the disappearance of some movement which the aggregate had in relation to external things, and a continuance of those movements only which the different parts of the aggregate have in relation to each other. Thus, generalizing the process it becomes clear that all forms of equilibration are intrinsically the same; since in every aggregate, it is the centre of gravity only that loses its motion: the constituents always retaining some motion with respect to each other—the motion of molecules if none else. (Spencer 1892 [1864], p. 488)

Notice that his model has all the basic elements of classical equilibrium. The parts of the system are all interrelated, as if held by rubber bands. Any external force which disrupts the equilibrium by moving one component out of position would put strains on the other components (like stretching their rubber bands), thus causing them to pull the displaced part back into position as soon as the stress ceases, and thus restoring equilibrium. Spencer's description of the internal workings of the equilibrium system is very similar to that of Le Chatelier's principle (1888) formulated sometime after the publication of Spencer's work in *First Principles*, and also to Cannon's (1929) concept of homeostasis, formulated approximately seventy years later.

Spencer conceived of a series of evolutionary adaptations, each resulting in a progressively higher level of moving equilibrium. The ultimate or equilibrium state would be utopia (Spencer 1892 [1864]). Thus, we see Spencer attempting to reconcile the synchronic equilibrium model with diachronic analysis a full century before attempts to reconcile functionalism and conflict theory by van den Berghe (1963) (see Ritzer 1983, pp. 232–34).

Spencer apparently believed his theory was rigorous. He was devastated to be told by Professor Tyndall at a dinner party that his "ultimate equilibration" was maximum entropy according to the second law (Duncan 1908, vol. I, p. 136). His equilibrium was not integration but devastation. How could this be reconciled so that his basic model could be saved? This is the Spencerian dilemma. He was devastated by this awful theoretical news. He was still writing *First Principles* at the time (1850s), and tried to reconcile the dilemma. He acted as though he had solved the problem in *First Principles* (1892 [1864]), but the matter remains starkly unresolved (see Russett 1966; Spencer 1892; Bailey 1984a). Unfortunately, some of his successors failed to learn from history, perhaps because of claiming that "Spencer is dead," as quoted by Parsons (1937 p. 3). Thus, they were doomed to repeat Spencer's steps, and to also become mired in the Spencerian dilemma by thinking that thermodynamic equilibrium meant integration (instead of the opposite).

Gibbs (1882)

J. Willard Gibbs is probably the best-known American-born scientist of the nineteenth century. Although little acclaimed during his lifetime (Wheeler 1951), his work was widely acclaimed after his death and, though largely mathematical, it had important empirical applications. Gibbs's chief contribution was to move equilibrium from the relatively simple heat baths and gas molecules of the thermodynamic theorists to the more complex world of physical chemistry. He moved equilibrium from the homogeneous to the nonhomogeneous. The bulk of his acclaim comes from his famous paper "On the Equilibrium of Nonhomogeneous Substances." The work was brilliant and of enduring value, but poses a trap for the unwary. One well-known contribution is the "prime equation" (Wheeler 1951)

$$dE = TdS - PdV. \tag{3.1}$$

This is essentially a mathematical combination of the first and second laws of thermodynamics. It can be used to generate

$$dS = q/T \tag{3.2}$$

where dS = change in entropy; q = heat added to the system; and T = absolute temperature of the system.

The problem with it is that it says that entropy increase (dS) is directly proportional to the amount of heat added (as a ratio of the absolute temperature). In the general case, *this is not true*. It is true *only for the equilibrium state*. Gibbs assumed this, of course, because his work was generally confined to equilibrium, so it caused no confusion. However, in some books (see Lewis and Randall

1923, p. 133) one may read several pages without seeing the statement that equation 3.2 holds only for the equilibrium case. Even then the reader may not realize that this means that

$$dS = 0 \qquad (3.3)$$

(see Lewis and Randall 1923). This means that *empirically speaking,* the analysis of equations 3.1 and 3.2 is a hypothetical analysis or an imaginary exercise. Applying equation 3.3 to equation 3.2, we see that

$$dS = q/T,$$

but only when

$$dS = 0,$$

so that

$$0 = q/T = 0.$$

We are in effect analyzing nothing at all, and it has just that much significance empirically (although it is theoretically very significant).

Referring to figure 3.1, we see that the problem lies *not* in the physical analysis or the prime equation as devised at level X'' (a mathematical model). Equilibrium has proven invaluable as a model. It serves at least three functions as a mathematical model: (1) it serves as a mathematical criterion point; (2) it allows the deduction of an exhaustive set of mathematical relationships (all relationships among all variables—entropy and temperature, volume and pressure, etc.); and (3) it simplifies the complexity of the mathematical analysis, much like the linearity assumption or recursiveness assumption greatly simplify the mathematics of path analysis for sociologists (see Duncan 1966).

Within level X'' (the model level), the analysis is precise, elegant, and valuable. It is only when it is misplaced from level X'' (model) to level X' (empirical) of figure 3.1 (along path c) that problems occur, and these are consequential. This sort of displacement, although common, is *entirely illegitimate,* because by definition $dS = 0$ and $S =$ maximum. This *cannot be found empirically,* as it is system death. Analysis which has illegitimately displaced equation 3.2 to the empirical case, such as that by Odum (1983), has not surprisingly reached logically contradictory results in which both low entropy and high entropy simultaneously are said to represent "order," and this is impossible. Odum (the son of regional sociologist Howard Odum) thus concludes that the central concept of "order" should be abandoned (Odum 1983, pp. 315–16). It is actually the illegitimate displacement of the equilibrium model along path c that should be abandoned, and not the crucial notion of order.

Odum's plight offers a good example of the disciplinary pitfalls of what some ethnomethodologists call shared understanding (Garfinkel 1967). If one reads the physical chemistry book from the first page on, or is socialized into the shared understanding that $dS = 0$, then the misapplication (path c of figure 3.1) will not occur. Odum apparently did not share the basic assumption ($dS = 0$), or had forgotten it. There is no way around the problem. One might think that it can be escaped by looking for cases of $dS = 0$ where S is below the maximum, and thus empirically possible ($S =$ maximum is empirical death). This neglects the fact that dS *cannot* be 0 if one wishes to analyze both change in S (dS) and added heat (q) as Odum (1983) does. Any way you look at it is illegitimate.

Pareto

Pareto was a sophisticated student of thermodynamics who was familiar with both the first and second laws, and adroitly avoided the "Spencerian dilemma." He was enamored of Gibbs, aware of the work of Clausius, and probably aware of Boltzmann's work on statistical entropy. I will present a brief synopsis of the context for Pareto's work. Clausius (1850) had studied thermodynamics on the empirical level (X'), modeling it mathematically on the indicator level (X''). Clausius had hardly mentioned equilibrium, but had studied entropy exhaustively. Gibbs (1874–1877) had worked primarily with the model level (X''), extending the concept of equilibrium from homogeneous to nonhomogeneous substances. His primary focus was on equilibrium, but he had used the entropy concept where it was useful (as in the prime equation). We also realize that perceptually (X) the connotations of balance, harmony, and even justice (e.g., discrepancies or lack of balance will be righted) that are commonly attributed to equilibrium are much more pleasant than the destructive gloom of entropy.

What concept would we expect Pareto to emphasize? Equilibrium of course, and equilibrium it was. Although his concept of equilibrium was sophisticated, and sound (he learned it as a student of thermodynamics), he was in a curious position compared to other equilibrium theorists. The classical thermodynamicists generally derived their models (X'') from empirical observation (X'), though to be sure they then derived hypothetical conditions and used assumptions not found empirically. Nevertheless, their work was from X' to X'' (path c of figure 3.1). Gibbs worked primarily with X'' although his work was later applied (from X'' to X', path c). The later homeostatic theorists such as Cannon (1929) also did exhaustive empirical studies (X') of homeostasis in living organisms, and used these to derive the model (X'').

Thus, while both thermodynamic equilibrium theory before him and biological homeostatic theory after him were "grounded theories" (Glaser and Strauss 1966) which proceeded from X' to X'' and could be termed inductive, Pareto began with the model (X'') of equilibrium that he borrowed from the thermodynamic theorists. As we have already seen, he knew that equilibrium was only a model (X''), and he knew how models were used to generate theory in thermodynamics (e.g., by assuming that a small amount of heat could be added through the boundaries of an isolated system, although definitionally this was impossible). In a way, thermodynamics set a bad example for Pareto to follow by allowing the utilization of assumptions which not only violated the definition of an isolated system, but also contradicted empirical reality (such as the use of "unnatural systems" such as heat flowing from cooler to hotter substances, which is empirically impossible). This gave Pareto a sort of license that was validated by physical science. I leave it to the reader to judge whether this is positive positivism of the sort we seek. Pareto was somewhat reluctant to admit that he used equilibrium empirically, but this is clear, as in the quote already provided (Pareto 1935, p. 1436). In that passage he refers, for example, to short wars in rich countries. This is clearly an empirical reference.

Russett (1966) also comments on this. Pareto took equilibrium for granted. He reasoned that if the forces responsible for bringing the equilibrium state about originally were still in force (and he assumed that they were), then if equilibrium were disturbed (from outside), the forces still in play would soon reestablish it. Pareto's equilibrium exemplifies three unorthodox and perhaps untenable extensions of prior usage:

1. He used an isolated-system concept for open systems (societies).
2. He applied a theoretical model (X'') to new empirical situations (X'), thus *reversing* the direction of flow along path c (from X' to X'') taken by the original thermodynamic theorists during their inductive development of the model of equilibrium from the empirical observation of heat (water) baths.
3. He used empirical boundary openings ("short" and "small" deviations) as analogues of theoretical or hypothetical boundary deviations in physics.

All of these unorthodox extensions resulted in an untried equilibrium concept that had not earned the positive connotations of perceptional stability (X) and scientific respectability (because of so many deviations from the original) that its adherents coveted.

Depending upon how one counts, Pareto made as many as six changes from classical thermodynamic equilibrium. He extended it empirically (1); he extended it theoretically (2); he shifted from model (X'') to data (X'), instead of from data (X') to model (X'') as had previous theorists (3); he used the empirical disturbances in place of theoretical (hypothetical) ones (4); he used a largely verbal analysis instead of a mathematical one (5); and he analyzed an empirical phenomenon (society) for which the connotations of equilibrium as stability assumed the role of value judgments (X'') (6).

In terms of value judgments, it makes little difference when dealing with gas molecules whether one is discussing entropy or equilibrium. In thermodynamics, the analysis was driven by the indicator level (X'') in terms of modeling considerations, or alternatively by the empirical level (X') in terms of empirical concerns. In neither case were value judgments a guiding force. In contrast, in sociology, the connotations of equilibrium such as "stability," raise the specter of the use of a scientific concept being directed not primarily by its theoretical, mathematical (X''), or empirical properties (X'), but by its conceptual connotations (X) in the form of value judgments such as balance and harmony. Value judgments thus raise their head in a significant way in a supposedly "positivistic" analysis (not to mention the ever-present danger of reification).

Le Chatelier

Le Chatelier's principle (1888) was somewhat revered in science and still is well known (see the analysis in Miller 1978). This principle, as translated by Lotka (1925, p. 281), is as follows:

> Every system in chemical equilibrium, under the influence of a change of every single one of the factors of equilibrium, undergoes a transformation in such direction that, if this transformation took place alone, it would produce a change in the opposite direction of the factor in question.
>
> *The factors of equilibrium are temperature, pressure, and electromotive force,* corresponding to three forms of energy— heat, electricity, and mechanical energy. (italics in the original)

The similarity to Spencer's (1892 [1864], p. 488) earlier systems formulation, cited above, is striking. Although always the positivist, Lotka himself warned against extending Le Chatelier's principle beyond the specific applications for which he formulated it. Lotka (1925, p. 289) writes:

> On the whole, so far, it must be said that the result of a careful analysis of the principle of Le Chatelier yields negative results, so far as practical application to biological systems is concerned. The chief conclusion is that great caution must be exercised in employing this principle.

In spite of this caveat, the appeal of the principle seems irresistible for some social equilibrium theorists. It implies that whenever equilibrium is threatened, internal forces in the system will react so as to reestablish it. This is what social equilibrium theorists wanted to hear, and many of their formulations mirror it (see Russett 1966), though relatively few specifically quote Le Chatelier.

The dangers of extending this principle even to other physical phenomena other than the chemical equilibria for which it was formulated, let alone extending it to (unmeasurable) social phenomena, are made clear by cursory perusal of the properties of the forces that Le Chatelier is analyzing. Sociologists of the 1960s were spoon-fed the merits of deductive theorizing by Zetterberg (1965),

Homans (1964), and others. It may come as a shock to some of these would-be positivists to see that even the simple deductive syllogism that was touted to sociologists does not hold if extensive and intensive thermodynamic variables are mixed. For example, consider the following simple syllogism.

Axiom 1:   The greater the mass (extensive property), the greater the temperature (intensive property).

Axiom 2:   The greater the temperature, the greater the pressure (intensive property (i.e., the type given in Le Chatelier's principle).

Therefore

Theorem 1:   The greater the mass, the greater the pressure.

Even if the original two axioms could be assumed to be true, the deduced theorem would not necessarily follow, as is usually true for syllogisms, simply because extensive and intensive properties are mixed, and they have different qualities (see Lewis and Randall 1923, p. 13).

The lesson for extending thermodynamic concepts to social phenomena *which are not even operationalized* seems clear. One could easily commit all sorts of logical fallacies *and not even know it*. This caveat holds for our analysis of social entropy (Bailey 1990) as well. The difference is that we are using physical entropy, legitimately applied to physical phenomena (a la Comte) as a foundation for the analysis of social entropy. The latter is predominantly statistical, and is based on information-theory more than thermodynamics (although ultimately the second law still holds— deriving from the foundation of physical entropy).

Lotka

Lotka (1925) also championed equilibrium, and his work was read fairly widely by sociologists. His work is a good representation of his era.

Lotka offers a mathematical definition of equilibrium that is analogous to the definition in physics of a static state of equilibrium. For a set of variables X_i comprising a system, equilibrium exists when $dX_i/dt = 0$. This means that each variable remains constant in value over time (there is no change over time).

Lotka also uses the term "moving equilibrium extensively," but in a somewhat different fashion than does Spencer. He says that moving equilibrium exists when the variables are not constant, but are changing slowly. However, he assumes that the change is slow enough, that the equations for static equilibrium can be used for moving equilibrium. Lotka (1925, p. 260) acknowledges that, "Strictly speaking, this involves a contradiction. For if the velocities F are zero, the variable X cannot be changing." Lotka says further that this approximation of the values of X_i "represents a first approximation which is not in all cases free from significant error."

It is understandable that an evolutionary theorist would want to utilize a concept of moving equilibrium. Unfortunately, Lotka never offered a moving equilibrium model distinct from a static equilibrium model. To use the equations of static equilibrium to analyze moving equilibrium seems inherently unfruitful (not only empirically, but heuristically as well), and an inadequate way to analyze evolution. Lotka was doubtlessly motivated to use the concept of moving equilibrium by Spencer. He says (1925, p. 262), "Moving equilibria play an important role in evolutionary processes of the most varied type, as emphasized almost *ad nauseam* by Herbert Spencer." However, it is clear that Lotka's moving equilibrium model was simply a revised static equilibrium model and is thus similar to Spencer's first and second forms of equilibrium. It can be compared with Spencer's fourth form (independent moving equilibrium) which is characterized by cyclical motion, but should not be confused with Spencer's major model, his third-form dependent moving equilibrium which is similar to the homeostatic model developed later by Cannon.

His care in not carelessly extending the concept of equilibrium, as exemplified by his admonition regarding Le Chatelier's principle is admirable. However, he clearly was biased toward equilibrium, and recognized its failings with regard to the analysis of change. He attempted at length to develop a "moving equilibrium" model mathematically. The end result (although he, like Spencer, was reluctant to admit it) was failure. It is clear that the only way he can make the differential equations of moving equilibrium valid is to say $dx/dt = 0$. In other words, the equations improve as dx/dt approaches 0, and really hold when $dx/dt = 0$. But in that case (like in the prime equation), we have contradicted ourselves and are back to the static equilibrium model, as the case of $dx/dt = 0$ represents *no* change over time. In other words, moving equilibrium works *only* when there is no movement (as in the case of maximum entropy). Ironically, moving equilibrium is clearly shown by Lotka to be a fiction, despite his vigorous efforts to prove the opposite.

Henderson

L. J. Henderson was a Harvard physiologist who was a great admirer of the equilibrium theories of both Pareto and Gibbs. He wrote a book whose title implies that it is solely concerned with Pareto (*Pareto's General Sociology,* Henderson 1935), but the latter portion of the book deals with the work of J. Willard Gibbs.

Henderson wrote an equilibrium theory about blood (Henderson 1928) using a closed systems model of equilibrium. He apparently was unaware that the isolated-system concept of equilibrium could not be legitimately applied to social systems, and he advocated equilibrium strongly to those he influenced around Harvard, including Parsons, Homans, and James Grier Miller. As a "true believer" he probably as much as anybody was responsible for promoting the notion of equilibrium as a "given" that no one would think to question.

Samuelson

Later to win the Nobel Prize, Paul Samuelson was a Harvard Fellow along with George Homans and James Grier Miller in the days when equilibrium was taken for granted and reigned supreme. Samuelson used his tenure as a Harvard Fellow in part to develop an equilibrium analysis of economics (1983 [1947]). This, however, was a mathematical model (X'') and not a data analysis (X'). As had Lotka before him (though without reference to him), Samuelson also attempted to write the differential equations for moving equilibrium, but to no avail. The conclusion is always the same—equilibrium is synchronic and "movement" is diachronic. The two do not meet mathematically unless one lets dx/dt approach zero so closely that for all practical purposes movement has ceased. In other words, "moving equilibrium" works only when there is no movement—a total contradiction.

Other Forms of Physical Equilibrium

In addition to thermodynamic equilibrium in isolated systems, several other forms of equilibrium are defined in physics. All of these can be found empirically (X'), and *all* are subsumed under thermodynamic equilibrium and the second law. That is, if entropy is maximized (and thus $dS = 0$), then these other forms will also be in equilibrium.

Stable Equilibrium.   This is a form where a physical object is at rest with its center of gravity at its lowest point. For example, a book lying flat on the table is in "stable equilibrium." There is no available energy to be dissipated in this system, and thus disturbances of the equilibrium cannot come from within the system. If disturbed by an external force (such as someone's hand turning the book up on one corner), the kinetic energy is generated by raising the center of gravity. As soon as the external force is removed, the book will fall to its lowest center of gravity, thus expending the energy and reestablishing its stable equilibrium. Two other forms of physical equilibrium are similar.

Static Equilibrium.   Static equilibrium is attained when an object in linear motion comes to rest. This was Spencer's (1892 [1864]) first form of equilibrium.

Rotational Equilibrium.   Rotational equilibrium is established when two torques are balanced (as in a spinning top). This was Spencer's first form of moving equilibrium.

Other Equilibrium Theorists (Nonfunctional)

There were a number of other sociologists who used the concept of equilibrium, including Small (Small and Vincent 1894) and even a number of social psychologists and micro theorists. Most of these usages were verbal expositions quite similar to the one we have already examined. For further discussion see the excellent analysis by Russett (1966).

Other Equilibrium Theorists (Functional)

Early functionalist theorists in anthropology were also equilibrium theorists of a sort, but they did not unduly exaggerate the concept. Their emphasis was on listing the components that were functional for a society, and for identifying the function or functions that each component served. They emphasized part-whole analysis—the function that each part served in the functioning of the whole society. Sometimes the function was to maintain equilibrium, other times to merely maintain "survival" or a subsistence state, particularly in the more primitive societies.

HOMEOSTASIS

One person at Harvard who *did* question equilibrium as a generic concept applicable to most everything (including social phenomena) was another physiologist, Cannon (1929, 1932). Like Henderson he turned to an interest in sociology later in his career (the last chapter of *The Wisdom of the Body* is devoted to the analysis of homeostasis in organizations). But the two differ widely in their approach to equilibrium. Henderson, following Pareto, applied the closed system equilibrium model to society even though the latter, an open system, did not meet the basic assumptions, and therefore its application was in doubt. Pareto, Henderson, and later Parsons, acted as though equilibrium was generic (or "robust" in modern statistical jargon), and thus applicable (perhaps with modification) to almost any phenomena, including social phenomena.

Cannon knew better. He developed his model of homeostasis inductively, from intense study of living organisms, including humans. He was one of the relatively rare scientists who did not hesitate to use himself as a research subject, as in studies of homeostatic inundation where he and other subjects drank huge amounts of water to see if it would dangerously dilute the blood. He determined that because of effective homeostatic mechanisms in the body, there were no such dilutions of the blood.

Cannon taught the required physiology course for first-year medical students at Harvard Medical School, which James Grier Miller took in 1939. He required the students to take their own temperature hourly for a twenty-four-hour period. They discovered that temperature remained within a homeostatic range, but fluctuated, from roughly a high of 98.9 degrees Fahrenheit at 4 P.M., to a low of 98.2 or 98.3 degrees Fahrenheit at 4 A.M. (Cannon 1932).

The basic homeostatic model is rather precise. The body is an open system, which means that it is open to inputs of energy in the form of food, and is also affected by other stimuli (temperature, light, etc.). The body maintains certain key variables such as blood content and temperature in a range of similarity. This range is a constant, and can be analyzed synchronically, as it generally does not change over time. Cannon studied at least two basic empirical forms of homeostasis: homeostasis by inundation, and homeostasis by regulation. Homeostasis by inundation occurs as illustrated in the example of drinking a lot of water. The body does *not* allow dilution of the chemical balance of the blood. Rather, the excess water is inundated into areas of the body (such as the muscles and in fat cells near the skin) that normally do not hold liquid, at least in quantity. This inundation is not unlike the flooding of a lake upon the low-lying brushland around it. When the body is deprived of water, the opposite phenomenon takes place, and moisture needed for the blood content is pulled from the outlying body areas into the blood.

Homeostasis by inundation is apparently little known to sociologists. Homeostasis by regulation is much more familiar. Parsons (1961b, p. 339) discusses body temperature as an example of homeostasis by regulation. In this form, the body components work to regulate the variables so as to maintain their limits. For example, if the body gets too cold, it will contract or shut down in a way to use less heat. If it gets too hot, it will give off heat through perspiration.

Cannot could have used the term "equilibrium." After all, his Harvard colleague also studied and wrote about blood (Henderson 1928), and Henderson used the equilibrium concept. However, Cannon was aware that equilibrium theory was formulated *specifically* for isolated systems and *held only for those systems*. Living organisms were *not* isolated systems, but open systems (what we would call "regulated systems"). For theoretical clarity he coined the term "homeostasis" to apply to the equilibrium-like phenomena appearing in open systems such as living organisms. It was thus important to him that equilibrium and homeostasis not be merged, as theoretically they had quite different assumptions and implications.

There are a number of features of Cannon's homeostatic concept that must be clearly understood by social systems theorists.

1. Homeostasis is an empirically conceived phenomenon (X' level). Homeostatic theory is "grounded theory" or inductively

derived from exhaustive empirical analysis. It is not a verbal or mathematical model (X'') that was first transplanted from thermodynamics to sociology, then applied to empirical social phenomena (X') with the possible dangers of invalidity or reification. To reiterate, while sociological equilibrium went from X'' to X' (from thermodynamic theory to social empirical phenomena), homeostasis went from X' to X'' (from empirical observation to theoretical formulation).

2. The homeostatic level, while often treated as a specific value (e.g., body temperature of 98.7 degrees Fahrenheit), is actually generally a *range* of variation, perhaps with regular cyclical variation within this range. Thus, if one measures precisely enough, body temperature is seen to not be a *constant* at all, but a cyclical variant which has a fixed minimum and maximum. To really study body temperature precisely, one cannot use a synchronic analysis of a constant, but must use a diachronic analysis of a cycle.

3. Homeostasis only holds within a certain range of external disturbances. If key disturbances are too large (e.g., too little food energy, too little water, too little heat, too much heat), then homeostasis will not be effective and life will be threatened. If this were not so, humans could save money on winter heating bills by letting their internal homeostatic mechanisms deal with cold temperature. Since the body's homeostatic mechanisms only operate within a relatively narrow range of external variation, rooms have to be heated in extremely cold temperatures.

The notable thing about the social and biological equilibrium and homeostatic theory discussed until now is that *all* of it on the empirical level (X') *used Q-analysis of concrete systems*. This is true for Spencer, Pareto, the early functionalists, and Cannon. The mathematical models (X'' level) of course used R-analysis. Thus, as we turn to Parsons, we see a major turning point in equilibrium theory—from Q to R.

Parsons

Parsons was exposed to virtually all of the equilibrium concepts discussed in this chapter. His basic concept was that of dynamic or moving equilibrium, as in the concept of growth (allometry). But he also used the concepts of "static" or "stable" equilibrium (1951, p. 205; Parsons and Shils 1951, p. 107), which are not synonymous in physics as we have seen. He also used equilibrium as a synonym for "balance," as in a balance between social or social-psychological forces (Parsons 1951, p. 205). But his basic definition was cited in chapter 2. I will repeat it here for convenience:

> The most general and fundamental property of a system is the interdependence of parts or variables. Interdependence consists in the existence of determinate relationships among the parts or variables as contrasted with randomness of variability. In other words, interdependence is *order* in the relationship among the components which enter into a system. This order must have a tendency to self-maintenance, which is very generally expressed in the concept of equilibrium. It need not, however, be a static self-maintenance or a stable equilibrium. It may be an ordered process of change—a process following a determinate pattern rather than random variability relative to the starting point. This is called a moving equilibrium and is well exemplified by growth. (Parsons and Shils 1951, p. 107, italics in the original)

Parsons's systems theory focuses on three basic systems: the personality system, social system, and cultural system (which form a hierarchy of sorts). The personality system is basically psychological, dealing with individual personalities. The social system deals with relationships. The cultural system deals with values. These were the original three levels presented in *The Social System* (Parsons 1951). The organismic or biological level was added later. As I said in chapter 2, Parsons was very careful to guard against biological reductionism. This was one reason for his opposition to dependence upon concrete systems, and also the reason for his reluctance to include the biological system, lest it direct emphasis from the "true" objects of sociological study—social and cultural factors.

Each of these three main systems (personality, social, cultural) is a system in its own right. In addition the three are interrelated. Two basic difficulties that readers had with this formulation of three abstracted systems are: (1) How are the boundaries of each system determined and analyzed?; and (2) What are the links between systems?

Parsons paid some attention to the first problem, and apparently considered it nonproblematic in systems terms. He said that the boundary is what separates members from nonmembers. With *role* as the basic unit of analysis, whether one is part of the system depends upon whether one occupies the role of member or not. It seems clear, though, that one cannot really make the member/nonmember distinction in terms of roles without ultimately turning to an analysis of concrete objects (individuals) occupying those roles. The difference between membership and nonmembership roles (and thus the significance of the boundary) is somewhat meaningless if all roles are null. It is only when individual actors occupy roles, as either member (inside the system boundary) or as nonmember (outside the system boundary), that the boundary really has significance.

A similar condition exists with regard to the links between systems. As purely abstracted systems, the boundaries are purely analytical, and links between systems are analytical. It helps little to say that personality is "embedded" in the social system, or that the social role is "embedded" in the value (cultural) system, or that the systems (personality, social, cultural) are seen to form a hierarchy. In the final analysis, it is the concrete individual that forms the link between these three systems. Parsons admits this, but contends that the social role is still the superior systems unit for analytical purposes, saying:

> The social system is made up of the actions of individuals. The actions which constitute the social system are also the same actions which make up the personality systems of the individual actors. The two systems are, however, analytically discrete entities, despite this identity of their basic components.
>
> The difference lies in their *foci of organization* as systems and hence in the substantive functional problems of their operation as systems. The "individual" actor as a concrete system of action is not usually the most important unit of a social system. For most purposes *the conceptual unit of the social system is the role*. (Parsons and Shils 1951, p. 190, italics in the original)

He says further that:

> A social system consists in a plurality of individual actors *interacting* with each other in a situation which has at least a physical or environmental aspect. . . . (Parsons 1951, p. 56)

Despite this clear realization of the concrete system as the basic definition of the system, Parsons nevertheless chooses to disregard the individual as the basic systems unit in favor of the role or *status-role* complex (Ritzer 1983, p. 195). This may be perhaps just another example of Parsons's general trend over the course of his career from emphasis on action theory early in his career to the macro-theoretical emphasis on systems later in his career, reflecting a subsequent inability (in the view of some critics) to integrate or reconcile his action and structural-functional view (see Ritzer 1983, p. 182; Menzies 1977).

What is salient for my purposes at this point is the nature of the links between the personality, social, and cultural systems. Returning to table 2.1 (the score matrix), it is clear that by eschewing concrete systems analysis (although he demonstrated awareness of its ultimate validity), Parsons gained analytical freedom in his verbal theorizing. Such freedom may have been efficacious in allowing his thoughts to flow unfettered, and may well have been a boon to his creativity. The net methodological result, however, appears more like license than freedom, as it leads to a mix of units that are perhaps just so much analytical stew.

For example, in table 2.1, let us examine the basic unit of analysis (in concrete terms) of each of the four systems. The organismic system obviously has the individual organism as its unit of analysis, as does the personality system. For these two, the "objects" listed along the left margin can be individual persons. The social system is more problematic. The links between individual action and social structure are problems not only for readers of Parsons, but remain central issues not only in social systems theory but in all of social theory (see, for example, Giddens 1979, on agency and structure). One would theoretically use this same matrix to analyze the social system. However, it would have a huge N (perhaps several hundred million or even a billion individuals). Further, while it could represent additive (and interactive) properties of individual interrelationships (e.g., networks), it would have limitations as a tool for social systems analysis. Specifically, it could not be used to analyze any emergent properties of the social system that any systems theorists may believe exist. These include such phenomena as equilibrium and hierarchy. In order to analyze the latter we require a *new matrix* for table 2.1, with the *society* as the unit of analysis. Then we can analyze the properties of the system *sui generis* in Durkheim's terms.

The macro properties studied by Durkheim as social facts, can be, following Lazarsfeld (1958), divided into global macro properties, defined *only* for societies and operationalized without any knowledge of individual properties, and analytical macro properties that are aggregated from individual properties, such as the suicide rates studied by Durkheim. Notice that while the analytical macro properties of systems can be generated from table 2.1, *the global macro properties cannot be, as they are true emergents.* They require a new table, such as table 3.1, where the unit of analysis is now the whole society.

Thus, quantitative analysis of the Q-R distinction in tables 2.1 and 3.1 shows what was lost, probably irrevocably, in the verbal analysis. This is *that the basic unit of social systems theory is not either the object or the role, but is rather the society.* The relevance of this statement may not be apparent at first. The statement may seem truistic, tautological, or a case of belaboring the obvious or rediscovering the "wheel" of social fact that Durkheim analyzed so exhaustively. In fact, though, it is of fundamental importance as it unmasks the epistemological flaws inherent in verbal analysis, which cannot be easily decoded due to the lack of specificity.

To make this point more clearly, I can rephrase Parsons (1951, pp. 5–6). The basic unit of analysis of the social system is *neither* the individual actor, nor the unit act, nor the status, nor the role, nor the status-role complex (see Ritzer 1983, p. 195). The basic *unit of analysis* of the social systems is, and must be, that very *social system* itself, just as the basic unit of individual analysis is the individual. This truism is perhaps central to Parsons's inability to integrate action theory and functionalism. It represents a classic case of displacement of scope (see Wagner 1964) that is almost impossible to avoid in verbal theorizing due to its imprecision and the mercurial relationships between individual and society. A little

TABLE 3.1.
The Score (S) Matrix
for Societies

Objects			Variables			
(Societies)	1	2	3	·	·	·	N
Society 1	$Score_{11}$	$Score_{12}$	$Score_{13}$	·	·	·	$Score_{1N}$
Society 2	$Score_{21}$	$Score_{22}$	$Score_{23}$	·	·	·	$Score_{2N}$
Society 3	$Score_{31}$	$Score_{32}$	$Score_{33}$	·	·	·	$Score_{3N}$
·	·	·	·	·	·	·	·
·	·	·	·	·	·	·	·
·	·	·	·	·	·	·	·
O	$Score_{o1}$	$Score_{o2}$	$Score_{o3}$	·	·	·	$Score_{oN}$

reflection should suffice to show that if one permanently employs the individual as the systems unit one can never truly understand emergence. The true understanding of micro-macro links (or the "agency-structure question," see Giddens 1979) can only come from a dual analysis with *both* society and the individual (and intermediately, group and organization) simultaneously. The score matrix in tables 2.1 and 3.1 makes this clear.

Another way to make this slippery point is that systems analysis is lagged one level. Thus, of Miller's (1978) original seven levels, we need to work with a set or sample of N entities one level removed. This is because systems analysis (and statistical analysis) focuses on *relationships between units of analysis rather than on the units*. For example, to successfully understand society (Miller's level six), one has to go to level seven (the supranational system). If one focuses on level six as Parsons did, one cannot study relationships among societies. This explains several puzzling phenomena. It explains why systems analysis emphasizes meta-analysis (because it has to go another level higher), and it explains perhaps why Parsons and other systems theorists are often criticized for deriving sets of analytical or descriptive categories rather than constructing theories. This has not only to do with operationalism (level X'' to level X' of figure 3.1), but also with the epistemology of the Q-R distinction. As long as you verbally analyze a social system as a set of units, you cannot make the necessary theoretical leap to a set of systems. Thus, social systems theory focuses not on a plurality of individual actors as a system (this is a subsystem), but on a plurality of individual systems (what is often termed "cross-cultural analysis"). The basic unit of social systems theory is the society, *not* the individual. This point is difficult to grasp if one gets immersed in Parsons, but it must be held onto.

This confusion is not limited to Parsons, but is endemic in most verbal analyses, including perhaps at times, social entropy theory (Bailey 1990). It is lessened in Miller (1978) because living systems theory often works with several levels simultaneously, and emphasizes cross-level hypotheses, where a concept is studied on more than one level simultaneously. This will be discussed in more detail in chapter 5. Just remember, if one defines the social system as Parsons does, whether it is abstracted (with the role as the basic unit) or concrete (with the object as the basic unit), the unit used will lead to inadvertent displacement of scope or limit one to description (rather than theory) unless one goes to the next level up, and compares systems. The emergent properties are then clear, as the analysis is comparative and external, rather than internal.

ADDITIONAL EQUILIBRIUM MODELS

Homans

Parsons ironically verified his own theory of the internalization of concepts through socialization by becoming "oversocialized" with regard to equilibrium. In contrast, Homans was also socialized to internalize equilibrium. He studied Pareto under Henderson at Harvard, and wrote a book on Pareto (Homans and Curtis 1934), including of course the concept of equilibrium. But Homans's case dispels Parsons's teachings on the absolute power of internalization, as he soon dropped the concept of equilibrium. Early work by Homans includes equilibrium, but not his later work, as he began to doubt the efficacy of this concept (see Lopreato 1971).

Stinchcombe

I have noted wide dissatisfaction with social equilibrium, including social homeostasis. The latter has been confused or merged with thermodynamic equilibrium (Buckley 1967). Other charges are that the functionalist model is "weaker" than homeostasis, not showing the feedback mechanisms (Giddens 1979). Still others are unhappy with what they see as teleology, or the reversal of causal analysis in functionalism (see Dore 1967).

Stinchcombe (1968, pp. 80–98) has done a service by presenting a model of homeostasis as a causal loop. This illustrates feedback mechanisms, thus reducing the possibility of illegitimate teleology, tautology, or determinism. It also shows that "cause" and "function" are compatible. This fact is illustrated with Malinowski's (1948) analysis of magical ceremonies by canoeists prior to fishing expeditions, when they faced danger at sea. The threat of capsizing as a result of a storm (external

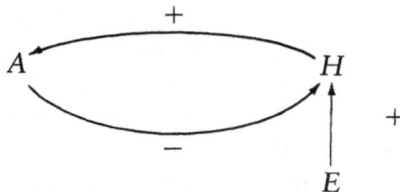

variable E) increases the anxiety level in the homeostatic variable (H). The increase in H leads to an increase in the rate of magical activity (A), which in turn decreases anxiety (H). The general model is that external activity affects the homeostatic variable, thus triggering the homeostatic balancing loop (one "minus" relationship balancing one "plus" relationship). For a somewhat different diagram of the same example see Stinchcombe (1968, p. 89). Causally, this can be written as a linear causal chain, à la path analysis (Duncan 1966), as

$$E \rightarrow H_1 \rightarrow A \rightarrow H_2.$$

Miller

James Grier Miller (1978) was also a Harvard Fellow, along with Homans and Samuelson, and was appointed by Parsons as the only clinician in the new Department of Social Relations at Harvard in 1946. Miller also was exposed to the view of equilibrium as a pervasive concept and a "given." He took the Pareto course from Henderson along with Homans, and also took Cannon's course as a first-year student in Harvard Medical School. His work on living systems theory represents a view intermediate to Parsons and Homans. His ideas germinated for nearly forty years before publication, so his view of equilibrium is less extreme than that of Parsons, and more modified by the study of biology and medicine. Nevertheless, he has not disassociated himself entirely from the equilibrium concept as Homans has done.

Miller's view of equilibrium is modified by his general systems perspective which includes knowledge of the role of entropy. He views living systems as systems which process energy and information in order to maintain appropriate levels of negentropy (negative entropy). He recognizes that the appropriate equilibrium model for living systems is homeostasis (as do the autopoietic theorists—see chapter 8). His requirements for homeostasis are minimal, saying that homeostasis consists of maintaining the level of at least one variable in the living system within a certain range. It is probable that most complex social systems meet this minimal definition. But while his definition of homeostasis may be minimal, Miller is adamant about retaining it, saying that "if this is not so my whole approach may be in jeopardy" (Miller 1978; see also chapters 5 and 7).

Other Views

There were a number of other functionalists—most of them students of Parsons—who were not really systems theorists. That is, they emphasized the study of institutions (parts) and the *function* each played for the whole. Their description of the whole system and the concept of equilibrium was secondary to their emphasis on the particular function. These functionalists included Davis (see Davis and Moore 1945), Aberle (Aberle et al. 1950) and Merton (1949). They were more inclined to stress the function as insuring "survival" of the society, or just the "functioning" of the society, rather than present a detailed model of equilibrium.

These functionalists were criticized on a number of grounds, including conservatism, tautology, and teleology, as already discussed (see Turner and Maryanski 1979). But methodologically, the biggest problem was dealing with the concept of survival. If a society survived, how could one say that it was due to the function of the particular institution without being tautological? The measure of "survival" was in a sense just another form of equilibrium, but it was superior in some ways. It was efficacious for the study of primitive societies in anthropological functionalism where it developed, as many of these seemed to stabilize near survival levels. Interestingly, the survival level is also much nearer the theoretical state of equilibrium (maximum entropy) in thermodynamics than is the concept of equilibrium as "order" or "integration." However, "survival" remains a dichotomous state with about the same methodological problems as equilibrium. For example, if the system "survives," then what system state is it in? Is it integrated or not? Entropy measurement will answer this. Utilization of the equilibrium and survival concepts will not.

Hempel (1959) addressed this measurement problem. Rather then mere survival, he proposed utilizing the concept of "normal" or "healthy" functioning—a direct carry-over from homeostasis in health research. He considered this more efficacious for social science. However, many sociologists are probably dubious. Although Hempel seems to feel the "normal" system state is easy to ascertain and measure, how can one do this for a pluralistic social system in which every public has its own notion of the "normal" or "healthy" state of the system?

SUMMING UP

The age of equilibrium in social systems theory is over. Equilibrium will always be with us, but it will not be a dominant concept. Social scientists, particularly systems theorists, of the late nineteenth and early twentieth centuries *had* to use equilibrium. It was an accident of birth. The concept was so pervasive, so compelling, and so unquestioned (at least at first), that there was little choice, particularly if one were recoiling from the heavily criticized evolutionary approach of social Darwinism.

However, they recoiled too far. In many cases, what verbal equilibrium theorists such as Parsons needed and wanted was primarily a concept of order, stability, or balance. Unfortunately, equilibrium was the wrong choice for myriad reasons. The way sociologists used equilibrium was unique in the scientific community. They seemed driven chiefly by its perceptual- or conceptual-level (X) connotations of balance, harmony, and stability and in retrospect these are seen as value judgments. Such perceptual-level (X) value judgments of equilibrium were generally absent in thermodynamics. Equilibrium was not a superior state in a value sense in thermodynamics. It did not connote stability, but system death or disintegration (maximum entropy) according to the second law. Equilibrium theorists in science were driven not by perceptions of harmony (X) so much as by empirical (X') and mathematical (X'') concerns. Sociological equilibrium theorists failed on both of these counts. They did not develop a consistent mathematical model of equilibrium (X'') *or* a satisfactory empirical analysis (X'), and thus because of their imprecise verbal analyses, were unable to keep the empirical (X') and heuristic (X'') realms distinct, leading to further problems (Russett 1966). Further, twentieth-century equilibrium sociological theorists reject Spencer ("who reads Spencer?"), and so failed to profit from history, and were doomed to once again become snared in the Spencerian dilemma.

Equilibrium is a weak concept in sociology. Instances of equilibrium that can be empirically established are generally very localized. For example, if you work in the central city and are deciding where to live, as you move farther from the center, you may encounter an "equilibrium point," where the cost of housing (cheaper as you move farther out) is balanced by the cost of commuting (more expensive as you move farther out). As local as this equilibrium point is, it is probably not the primary element in individual decision making. Most people probably do not settle at the economically rational equilibrium point, but rather base their residence decision on other variables (cultural amenities, family networks, etc.). There are other variables which may be forced into an artificial sort of temporary equilibrium, such as the value of the dollar. This is generally unsuccessful, and is accomplished only through tremendous expenditures of time, effort, and resources.

Another example is the phenomenon of lines in a bank or cafeteria. If several lines are long and one is short, there will likely be movement from the longer to the shorter which will appear to restore "balance," resulting in all lines of equal length. This is not "equilibrium" in the classic sense, but is rather an example of the statistical phenomenon of central tendency, where all lines are of roughly mean length, and dispersion is minimized. Even if this were interpreted as equilibrium, it is a rather local and trivial example, and not a proper basis for sound social theory.

This cannot be seen as an example of functionalist equilibrium where the part (individual) seeks to insure survival for the system. Rather the system (bank or cafeteria) monitors the lines, and opens new ones, not with a goal of restoring an "equilibrium" of equal-sized lines, but to minimize customer waiting time. Customers do not seek to restore equilibrium for the institution, but merely wish to get in a shorter line in order to reduce their individual waiting time (but by doing so they also establish lines of equal length).

Social systems theorists will continue to use equilibrium, but *properly*, in the spirit of positive positivism. This means not perceptually or conceptually (X) in terms of balance, harmony, or other positive cultural and social values, but mathematically (X''), as in whether a series of numbers continues to infinity, or instead reaches "equilibrium." When used in this manner, equilibrium serves as a *foundation* for social phenomena in the form of both thermodynamic equilibrium $(X''$ and $X')$, and homeostasis $(X''$ and $X')$. *These are not social equilibria,* but physical and biological bases or foundations for society (à la Comte).

I turn now to the field of general systems theory (GST) where equilibrium remains in use, but is not dominant, and does not block other avenues of analysis. Physical scientists have never understood the notion of social equilibrium, because the mathematics

(X'') was never worked out, and the variables could not be identified, let alone be measured in "centimeter/gram/second" measurement terms. This volume seeks to leave equilibrium in physics and homeostasis in biology where they belong, but to use both as underpinnings for sociology, as humans are biological organisms in a physical world. This way we will be free from the illegitimate analogies that have so long plagued systems science (see Lilienfeld 1975).

COUNTERPOINT

The reconstructive critique of equilibrium began in the preceding chapter with the discussion of blocking and is continued in more detail in this chapter. Here we see why and how equilibrium came to dominate sociology. Equilibrium was intellectually pervasive—it was in the air, and anyone who breathed consumed it. Thus, a sociology of knowledge analysis shows that utilization of equilibrium by sociologists during the late nineteenth and early twentieth century was almost inevitable. It is equally clear that physical equilibrium as a concept is obsolete in the new systems theory, and that the new systems theory is essentially an entropy and nonequilibrium approach.

Notice the clear parallel between the critique of equilibrium in this chapter and the preceding one, and the critique in neofunctionalism (Alexander and Colomy 1990). I noted that Parsons (and Pareto) merged the heuristic and empirical notions of equilibrium. Alexander and Colomy note the same thing, but speak in terms of "conflation," and rightly note that this merging by Parsons was more prevalent in later Parsonian cybernetic writing.

As for the methodological contributions of the new systems theory, the three-level model is particularly effective in demonstrating how Pareto changed the use of the equilibrium concept while transporting it from thermodynamics to sociology. The concept is entirely appropriate as used in thermodynamics, but not as used in sociology.

I agree with Collins, Giddens, and other critics that this equilibrium model of functionalism was flawed. It operated blindly, without clear explication of how equilibrium was achieved (Giddens 1979). It was hypostatized (Collins 1975). It was subject to teleology, tautology, and determinism. Notice too that it exhibited the problems with synchronicity (or "leaving time out") that were criticized by Giddens (1979). However, in fairness to functionalists, they probably were not aware of these pitfalls before constructing the model. We are fortunate in the new systems theory to benefit from this critique of functionalism and to build a systems model which escapes these flaws.

What evidence of metatheory is there in this chapter? Again, a number of types can be identified. Among the most prominent ones are the variant of M_U (internal-intellectual), through the focus on equilibrium and the three-level model. The third variant of M_U (external-intellectual—see Ritzer 1990a, p. 19) is also prominent in the sense of turning to other disciplines for tools. This is clear in the case of the multidisciplinary analysis of equilibrium.

What contributions are made in this chapter that are *not* found in mainstream theorizing, and thus give added breadth and richness to the mainstream? The chief one is the methodological critique of equilibrium, which neatly complements the standard theoretical critique. Another contribution is the use of the three-level model to show how Pareto subtly shifted the meaning of equilibrium in transporting it from thermodynamics to sociology. Another is the realization that equilibrium is not "the maintenance of order" and equilibrium is generally *not* a part of a definition of system.

CHAPTER 4

The Age of Entropy

If 1850 to 1950 was the age of equilibrium, 1950 to the present is just as surely the age of entropy, although perhaps without the glory and awareness of the preceding period. The concept of entropy is seen as a cornerstone or foundational concept in most of the current trends in contemporary systems theory. Entropy is central to the four most visible systems currents: nonequilibrium thermodynamics, cybernetics (including the newer developments of sociocybernetics and the "new cybernetics"), information theory, and general systems theory (GST). Also, social entropy theory and chaos theory depend heavily on entropy, and living systems theory (LST) and autopoietic theory can be discussed in terms of entropy. The first four of these approaches, as well as chaos theory, are discussed in this chapter, while SET and LST are discussed in chapters 5, 6, and 7, and autopoiesis in chapter 8.

The concept of entropy is so central to approaches such as cybernetics, general systems theory, information theory, and social entropy theory that one cannot read far in those literatures without encountering it. However, this should not imply that working with the concept of entropy is easy, or that it is always unanimously accepted. One problem is that the literature on entropy has developed primarily in thermodynamics (dealing with physical [generally heat] entropy), and also separately in information theory and cybernetics (dealing with statistical entropy). While I take the position that these two forms of entropy have a basic generic unity (at some higher level of abstraction), this position is not universally accepted, either by students of thermodynamics or of information. One reason for the remaining entropy gap is terminological confusion, with the statistical measure of information (H) being called by a variety of terms such as information, entropy, uncertainty, etc. This communications problem lies at the heart of the larger umbrella of general systems theory. It is somewhat ironic that while a major goal of GST is the integration of specialized subfields and

121

the standardization of their language, it has not been able to integrate the entropy literature within its own ranks, or even to standardize the basic entropy terminology.

The keystone of contemporary entropy theory is the school of nonequilibrium thermodynamics, particularly the entropy equation of Prigogine. Ilya Prigogine won the Noble Prize for his work on entropy in 1979. Nonequilibrium thermodynamics dates from the work of Onsager in 1931 (according to Prigogine and Stengers 1984, p. 137). As Prigogine and Stengers (1984, p. 138) say, "Equilibrium thermodynamics was an achievement of the nineteenth century, nonequilibrium thermodynamics was developed in the twentieth century, and Onsager's relations mark a crucial point in the shift of interest away from equilibrium toward nonequilibrium."

But while Onsager might have pioneered nonequilibrium thermodynamics in 1931, it has been the work of Prigogine over a period of more than forty years which has given the field importance and prominence (although sociologists are only now starting to see its importance and to accept it). Central works are his *Introduction to Thermodynamics of Irreversible Processes* (Prigogine 1955) and his culminating and more popularized (but exceedingly valuable) *Order Out of Chaos* (Prigogine and Stengers 1984).

Prigogine's work on linear nonequilibrium thermodynamics shows that even in nonequilibrium conditions, linear systems can move to a stationary state of minimal entropy production (remember that entropy production is zero [because $dS = 0$] at equilibrium by definition, as entropy has been maximized, and therefore cannot be increasing or decreasing).

Thus, when boundary conditions prevent the system from going to a state of equilibrium, it goes to a state of minimum entropy production. In this stationary state, entropy processes do not change, and so become "time invariant," and somewhat independent of the initial conditions of the system from which they evolved. As Prigogine and Stengers (1984, p. 139) say, "Whatever the initial conditions, the system will finally reach the state determined by the imposed boundary conditions. As a result, the reaction of such a system to any change in its boundary conditions is entirely predictable."

But while Prigogine stresses the principle of minimum entropy production, especially in his later work, the achievement that has had the most impact on general systems theory and on sociological systems theory is the Prigogine entropy equation:

$$dS_t = dS_i + dS_e \qquad (4.1)$$

where dS_t = total entropy change in the system; dS_i = change in internal entropy (entropy produced within the system); and dS_e = change in external entropy (entropy export from outside the system).

In living and other open systems (including all human social systems), internal entropy buildup (S_i) can be balanced by negative entropy (negentropy) (dS_e) through the importation of energy and information into the system from the environment. The tremendous significance of this equation is that it shows how open systems such as social systems (apparently) violate the second law of thermodynamics which dictates increase to maximum entropy (equilibrium) for all isolated systems.

In the case of equilibrium or a stationary state (minimal entropy production) of a linear nonequilibrium system, $dS = 0$. This implies that $dS_e = -dS_i < 0$. In this case the flow of negative entropy (negentropy) or dS_e being exported into the system from the environment is *matched* by the internal entropy production (dS_i) so that total entropy is zero. When this occurs, no growth of organization or complexity is possible.

However, in living open systems such as human groups, negative entropy or negentropy from outside the system *exceeds* the internal production of entropy. Negentropy can take the form of energy such as food and fuel (or information). In such a case, the system can grow in organizational complexity, and become *more* organized rather than less, in apparent violation of the second law. The difference is that while the internal entropy production (dS_i) is still obeying the second law, and the system is thus internally "running down," burning up energy, and becoming less organized and complex, this is being *more than offset* by the energy and information being brought in from the environment.

Equation 4.1 is in a sense the foundation for general systems theory and for social system theory. Without it these fields would be still mired in the apparent contradiction of the second law, and have the same problems with equilibrium as did Pareto, Spencer, and Parsons. With it, one sees that the second law holds for isolated systems only, but that the growth in organizational complexity of social systems is also easily explained.

Thus, Prigogine's nonequilibrium theory is foundational for most of contemporary social systems analysis. While not eschewing appropriate use of equilibrium, it concerns itself with the many instances of nonequilibrium systems behavior, which are the rule rather than the exception for sociological systems theory. Prigogine's equation (4.1) is thus to contemporary social systems theory what Gibbs's prime equation was to nineteenth-century thermodynamics (see Bailey 1987).

Cybernetics and Sociocybernetics

By the time *The Social System* was published (Parsons 1951) the age of information had already begun. The seeds of a new systems movement were already germinating. General systems theory and cybernetics date primarily from the 1940s, although Bertalanffy (1967) wrote manuscripts on GST during the late 1930s which were destroyed in the war, thus delaying publication of this movement by a decade or so. Although drawing on thermodynamics as did the social equilibrium theorists before them, general systems theorists received major impetus also from developments in information processing, principally during the 1940s and 1950s. This information emphasis was set upon a foundation formed by classical thermodynamics, with some continuities and some discontinuities, as might be expected. The revolution in information analysis roughly coincided with the development of computers.

Norbert Wiener, a former mathematics prodigy, published *Cybernetics* in 1948. This perspective is at the heart of what is now broadly called "control engineering." It emphasizes the control of systems through the monitoring of flows, principally energy and information. This perspective is responsible for popularizing a number of concepts now thoroughly ingrained in public parlance, as well as in the academic language. The most famous of these is the "feedback loop," often now shortened merely to "feedback," probably with resultant change in meaning. Other famous terms are "servomechanism," "steady state," and "black box."

In a sense, cybernetics applied an analogue of the homeostatic biological model (developed for living organisms) to nonliving systems, principally machines or energy or information systems (such as circuits). But cybernetics is much more than just the analysis of loops in electrical circuits, long a staple of physics. Rather, cybernetics analyzes control of a system which has a number of distinct parts. This control often results in attainment of a "steady state" somewhat analogous to the homeostatic state of living systems. The quintessential example of the elementary cybernetic system is the room thermostat, clearly parallel to the quintessential homeostatic example—regulation of temperature in humans, à la Cannon. But while the parallel between cybernetic temperature control in a room and homeostatic temperature control in the body is obvious, somewhat less obvious is the rather striking parallel between the achievement of a steady state in the cybernetic system, and Spencer's (1892 [1864]) form 3 of moving equilibrium, which features analysis of the steam engine.

In the room-thermostat example, the system is clearly a nonliving, but open system. It is open because it allows energy in the form of heat to be added to the system. However, heat is not added constantly, or at a steady rate. Rather, the system is *controlled* (or "regulated" in our terms) so that heat is added only when needed to maintain the temperature of the room at a previously set level (e.g., 70 degrees Fahrenheit). The system consists essentially of an energy source, which is the heater or fuel burner (Spencer emphasized this in his analysis of the steam engine in type 2 moving equilibrium), the servomechanism (thermostat), and the wire loops (electrical circuits) which connect all parts of the system, including the heat source and servomechanism.

The servomechanism is the monitor of system levels. It is preset by humans, and contains a thermometer which can monitor the room temperature. The servomechanism also includes a control for turning the heat on and off. In the simple room thermostat this is a relatively rudimentary device containing a strip of metal which contracts as it cools and expands as it is heated. It is constructed so that when the room temperature as measured by the thermometer falls below the level set on the thermostat, the metal strip will make contact with another piece of metal in the thermostat, thus closing the electrical circuit and allowing the heat to come on. When the temperature rises above the desired level, the circuit is broken, and the heat is off until the next time the thermometer falls below the preset level. This is diagrammed in figure 4.1. The feed-

FIGURE 4.1.

Components of a Cybernetic System, Including Feedback Loops

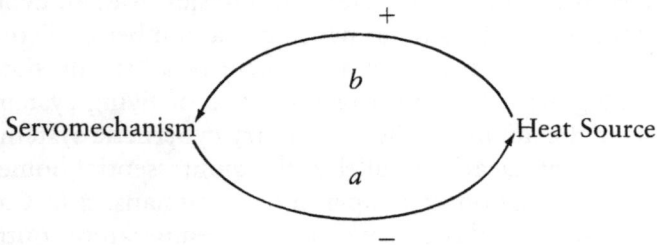

back mechanism is clear. A decrease in room temperature leads to an increase in activity of the heater (an inverse or minus relationship along path *a*), while an increase in the activity of the heater subsequently leads to an increase in room temperature (a positive or plus relationship along path *b*).

The term "feedback" refers to the whole circular loop. Information flows in both directions—to the heat source, turning it on, to the servomechanism which monitors temperature, and back to the heat source. The balance between the loops (+ and −) maintains a relatively constant ratio:

$$q/T \tag{4.2}$$

where q = heat added and T = room temperature. This is a fundamental ratio in thermodynamics which was first encountered in the prime equation in chapter 3.

The balance between the positive and negative relationships leads to a constancy which is clearly analogous or parallel to the condition of homeostasis in living organisms. In order to distinguish this general mechanical or nonliving condition from homeostatis, the cybernetic constancy is generally referred to as a "steady state." The steady state also has a clear statistical operationalization. In a stochastic process, a steady state occurs when

$$pt_1 = pt_2 \tag{4.3}$$

or when change in probabilities ceases. This is generally interpreted in causal terms as being a situation where no major cause of the probabilities can be determined, and the steady state may result from random processes. In causal terms such randomization can represent no one or few major causes, but rather a large number of offsetting small "causes," each representing a small amount of explained variance. This stochastic steady state is the statistical or probabilistic equivalent of mathematical equilibrium in a set of differential equations, representing a set of continuous variables. A steady state of a stochastic process can be approximated empirically by a distribution which grows progressively for a time and then settles into essentially a constant or time-invariant relationship. When graphed, this appears as an exponential or *J*-curve. Examples include the distribution of scientific citations (Simon 1955), and the "rank-size rule of city-size distributions" (Berry and Garrison 1962).

Although the term "servomechanism" is used sparingly by the lay public, the terms "steady state" and "feedback" have had explosive growth. Steady state became a favorite term of bureaucrats forced to hold budgets constant during the late 1960s and early 1970s. Feedback is found everywhere, including the adjectives positive feedback and negative feedback.

Unfortunately, the terms positive and negative feedback are often used in verbal discourse quite carelessly, as is the case with equilibrium. As with equilibrium, they often become value-laden terms, driven by their perceptual (X) connotations rather than by either their use in mathematical cybernetic models (X'') or in actual empirical (X') flows of energy or information. Thus, members of an organization will present a proposal to see whether it will invoke "positive feedback" (approval) or "negative feedback" (disapproval). Once again, as with equilibrium, the use of the verbal value-laden term bears little resemblance to its original mathematical specification. "Negative feedback" in a cybernetic model is *not* negative reaction, disagreement, or disapproval. Rather it is a negative contraposed with a positive in such a way as to rectify deviation from the steady state. If anything, this is "positive," as it makes the model operate efficaciously, rather than continuing to increase toward infinity.

The importance of cybernetics for social systems theory lies not in the notion of steady state or even feedback, but rather in the emphasis on rational control. The homeostatic system can be graphed in a form very much like the cybernetic system of figure 4.1, as in figure 4.2. The format of figure 4.2 is about the same as in figure 4.1. The difference is that in figure 4.1 *we know why and how control is exerted and why and how the system works: because it was designed that way by human actors.*

FIGURE 4.2.
An Illustration of a Homeostatic Loop

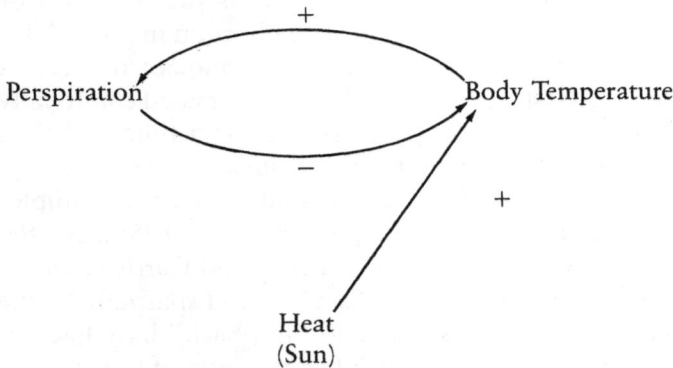

In functionalism and homeostatis we were often reliant upon notions such as evolutionary adaptation, survival, function, manifest function, latent function, etc., to explain the model (see Turner and Maryanski 1979, and Merton 1949). Obviously, this engendered criticisms involving determinism, teleology, etc. In the cybernetic model, humans are in control. The feedback works to maintain the steady state because they designed it that way, and they control and maintain it. Emphasis is on the servomechanism, notably lacking in homeostatic models, which often appear "automatic." If you walk outside your eye involuntarily and "automatically" adjusts to the sunlight without your conscious and voluntary control. This seems deterministic, mysterious, and to be explained chiefly as the result of years of evolutionary adaptation, where nonadaptive eyes failed to survive. In contrast, in cybernetic models, humans are clearly in control. This is a perfect vehicle for Parsons's action theory—goals, motivation, expectations, and so forth all fit right in.

Control.   Control is the key word here. The social equilibrium theorists relished the connotation of stability in equilibrium theory, and sought to buttress this by equating equilibrium with order, in direct contradiction to thermodynamic usage. However, while emphasizing order and stability, they lost the emphasis on control.

This was potentially ruinous, because along with "order," control is a crucial and pervasive word in scholarship (see Gibbs 1989).

Control is at the heart of positivistic experimentation. The "controlled experiment" is the essence of science, emphasizing control by the experimenter, control over the independent variable, control over the dependent variable, control over the experimental setting, control of extraneous variation, and so forth. Control is perhaps no less vital to humanistic sociology and verbal theorizing, forming the essence of concepts such as power and freedom. The classic definition of power is the ability to *control* the actions of another, even without his or her approval. As another example, loss of control is a definitional factor of alienation.

While emphasizing order and stability, social equilibrium theorists progressively lost touch with the notion of control. Beginning with Pareto, equilibrium would be reestablished if disturbed from outside, because the same forces that originally formed the equilibrium would still be functioning, and would reinstitute it whenever it was disturbed. Discussion of individual control is lacking or minimized. Under Parsons, equilibrium was true *by definition* (Parsons and Shils 1951, p. 107). This implies virtually no individual control, as equilibrium is defined as existing regardless of individual action. This was, if anything, exacerbated under homeostatic models, featuring "automatic" responses to restore equilibrium in the body, such as the pupil of the eye adjusting to sunlight, without the conscious control, permission, or even knowledge of the individual. Thus, in social equilibrium theory, the perception of stability (though counterfeit), was comforting, but this was offset by a discomfiting loss of control. There is no question that Parsons *did* discuss mechanisms through which equilibrium would be constructed (and reconstructed if disturbed—see Parsons 1951; Parsons and Shils 1951, p. 107). However, it is also clear that these mechanisms were almost illustrative or secondary, rather than central to the model, since equilibrium already was a condition of the society by definition (Parsons and Shils 1951, p. 107).

Cybernetics restored the desired emphasis on control. The model was "automatic" only in the sense that humans purposefully designed it to be automatic. Its goal or "purpose" was designed and controlled by humans—there was no tautology or teleology. The feedback loops, though nonrecursive or symmetrical, could be understood by and accepted by causal theorists. Further, the loops are compatible with other scientific developments. They lend themselves to diagrams of the form used in computer flow sheets, and they also are amenable to mathematics such as graph theory. Adaptations of the developments can be seen in sociology today, in graphs such as Stinchcombe's (1968, p. 89), and in the graphs (though recursive) of path analysis (Duncan 1966). Furthermore, the feedback *did not have to lead to stability*. Positive loops, as well as negative loops could be constructed, as could multiples of intersecting loops (perhaps with some positive and some negative loops, leading to an overall balance).

Cybernetics also showed that achieving "equilibrium" is not easy. Maintaining a constant temperature in a room requires construction of the room, insulation, fuel, maintenance of the system, and so forth. Wild swings in outside temperature require changes in the amount of fuel provided, and may require humans to change the thermostat setting. In an open system, even the maintenance of a simple empirical "equilibrium" such as a constant temperature requires constant maintenance and monitoring, and is certainly not "automatic," even though the thermostat functions "automatically."

In Europe there is little distinction between the terms "cybernetics" and "systems theory." Thus, the systems movement is synonymous with the cybernetics movement. In America, the two are distinct, and separate societies exist (the American Cybernetics Society and the International Society for Systems Science), although with considerable overlap in membership.

Parsons's systems theory was largely developed before cybernetics became popularized. However, he did incorporate it to some degree. His best-known cybernetic formulation is his so-called "cybernetic hierarchy of control" (Parsons 1966) which visualizes a hierarchy from environment on the bottom (as a base) to information on the top. The levels of analysis are formulated in terms of this hierarchy, with the biological or organismic level on the bottom, then the personality level, then the social level, and lastly the cultural level. This hierarchy signifies that the biological organism is most concerned with energy processing, while culture and society are predominantly concerned with information processing. Parsons's abstracted systems model is compatible with the study of symbol processing on the information level, and de-emphasizes environmental analysis and the study of energy processing.

One of the principal contributors to cybernetics is Norbert Wiener (1961, 1950). Wiener reports that the notion of developing a statistical theory of the amount of information occurred about the time to himself, to Shannon, and to the statistician R. A. Fisher (Wiener 1961, p. 10). He says:

> The notion of the amount of information attaches itself very naturally to a classical notion in statistical mechanics: that of *entropy*. Just as the amount of information in a system is a measure of its degree of organization, so the entropy of a system is a measure of its degree of disorganization; and the one is simply the negative of the other. (Wiener 1961, p. 11, italics in the original)

Wiener says further that a group of scientists dealing with these problems in about 1943 were hampered by the absence of a common terminology, and by the lack of unity of the literature. They decided that existing terminology was not sufficiently neutral (see the discussion of such matters in chapter 5), and were forced to coin the Greek term *cybernetics* (from *steersman*) to fill the gap (Wiener 1961, p. 11).

In dealing with problems of feedback and control, Wiener is dealing mathematically with problems of nonlinear circuits. He uses the concept of a "black box" to represent an unanalyzed nonlinear system, while a "white box" is a known structure. Wiener defines cybernetics to include both nonliving systems (machines) and living systems (animals and humans). In the second edition of *Cybernetics* (Wiener 1961), he discusses not only society, but also "self-reproducing machines." Although autopoietic theory is strictly limited to living organisms, this does provide some common point of reference between cybernetics and autopoiesis, as does the notion of steady state (nonliving systems) and homeostasis (living systems).

Another important developer of cybernetics is Ashby (1954, 1956). Like Wiener, he is concerned with feedback, entropy, the black box, and control. Perhaps Ashby's most notable contribution is his "law of requisite variety" (Ashby 1956, pp. 202–218). Since the law of requisite variety can be viewed in a parallel fashion to autopoietic processes, it is treated again in the discussion of autopoiesis in chapter 8. In any controlled cybernetic system there are a certain number of alternative control actions (leading to a certain number of system states). The law of requisite variety says that the number of control actions must be at least equal to the number of spontaneous fluctuations to be corrected, if the control system is to function effectively. Specifically, "*Only variety in R's [the regulator's] moves can force down the variety in the outcomes*" (Ashby 1956, p. 206, italics in the original).

In other words, the law of requisite variety says "*only variety can destroy variety,*" or if R is the regulator and D is the disturbance, "*Only variety in R can force down the variety due to D*" (Ashby 1956, p. 207, italics in the original). Thus, there is a close connection to entropy (see Geyer and van der Zouwen 1986, p. 216; and Aulin 1986) because disturbance in a system results in disorganization (entropy) while control or regulation results in organization (negentropy or negative entropy). In fact, if $H(D)$ is the variety or entropy of the environmental disturbance D, and $H(R)$ is the variety of the regulation, and K is a constant, then

$$H(Y) = H(D) + H_D(R) - H(R) - K \qquad (4.4)$$

(Aulin 1986, p. 109). Aulin (1986, p. 110) shows how entropy is decreased (and order is created) in a "hierarchy of regulation and control." This is an extension of Ashby's work, and an example of the "new cybernetics" or "new sociocybernetics" discussed later in this chapter.

Another author whose work has a cybernetic flavor is Ackoff (Ackoff and Emery 1972; Ackoff 1974). Although I might just as well classify Ackoff as a systems theorist, his work falls within the cybernetic camp due to his emphasis on control. His principal work is on purposeful systems, and is quite reminiscent of the newer work in sociocybernetics and self-steering.

There is a direct connection between the work of Ashby and Ackoff in Ackoff's use of variety. He says that in designing socio-psychological systems whose objectives are to change the values and other psychological properties of at least one subset of its members, there will be *variety reduction* (entropy reduction). Specifically:

> To the degree that their members are instrumental to these systems' achieving their social purposes, there will be a tendency toward *variety reduction*: the very behavior that would manifest

the higher values of these systems will tend to be inhibited. (Ackoff and Emery 1972, p. 216, italics in the original)

Ackoff claims that the "systems age" is preoccupied with purposeful systems that can display choice of both means and ends (Ackoff 1974, p. 18). There are three central problems in the management and control of purposeful systems (Ackoff 1974, p. 18): the effectiveness of their own purposes, the purposes of their parts, and the purposes of the larger systems of which they are a part. These can respectively be labeled the self-control, humanization, and environmentalization problems. Among the applied social problems that Ackoff (1974) deals with are health, narcotics, aging, the environment, and transportation.

The "First" and "Second" Cybernetics

Classical cybernetics emphasizes, as has been seen, deviation-minimizing loops. These are loops utilizing "negative feedback" so that every deviation is nullified by an equal and opposite change (every "plus" in the feedback loop is matched by a "minus"). The processes of equilibrium, homeostasis, and negative feedback (steady state) are termed "morphostatic" process by Buckley (1967, pp. 58–59).

In addition to morphostatic processes, Maruyama (1963) has conceived the notion of morphogenetic or structure-building processes. An example is a feedback loop in which deviations are amplified instead of minimized. This is an example of positive feedback ("the second cybernetics") rather than negative feedback ("the first cybernetics"). As Maruyama says:

> By focusing on the deviation-counteracting aspects of the mutual causal relationships . . . cyberneticians paid less attention to the systems in which the mutual causal effects are deviation-amplifying. Such systems are ubiquitous: accumulations of capital in industry, evolution of living organisms, the rise of cultures of various types, interpersonal processes which produce mental illness, international conflicts, and the processes that are loosely termed as 'vicious circles' and 'compound interests': in short, all processes of mutual casual relationships that amplify an insignificant or accidental initial kick, build up deviation and diverge from the initial condition. (1963, p. 164)

Maruyama says that the process of organization can be explained in terms of deviation-amplifying mutual causal processes. Deviation-amplifying feedback loops amplify complexity and thus increase organization.

It is interesting to note (as discussed in chapter 2) that Archer (1985) prefers morphogenetic theory to the theory of structuration espoused by Giddens (1979). Archer (1985, p. 61) says, in comparing morphogenetic theory to structuration theory, that "This perspective has an even better claim than the former to call itself a 'non-functionalist manifesto'. . . ." Archer rejects Giddens's notion of dualism in structuration theory for the notion of *analytical dualism* in general systems theory and morphogenetic theory, which allows the analytical separation of structure and interaction over time, thus permitting theorizing about temporal structuring in a way that Giddens's structuration approach does not allow (Archer 1985, p. 82). She concludes:

> The morphogenetic perspective is not only concerned with the identification and elaboration of social structures, it is preoccupied above all with the specification of the mechanisms involved. (Archer 1985, p. 84)

The mechanisms referred to are feedback mechanisms, both positive and negative.

Parsons was also concerned with cybernetic theory. Alexander and Colomy (1990) point out that he had more trouble maintaining the distinction between equilibrium as heuristic and as empirical entity in his later cybernetic work. Further, Fararo (1989) has recently presented a formalization of Parsons's action theory. Included is a general presentation of a cybernetic model in the form of a general negative feedback control system and a two-level cybernetic hierarchy (Fararo 1989, pp. 165–173), as well as an application of the general cybernetic feedback model to Parsons's work. This results in a mathematical treatment of action theory (see Fararo 1989, pp. 169–195).

Another notable development in social cybernetics is the work of Pask (1975). Pask applies cybernetics in a statistical and computerized approach to learning and communication. He is concerned with conventional cybernetic concepts such as regulation, but the gist of his approach is the application to language and conversational theory. He uses the entropy measure (H) but terms it uncertainty rather than entropy (Pask 1975, p. 125). He also utilizes "fuzzy" algorithms (Pask 1975, p. 466).

Information Theory

Information theory began about the same time as cybernetics. Shannon and Weaver's *A Mathematical Theory of Communication* was published in 1949. This book is a classic, and is generally regarded as the cornerstone of modern information theory. Wiener (1948) was also instrumental in developing this perspective as well as the related perspective of cybernetics.

The symbol H is used widely in information theory. It denotes information. However, as shown later, this interpretation is problematic. The symbol H is also used to represent entropy and uncertainty. The basic equation of information theory is H, written either as equation 4.5 or equation 4.6 (they are mathematically equivalent).

$$H = - \sum_{i=1}^{K} p_i \log p_i \tag{4.5}$$

$$H = \sum_{i=1}^{K} p_i \log 1/p_i \tag{4.6}$$

where K = the number of categories in the variable; and p_i = the probability of occurrence of each category. This is illustrated in table 4.1. Table 4.1 shows the *maximum* value of H in table 4.1a and the *minimum* value of H in table 4.1b. The *maximum* value for a nominal or categorical variable illustrated in such a table is always log K, where K is the number of categories. The *minimum* value is always zero. Thus, H varies, for a table of K categories (where K is a constant greater than zero) between a maximum of log K and a minimum of zero. Or,

$$\log K \geq H \geq 0. \tag{4.7}$$

This is the formula for *univariate H*. It was developed in the Bell Telephone Laboratories and named H after Harry Hartley, who worked on information theory at Bell for many years. The major original presentation of this work was authored by Shannon, with an extended foreword by Warren Weaver, and is thus generally catalogued as Shannon and Weaver (1949). Hall and

TABLE 4.1
Illustration of Maximum and Minimum H values

K_1	K_2	K_3	K_4	K_k	
$p_1 =$ 0.20	$p_2 =$ 0.20	$p_3 =$ 0.20	$p_4 =$ 0.20	$p_k =$ 0.20	a

A H = Maximum = log K

K_1	K_2	K_3	K_4	K_k	
$p_1 =$ 0.0	$p_2 =$ 0.0	$p_3 =$ 100.0	$p_4 =$ 0.0	$p_k =$ 0.0	b

B H = Minimum = Zero

Fagen (1956), who authored the definition of system presented in chapter 1, also worked in the Bell Laboratories.

The maximum value of H is not a constant bound, but varies with the number of categories. As we shall see later, this is some-times considered a negative feature in an interpretative sense, but sometimes a positive one. The basic negative argument is that it is difficult to compare H for variables with different numbers of cate-gories. This can be easily rectified by dividing the computed value of H by log K. When this is done,

$$H' = H/\log K. \qquad (4.8)$$

H' will vary between a minimum of zero and a maximum of 1.0.

Another prominent feature of H is that its maximum value is attained when each of the K categories has an equal probability of occurrence. That is, maximum H would result in a five-category table when each category had a cell frequency of $N/5$ (where N is sample size), or each cell of a ten-category table had a cell frequen-cy of $N/10$, etc. Thus, maximum H occurs *if and only if* the frequency of every cell in the table is N/K, where N = sample size and K = number of categories. At the other end of the spectrum, minimum H (zero) *only* occurs when all N cases are in a single cell, and all other cells are null. Thus, H = zero, if all N cases occupy a single *one* of the K cells, and the other $K - 1$ cells are null. In every case, $1 \geq p_i > 0$, and $\Sigma\, p_i = 1.0$.

How can one best characterize maximum H? It is clearly a *uniform distribution*. There seems to be agreement on that. It also can be characterized as the "most probable" state in the statistical sense, or the case of "no order." There is relative agreement on this. Also, many scholars characterize it as a *random* distribution, as the case where all K categories have equal p_i seems to fit the basic definition of randomness. Not all scholars are comfortable with this, however. Part of the problem seems to lie in the interpretation of what is being randomized.

There are two possibilities. Students of sampling think that random selection would mean that each of the N cases would have an equal probability of selection. Since they are all selected, regardless of the computed value of H, this does not quite seem to fit the definition of random selection as these sampling theorists are accustomed to it, either in sampling theory for surveys or for experimental selection. That is, all N cases will be used whether H is maximum, minimum, or intermediate.

Another interpretation (and probably the dominant one) is that when H is maximized, the cells or categories are randomized, as each cell has an equal probability of selection in the sense that there is no ordering among the cells—each gets an equal percentage f_i/N or equal frequency N/K. There is also no ordering among the cells in minimum H, as any one of K cells can contain all N cases, and any set of $K - 1$ cells can thus be null. The issue is not epistemologically serious, and is in a sense a false problem. It can lead to interpretative confusion, however. The answer to this point is that the K categories are of a *variable*, and it is the variable that is displaying a random distribution. There is no predictability or explanatory power. This is the most probable statistical state of the distribution.

The answer to our randomization puzzle is made clear by reexamining our Q-R distinction. The confusion over what is being randomized or distributed is a Q-R confusion. The answer depends upon what the categories represent, and thus how they are labeled. If they represent quantities of income (abstracted system) for example, this is clearly R-analysis. If they represent fifths of the population of objects (concrete system), this is clearly Q-analysis.

Remember, though, that in table 2.1, there is only *one* set of internal data in the table, and Q and R are thus two sides of the same coin, like heads and tails. If the sample size is N, then *objects* are distributed across variables, and the analysis is R. If the sample size is M, then *variables* are distributed across objects, and the analysis is Q. The calculation of H generally involves both a set of objects and a set of variable categories (see Theil 1967). If the categories are not properly labeled, the result is interpretative confusion.

For our purposes, let us assume that the K categories in table 4.1 represent some variable (e.g., occupation). In this case, maximum H represents randomization, or *lack* of order, in the univariate sense. We have no basis for predicting one occupation over another (or conversely, in terms of persons, we cannot predict which occupation a person will have). This is clearly *random,* and thus represents maximum entropy ("equilibrium").

Other issues regarding H include the base of the logarithm to be used, and the direction in which it runs. Much of the original work on H used base 2. This binary arithmetic has the advantage of interpretation in terms of bits, and thus is compatible with many computers. One difficulty with binary logic is that tables of logarithms of base 2 are not always readily available. Fortunately an excellent one is provided by Theil (1967). Either e or 10 or any base can be used. The choice of base is quite immaterial (as long as they are not mixed in the same analysis, of course), as we can translate from one to the other (see Theil 1967; McFarland 1969).

The direction that H runs is also a topic for discussion. Notice that randomization or lack of order is not represented by the *minimum* value of that statistic as in correlation coefficients, but by the *maximum* value. Correlation coefficients for categorical data ("nonparametric measures") have minimum (generally zero) values for randomization, connoting no relationship, and increase to some maximum (generally, but not always, 1.0) connoting perfect correlation.

Statisticians accustomed to the direction of these correlation coefficients may think of H as backwards. This is somewhat like saying that the term "prejudice" is "backwards" because its maximum values are "negatives." The truth is, H should simply be interpreted as belonging to the general class of distance measures (see Sokal and Sneath 1963; Sneath and Sokal 1973; Bailey 1975). There are a number of these in the statistical literature, with many applications in numerical taxonomy. The most familiar is the Euclidian Squared Distance, but the Mahalanobis D^2 is also widely known and used. When distance measures are applied to two points in geometric (e.g., Euclidian) space, maximum distance indicates maximum separation of points in space, and minimum correlation between them. Minimum distance (zero distance) indicates *no* separation, e.g., the two points are identical. Although H is not customarily interpreted in terms of distance, it can be visualized in a very similar fashion. Minimum H means all points are identical in the space, while maximum H denotes maximum distance among all graphed points. Whether the points represent objects (e.g., persons) or variables depends on whether the analysis is Q or R (see Bailey 1975).

If one is uncomfortable with the direction of H, it is simple enough to reverse it. I have already said that the maximum value of H depends on the number of categories (K). Some people like this feature in certain applications. For example, in computing the division of labor with H, it can be argued that a larger number of categories (occupations) in a larger society rightly indicates a larger degree of complexity than in a small society. However, for researchers who desire a measure more in conformance with standard measures of correlation, a common practice is to combine both division of H by maximum H (log K), which standardizes the maximum at 1.0 (as in equation 4.8), and to reverse its direction,

$$H'' = 1 - H/\max H = 1 - H/\log K. \qquad (4.9)$$

H'' is called the coefficient of redundancy (Rothstein 1958). Now H'' varies like the usual correlation coefficient. Thus, H'' is zero when H is maximum, and 1.0 when H is minimum (zero) (see Bailey 1985). I should also note in passing that while H is clearly affected by the number of categories, this is of course true for *all* categorical or nonparametric statistics. They all utilize cell frequencies or probabilities as their bases for computation, and changing the category boundaries changes these frequencies or probabilities. This is unavoidable, but is clearly not a weakness specific to H. Another issue with H is lack of a clearly specified underlying probability distribution. See Theil (1967) for a detailed discussion of this point.

The largest controversy surrounding H has been over what to call it. It might have been better to simply call it H or Hartley (after Harry), and leave it at that. Shannon, though his book title indicates that he is studying *communication* and not information, called H "information." As James Grier Miller (1987) says, "H is a measure of information because Shannon defined it that way". (personal communication) This position, if taken in a "strict operationist" sense, merely says that information is whatever H measures. Miller (1978) is one of the authors who says that H measures information.

There is a clear interpretative problem with this. From an extreme operationalist ("strict operationalist") position, Shannon may operationally define information any way he wishes, especially since the specific meaning of the term is somewhat unclear. However, this commits us to a position in which maximum information (H) tells us nothing (as it constitutes randomization, or lack of order), while minimum information (H) provides complete predictability and order. Does one really want to cling to an interpretation of H as information when it (H) is clearly the reverse? How can one utilize an operational definition of information in which maximum information is complete ignorance (maximum disorder, and no explanatory or predictive power), while minimum information is complete order, complete explanation, and perfect prediction?

Although some social scientists may think that natural scientists and mathematicians are largely immune to "value-laden" problems, and are basically "value-free," the inescapable conclusion here is that "information" is the favorite perceptual (X) interpretation of H. Actually, H is *clearly* a measure of entropy. This has been recognized from the start. It is almost identical to Boltzmann's statistical entropy formula

$$S = -K \sum_{i=1}^{N} p_i \log p_i \qquad (4.10)$$

except for Boltzmann's constant (see Olsson 1967). But entropy is not a pleasing interpretation. The interpretation of H poses a classic paradigm anomaly in the words of Kuhn (1962). Some scholars *want* it to signify information, but it does *not*. What can they do? One possibility is to attempt to explain away the obvious anomaly, or at least to make it seem more apparent than real. This is often de-emphasized (as in a footnote). One example from a footnote is as follows:

> The use of the term "information" for a quantity that measures uncertainty may seem confusing and deserves some explanation. The amount of information gained, in learning the outcome, is the amount by which the prior uncertainty is reduced. But learning the outcome reduces the uncertainty to zero, making the amount of information gained equal to the amount of prior uncertainty. Thus, the same quantity measures both the uncertainty existing before the outcome is known and the information gained in learning the outcome. (McFarland 1969, p. 45)

While the explanation seems plausible, it is clearly a legerdemain solution which does not resolve the anomaly, but only appears to.

The explanatory attempts are vacuous. The "outcome" that is learned is whatever the value of H is, and "uncertainty" is not reduced to zero, but remains whatever the H value is, as uncertainty is the difference between H and minimum H (zero), while *information* is the gain from maximum H. That is, *information gain* is the difference between two H values. For example, maximum H is the minimum available information, and H is departure from this. Thus, information gained is

$$\text{Max } H - H. \qquad (4.11)$$

In reality, H is entropy. Information is negentropy. This cannot merely be defined as $-H$ because there *are no* negative H values by this statistical formula, so $-H$ is just the mirror image of H (see Odum 1983). This situation is complex, involving in part synchronic and diachronic considerations. For further discussion of this issue see Bailey 1987.

The attempts to explain away the anomalous interpretation of H appear somewhat convincing on the surface, and have apparently satisfied some readers (probably those most committed to interpreting H as information). However, many remained unsatisfied, and so apparently began to deal with this problem by the relabeling and multiple labeling of H. As a result, H now is interpreted variously as information, entropy, uncertainty, and surprisal (see McFarland 1969; Bailey 1983). Moreover, authors have followed suit in some cases by even changing symbols. Thus, the left side of the H equation (equations 4.5 and 4.6) is sometimes presented as U (uncertainty), S (entropy), I (information), or even no symbol at all, with just the right side of the equation given (Galtung 1980). This is discussed in more detail in chapter 6. For now I conclude that H is a measure of entropy. It can also be interpreted as uncertainty or surprisal, but *not* as information, unless one's intention is to reverse the standard meaning of that term.

The equation that we have been discussing is for univariate categorical H. The formula is easily extended to the multivariate form

$$H(XY) = - \sum_{i=1}^{K} \sum_{j=1}^{L} p_{ij} \log p_{ij} \qquad (4.12)$$

Similarly, conditional H can be written, either for predicting values of y when x is known,

$$H(Y|X) = - \sum_{j=1}^{L} p_{j|i} \log p_{j|i} \qquad (4.13)$$

or for predicting values of x when y is known

$$H(X|Y) = - \sum_{i=1}^{K} p_{i|j} \log p_{i|j} \qquad (4.14)$$

The multiple H (equation 4.12) is analogous to multiple R (the correlation coefficient). It shows how two or more variables vary together. Although computation is limited by computer capacity, mathematical extension of the formula to almost any number of variables is simple. The conditional entropy measures (equations 4.13 and 4.14) are analogous to partial correlation. They show entropy in one variable when the value of the other variable is known. In addition to the categorical equations, the entropy measure for continuous variables is an easy extension (Theil 1967).

$$H = \sum_{i=1}^{N} X_i \log 1/X_i \qquad (4.15)$$

Despite the interpretative confusion, information theory has had widespread application and impact in a large variety of fields. The *H* measure has been applied in sociology for over twenty years (see Bailey 1983 for a review), but not in a particularly comprehensive or cumulative form. There is finally a Sage quantitative series monograph on information theory (Krippendorff 1986).

GENERAL SYSTEMS THEORY

General systems theory (GST) is generally dated from the early 1950s. The Society for General Systems Research (SGSR) was founded in 1954. It has recently undergone two name changes, first to the International Society for General Systems Research (ISGSR), and then to the current International Society for Systems Science (ISSS). Since 1956 the society has published the *General Systems Yearbook*. Miller (1955) wrote a pioneering article on the general systems approach. Other pioneering statements were by Bertalanffy (1962), and Boulding (1956). Among the founders of SGSR were Miller, Rapoport, and Mead.

There are a number of significant contributors to early cybernetics and GST in addition to those already mentioned. Some that come to mind are von Neumann (1958) and his theory of "automata"; Deutsch's (1951) seminal comparisons of feedback and equilibrium (see Buckley 1967, pp. 55–56); the work of Rapoport (1956) on organizations, systems, and game theory; the cybernetic work of Vickers (1959) including his work on stress; the work of Simon on organizations and steady states (1964); and the good introductions to systems by Berrien (1968) and Churchman (1968). Also notable are von Foerster et al. (1947–1957), Easton (1965), and also the work of Seyle (1956) on stress. Also some good early references on entropy are Schrodinger (1945) and Brillouin (1949, 1956). See also Bertalanffy (1956) and Foster, Rapoport, and Trucco (1957). Another important early contribution is Boulding's work on organization (1956), as well as the work of Rothstein (1958) on organization as complexity, with a discussion of entropy.

An interesting aspect of large-scale systems that deserves comment is the occasional appearance of a steady state, for example in the case of the rank-size rule of city-size distribution (Berry 1964).

Here, cities attain, over time, a size-ranking pattern that appears rather constant, not changing for a long period of time. This can be seen as a steady state of a stochastic distribution. In this statistical form, the pattern is the statistically "most probable" case. In causal terms, this means that no one single cause or even a few major causes can be discovered, but rather that there are a whole host of very small "causes." This steady-state phenomena is the open system analogue of closed system equilibrium or of homeostasis. It appears not only in distributions of cities, but also in other distributions such as word frequencies and other phenomena (see Margaleff 1958; Berry and Garrison 1962; Simon 1955). In the steady state, entropy is maximized and thus entropy production is minimized (or is zero), as in equilibrium (see Olsson 1967; Berry 1964).

Another important current in early GST that deserves mention is systems philosophy. A major systems philosopher is Laszlo. A review of Laszlo's (1972, pp. 153–54, 163) discussion of ontology discloses three elements. These are physical systems, mental (cognitive) systems, and systems models. The physical and cognitive systems have separate models which are isomorphic with each other. Thus, changing from a physical system to a cognitive system does not necessitate changes in theory. The content of the model changes, but its form remains invariant. This leads to the notion of the natural-cognitive system, and the well-known notion of bi-perspectivism. From the standpoint of bi-perspectivism, cognitive systems can be viewed externally and concrete systems can be viewed internally, and thus the overall system is a natural-cognitive, or bi-perspective system.

Also important is the work of Jantsch (1975). He distinguishes three "interference systems" (physical milieu, culture, civilization), and three "transformer systems" (institutions, instrumentalities, and iconological systems). He also includes time in his model. He discusses self-organizing behavior, as well as a system approach to planning. For further discussion see Jantsch (1975).

It is somewhat ironic that systems theory in sociology was nearing its centennial as this "new" field was emerging. The concept of "system" was not new in sociology. It was used by Comte, and found throughout the nineteenth century. Thermodynamic researchers conceptualized the isolated system, as has been seen, and Spencer and Pareto used the concept of system. It is not surprising that sociologists have used the concept almost since the beginning of the field, as despite the criticisms and its problems of application, the concept is a "natural." There is no question that society meets the definition of system.

Thus, the system concept promises to be a very important tool in the sociologist's conceptual armory, not only for linking sociology to other disciplines, but also for applications within sociology. It is in the latter realm where there is a long history with a great deal of success, marred only by a few anomalies such as the problems with equilibrium. If equilibrium is handled correctly, the field's promise can likely be fulfilled. Thus, it is not precisely accurate to say that the systems concept was largely unknown before the 1930s, when Henderson and Cannon popularized it at Harvard, to be learned by Parsons, Homans, James Grier Miller, and others.

However, it is true that the *general* systems approach as a means of interdisciplinary integration for all systems was generally unknown before the 1930s. Bertalanffy worked on theoretical biology in Europe in the 1920s and 1930s, another approach which, though widely accepted today, was then brand new and encountering a lot of resistance. Bertalanffy also worked on general systems theory pieces during the 1930s, but as mentioned earlier, their publication was hindered by World War II, and some manuscripts were destroyed. It was only after World War II that general systems publications began appearing in any significant numbers. In fact, when Miller's interdisciplinary group of scientists began regular meetings in 1947, their goal was interdisciplinary integration. However, at the time they had not even decided on a label for their endeavors, and did not choose the term "systems theory" until 1952.

Although the systems concept had been in use for at least a century, one reason for delay in the general or interdisciplinary approach was the dominance of another interdisciplinary paradigm—the notion of a "field." The field paradigm was dominant during the late nineteenth and early twentieth centuries, and is still used today, although in some ways the prominence of the systems approach has lessened emphasis on field theory. A "field" is a set of forces operating strongly in a particular region, then extending out in space in all directions from that region. The field generally contains a set of forces, or dimensions. These may be energy forces in physics, for example. Forces are often illustrated in terms of vectors. This force field is ideal for analyzing concepts such as static equilibrium or rotational equilibrium, which occur when vector forces in one direction are balanced by vector forces in the other direction.

Such a vector field is similar to the notion of system in that relationships can be analyzed in a particular area, and the notion is sufficiently general to be applied in a number of disciplines. But the notion of "field" differs from "system" in some significant ways also. Firstly, a field is quite homogeneous, while a system can be composed of relatively incongruous parts (such as the wires, thermostat, and heater of the cybernetic system). Secondly, a field generally does not have sharply demarcated boundaries, but simply fades gradually off into space, at some point losing its forces, and becoming progressively weaker until it is no longer operative. Field theory is perfect for the localized, homogeneous case. Here there is probably no parallel for the analysis of vector forces. The analysis of local optima or equilibrium under these conditions (where possible) is probably quite sound. However, the systems approach is decidedly superior for the analysis of relationships between more heterogeneous components, and for the analysis of multiple systems, where boundary determination is critical. Similarly, the concept of equilibrium often suffices analytically in the simpler homogeneous case, but entropy must subsume it in the more complex case (see Pickler 1954, 1955). Although field theory was most widely applied in natural science, it also had some social science applications, with the field theory of Lewin (1936) being the best known.

The Beginning: 1950 to 1970

The stated goal of general systems theory was clear enough. As stated in the statement of purpose of the fledgling Society for General Systems Research (1954), the goal was to further interdisciplinary integration via the systems perspective, chiefly through searching for "isomorphisms" or similarities between seemingly diverse researches (see Turner 1991, p. 130).

While the purpose was noble, the goal was somewhat formidable for a number of reasons. Two reasons predominate. One was the sheer number of disciplines to be integrated, their sometimes extreme degree of specialization and isolation from one another, and the frequent antagonism between disciplines. Researchers in many disciplines view the GST movement as a matter of boundary encroachment, and thus as a threat to their discipline. They feel that their "turf" is being invaded by interlopers who, if they are not charlatans, may be hopelessly naive. They may not feel integration of their field with others is possible or even desirable. They may feel that this is the "age of specialization" and that the most significant advances come from specialized and single-minded pursuit of their specific research goals. They may feel that integrating their approach with some other that is weaker (like social science) or even antagonistic can only weaken it or destroy progress, or at best result in an overly broad, superficial approach. Again, there is probably an underlying fear of loss of control—a problem that has plagued equilibrium theorists in sociology for over one hundred years.

The second problem was not how to overcome the resistance of existing specializations, but rather how to build upon the legacy of past general approaches—including systems approaches and other general approaches such as information theory. Ironically, the apparently competitive "field" perspective has proven easier for GST to deal with than some of its systems ancestors. Being more localized, the field approach has essentially been submerged by the broader GST approach, very much like a rock in a lake being submerged by rising water. It is still there, but is rarely visible. When it is, it generally can be quite useful in local analysis, but it is almost wholly subsumed by GST, posing no current threat to the systems approach, and thus no "anomaly" in the sense of Kuhn (1962).

In Kuhn's formation (Kuhn 1962), science is *not* a continuously accumulative process. Rather, "normal science" is seen as a period of rather peaceful accumulation of knowledge in a paradigm. This cumulation may be threatened by the occurrence of "anomalies" or contradictions which seem to contradict conventional wisdom, or at the very least to be ambiguities that need to be clarified. When anomalies arise in a paradigm, they can be handled in a number of ways. Probably the most common approach is to simply ignore the anomaly. If the anomaly is small, or the adherents wish to ignore it and are very powerful and totally in control of the paradigm, including all publishing outlets, then this approach may work. It is risky however, as it allows the anomaly to grow and become more evident, perhaps to the extent that the paradigm is threatened.

If the anomaly cannot be ignored, another tact is to explain it away. This is the approach that was used with the H anomaly. Paradigm adherents were faced with a blatant anomaly when they tried to characterize randomization as maximum information, in direct contradiction to all extant connotations of the term. They attempted to explain this anomaly away through a verbal explanation which seemed plausible on the face of it. In reality, however, it did not deal with the problem, and the anomaly remains (see McFarland 1969). While paradigm adherents have attempted to show that the anomaly is more apparent than real, in reality the anomaly is *real,* and it is their explanation which is seen to be more apparent than real.

Another example of a real anomaly that failed to yield to verbal explanation is the Spencerian dilemma (Bailey 1990). Spencer learned in the 1850s that equilibrium, according to the second law, meant *disintegration* not *integration*—the exact opposite of the result he desired. He attempted for years to rewrite this, but was never able to (Spencer 1892 [1864]; Russett 1966).

A third but related approach to the anomaly is to simply re-label or redefine it. To a certain extent this is what Spencer did. This is also how Parsons dealt with the problem. Equilibrium clearly is maximum entropy or "disorder" under the second law. Yet Parsons *defines* equilibrium as order (Parsons and Shils 1951, p. 107), thus reversing the standard meaning. This relabeling or redefining is in a sense a denial of reality. This denial must then be defended by adherents (for example, by saying that they are referring to social equilibrium, not physical equilibrium, and so its meaning is different). The question still remains of why one would choose a meaning that *directly contradicts* a meaning in a different discipline. The information theorists also did this. The maximum value of H turned out to be maximum entropy when they wanted it to be maximum information—so they simply relabeled it and tried to explain it and defend it. In this case the inconsistencies led to further relabeling and redefinition so that the same statistic has multiple names and labels, some contradictory. This is the kind of paradigmatic nightmare that systems theorists wish to avoid at all costs, but it was dropped squarely in their laps. Ironically, but as a reading of Kuhn (1962) might lead one to predict, GST has not proven necessarily better equipped to deal with this anomaly than any other paradigm. This may be a matter of priorities. General systems theory has concentrated on the second-law anomaly, and has only recently become concerned with confusion over H.

The Second Law

There have been two chief mainstream approaches within GST. While both are theoretical, one is more deductive and directly concerned with anomaly reconciliation, while the other is more inductive and definitional, and concentrates on constructing inter-disciplinary links from the ground up. This second approach is epitomized by Miller (1978; Miller and Miller 1992) and his inter-disciplinary group which formulated 20 subsystems common to all living systems, and dealing with energy or information flows. I shall discuss this approach in detail in the next chapter (chapter 5).

The first approach is exemplified by the work of Bertalanffy (1968). *General System Theory* (not *systems* theory but *system* theory) is difficult to generalize because it is so inclusive. Bertalanffy seems reluctant to exclude any general approach, and so encompasses a range of analyses—from mathematical and conceptual systems to systems philosophy and methodology, to applications in all fields from science to social science to humanities.

The work of Bertalanffy (1968) and others among the founders of GST represents this Catholic view, but also has a clear initial focus on reconciliation of the dilemma surrounding the second law. While social equilibrium theorists from Spencer to Parsons really believed (at least initially) that equilibrium meant order rather than disorder, the early general systems theorists suffered under no such illusion. They *knew* that equilibrium was equated with maximum entropy under the second law, but the dilemma still remained. As stated by Clausius (1850), the second law applied to the entire world (*"Die Entropie der Welt strebt einem Maximum zu"*). The anomaly was clear, and was embodied in the following contradiction—how could the organizational complexity that clearly existed in living systems, from biological organisms to social groups, be explained *without contradicting the second law?* Since they could not conclude that equilibrium (maximum entropy) meant order, as Parsons and others had done, how could they explain order in organizations?

It was clear that equilibrium was maximum entropy. To equate equilibrium with order would thus just be another contradiction, and no solution to the problem. Compounding one anomaly with another à la Parsons gets one nowhere, but simply leads to criticisms and challenges as the anomaly becomes more and more apparent. Rather, they had to either discard the second law, discard the empirical evidence of organizational complexity, or find a way to resolve the contradiction between these two. Since evidence for both the second law and organizational complexity was overwhelming, the obvious strategy was to find some reconciliatory model.

Fortunately, by the time GST was born, the tools for doing this were available. They were explained in detail in *General System Theory* (Bertalanffy 1968). The timing seemed to be right for this book. It was published seventeen years after Parsons (1951). Thus, readers knew of systems theory, but were aware of the mounting criticisms of functionalism during the 1960s, and curious about this newer approach. Readers' interest may have also been piqued by Buckley's (1967) *Sociology and Modern Systems Theory.* Bertalanffy also drew attention to the title by using the singular "system" instead of the customary plural "systems."

Readers of this volume found emphasis on the resolution of the second-law dilemma. Bertalanffy handled the problem neatly. He introduced Prigogine's (1955) equation for entropy in open systems, which is the crux of the resolution. This is probably the first exposure that many readers had to Prigogine's equation, and its application was a good example of how GST can accomplish its integrative goals by resolving anomalies which hinder synthesis of two or more fields. In Prigogine's work, and Bertalanffy's GST application of it, the second law of thermodynamics and the organizational complexity of biology and social science fit together like hand and glove.

Briefly, what GST contributed was:

1. To alter the definition of isolated, closed, and open systems (presented earlier).

2. To show that social and biological systems are open systems.

3. To show (à la Prigogine) that entropy does not necessarily remain constant or increase in open systems as it does in isolated thermodynamic systems.

By the 1950s, the classical definition of isolated, closed, and open systems could be modified in two important ways. First, Einstein's work allowed the distinction between matter and energy to be more or less ignored when discussing the permeability of systems' boundaries. The classical distinction between isolated and closed systems was that isolated systems allowed transfers of neither matter nor energy across boundaries, while closed allowed transfers of energy only (and open allowed both energy and matter). Einstein's work negated this distinction, as energy could not be said to flow across boundaries without concomitant exchange of matter. Since Einstein, matter-energy can be hyphenated, and only one type of closed system need be designated. This system is closed against transfers of matter-energy across its boundaries. It could be called "isolated," but the term "closed" has been adopted in GST (see Miller 1978).

Thus, the first definitional change now common in GST was to drop the term "isolated system" (for all practical purposes) and to distinguish generally only between "closed" (no matter-energy transfers) and "open" (matter-energy transfers). However, in the 1950s, GST could benefit from advances not only in physics but also in information theory, cybernetics, and homeostasis. The next step was thus to enlarge the definition of closed and open systems to include openness to information flows as well. The current definitions are as follows:

closed system—a system whose boundaries *do not* allow transfer of matter-energy or information from the environment.
open system—a system whose boundaries *do* allow transfer of matter-energy or information from the environment.

Thus, while some subsystems deal only with information flows in living systems theory, and others deal only with matter-energy flows, the boundary is a dual subsystem, dealing both with matter-energy and information (see Miller 1978).

Once the basic definitions of open and closed systems were updated and clarified, GST was in a position to reconcile the Spencerian dilemma, or the second law, against organizational complexity. It did so by application of Prigogine's (1955) equation for entropy in open systems (equation 4.1). This equation shows that open systems such as social organizations do not have to increase in entropy over time, but still do not violate the second law of thermodynamics. This is because, while internal entropy increases, importation of resources into the system can lead the overall level of entropy to stay the same, or even decrease.

Thus, by importing energy and information in excess of energy and information loss, negative entropy can be added to the system in quantities great enough to offset internal entropy gain. Plants do this, animals to this, and social systems do this. The Spencerian dilemma has been reconciled successfully. Although primary credit in this effort must be given to Prigogine for developing the basic entropy equations, credit is also due GST for putting it all together and applying it widely to biological and social systems.

At first glance, the GST approach to the issue of social order may seem similar to that of the earlier functional equilibrium theorists. After all, sociologists have known about open systems for some time. However, the results of the two approaches are very different. Let us briefly compare Parsons's approach to social order with that of GST, specifically the work of Bertalanffy (1968). Some similarities are:

1. Both have similar definitions of system.
2. Both recognize open systems.
3. Both assume the existence of order in social groups.
4. Both are integrative approaches.

Some differences are:

1. Parsons defines order as equilibrium; Bertalanffy defines it as negentropy.
2. Parsons uses a purely verbal analysis; Bertalanffy combines verbal analysis with mathematics.
3. Parsons's approach violates the second law; Bertalanffy's does not.
4. Parsons assumes change will result in either a return to equilibrium or a new equilibrium; Bertalanffy does not.
5. Parsons uses an abstracted (R) system; Bertalanffy uses a concrete system (Q) that is amenable to R-analysis as well.
6. Parsons's approach emphasizes part/whole analysis; Bertalanffy emphasizes relationships among all parts of the system, as well as system/environment relations (see Luhmann 1982).

While the number of similarities in the two approaches roughly equals the number of differences, the ramifications and implications of the two approaches are dramatically different. Parsons's approach is still mired in the Spencerian dilemma, still exhibits anomalies, is not operationalizable, and does not handle change well. Bertalanffy's approach has broken through the Spencerian dilemma, does not exhibit anomalies, is operationalizable, and does handle change. Consider this latter point. There is only one condition under which the functionalist equilibrium model predicts change. That is when the function to be served changes, then the system must change. However, in most cases the function to be served is assumed to be constant and permanent, and so the institution serving this function is also constant and cannot be dramatically altered.

However, in the GST model, both internal entropy (and external entropy), are assumed to change. Notice that the model is a *change* model, as it is written as *dS*. Thus, while students of social change who are analyzing functionalism have to find some way to deal with the perennial threat of the return to equilibrium, no such onus is present in the GST model. Change is not only accommodated but assumed. The basic model is written in terms of *change of entropy* (*dS*) rather than entropy (*S*) (the symbol *d* signifies change). However, *S* is the continuous measure of entropy. The statistical measurement of entropy in its categorical or nominal form, e.g., as measured by *H*, is synchronic analysis rather than diachronic (change) analysis. Thus, the GST model can easily accommodate *either* change (diachronic analysis) or lack of change in the form of cross-sectional analysis (synchronic analysis), or equilibrium—it has no built-in blocks which preclude certain approaches to the study of temporal phenomena as equilibrium models do (see the discussion of blocks in chapter 2). Equilibrium analysis still exists in GST, but not was a way out of the Spencerian dilemma—study of entropy change (*dS*) serves this purpose.

A fundamental difference in perspective is evident between the social equilibrium approach and the GST approach. While some sociologists are unaccustomed to equating negentropy with order and may resist it, they accepted (or at least did until it was challenged) the invalid equation of equilibrium with order. General systems theorists accept the interpretation of negentropy with order, and in fact take it for granted. However, they reject the notion that equilibrium is order or organizational complexity. The GST approach benefits from modern developments, while social equilibrium is obsolete.

What would it take to revitalize Parsons? It is obvious that Parsons's approach cannot be made consistent with modern systems theory without discarding the equation of equilibrium with order. This is easily done by simply omitting a few lines in the definition of system (Parsons and Shils 1951, p. 107). Another task is to "ground" Parsons by setting his systems model on the foundation provided by the concrete systems model. This is easily done, because Parsons recognized at least twice (Parsons and Shils 1951, p. 190; Parsons 1967, p. 328) that the concrete model is the "ultimate" model. We cannot stop short of the "ultimate" if we want a successful model. Related changes are to emphasize boundary properties (which Parsons did, but not in the context of our other changes) and to conflate his analysis of information and energy.

That is, Parsons treated energy transmission and information flows on separate levels in the cybernetic hierarchy of control (Parsons 1966). They need to be analyzed together in a concrete model, so that both can be seen in their true and complementary relationship to entropy. Separation of information and energy is artificial and overlooks the fact that information transmission requires a physical carrier and an expenditure of energy (Miller 1978). Once these things and a few others are done, Parsons can be brought up-to-date and revitalized, thus paving the way for his many contributions to general systems theory (e.g., AGIL) to be recognized by GST and to receive proper credit.

Sociology and Modern Systems Theory

In the year before Bertalanffy's *General System Theory* was published (1968), Buckley's *Sociology and Modern Systems Theory* was also published (1967); the next year a companion volume, *Modern Systems Research for the Behavioral Sciences* (Buckley 1968), an edited book of significant articles on general systems theory, was published. These books were well received by readers, and were a valuable link between the sociological systems tradition and the more contemporary GST. Although the book was linked to "functionalism" (Ritzer 1975, p. 67), this was to be expected, as this approach was dominant and was the only familiarity many sociologists had with systems theory.

Buckley introduced cybernetic principles to sociologists, emphasizing concepts such as feedback much more than Parsons had. He presented the standard feedback loops where a positive and negative relationship in concert lead to a steady state (see figure 4.1). In addition, he drew upon the innovative work of Maruyama (1963) to show that loops could be other than homeostatic or stead state. For example, loops containing only "pluses" (positive relationships) are termed "deviation-amplifying" loops by Maruyama, as they do not lead to a steady state, but expand indefinitely. Loops which are "deviation minimizing" in contrast contain all "minuses" (negative relationships) and proceed continuously down in a negative spiral. Both can be seen as examples of what are colloquially termed "circular causation."

In addition to cybernetic work on feedback, Buckley (1967) also drew upon information theory. He introduced the basic concepts of information and entropy, and wrote about negentropy as organizational complexity. On the sociological side, Buckley discussed Parsons in great detail. While critical of equilibrium, and while noting Parsons's careless conflation of homeostasis and equilibrium, Buckley showed clear appreciation for Parsons's approach, as well as for GST. He discussed action theory at length. In fact, the book is better described as action theory than as functionalism. It is basically a critique of functionalism and a championing and elaboration of Parsons's earlier, more social-psychological action theory approach, involving individual actors, their socialization, motives, and goals.

Kuhn

Alfred Kuhn wrote significantly on the social system within the general systems rubric. Kuhn (1963, 1974) was an economist who adopted a general systems approach that utilized equilibrium in the economic tradition, but did not founder on it. As used in economics, equilibrium is a model (X'') and *not* an empirical entity (X'). As long as this usage is consistently maintained, the Spencerian dilemma is avoided.

Kuhn, like Buckley, had an appreciation of Parsons's work and elaborated upon it, but did not universally adopt all of it. Published before Miller (1978), Kuhn (1974) presented a distinction between types of systems which parallels Miller's abstracted and concrete distinction (see chapter 2). Kuhn (1974) distinguished in some detail between acting systems and pattern systems. Clearly influenced by Parsons, Kuhn's acting system is an empirical system (X') in which human action takes place. In that sense it is very similar to Miller's concrete system. Kuhn distinguishes this acting system from a pattern system. Again, the Parsonian influence on the labeling is apparent, although Kuhn's pattern system should not be confused with Parsons's "pattern variables" (Ritzer 1988, p. 28).

Kuhn's pattern system is best understood as a symbol system. In that sense it seems to be similar to Miller's (1978) conceptual system (level X''). However, the elements of a pattern system can also be roles or empirically occurring symbols (X'). In that sense the pattern system seems similar to an abstracted system. Although, as always, it is difficult to ascertain the exact states of sometimes vague verbal formulations, the pattern system seems to be an empirically occurring (X') system in terms of our three-level model, and an R-system as opposed to a Q-system.

The difference between acting and pattern systems is, to oversimplify, a process/structure and diachronic/synchronic distinction as well as a Q-R distinction. An acting system is composed of human actors who utilize symbols to structure their actions over time. The particular pattern and structure that they use to guide their actions is the pattern system. It is basically synchronic or "structural" as it cannot move or act, although it can be changed by the actors of the acting system.

Boulding

Another important contributor to the social systems approach in general systems theory is Kenneth Boulding, a well-known economist, and past president of the Society for General Systems Research. He is the author of a number of books, including *Ecodynamics* (1978). Though not an entropy theorist (he has remarked that the concept is "backward"), Boulding (1956) has made important contributions in the analysis of organizational complexity within GST.

1970 to 1994

The last two decades show a definite maturation of GST. The first volumes of the *General Systems Yearbook* were very seminal and exciting, with contributions by the founders such as Bertalanffy, Boulding, and others. During the last two decades, the *Yearbook* has become a reprint vehicle rather than a journal for original articles, and it seems fair to say that the articles reflect maturation of a field rather than the originality and excitement of the first volumes. Most of the new developments have sprung up independently and have been incorporated into GST. These include artificial intelligence (AI), fractal theory, duality theory, and hierarchy theory (see Troncale 1978).

While some observers such as Miller feel that a maturing systems theory must move past early emphasis on systems philosophy and metatheory to operationalization and quantitative formalization of theory, this has happened only partially. Recent volumes of the ISSS proceedings, as well as of the *Yearbook,* show work on applications, but also continuing emphasis on systems theory, systems philosophy, and systems methodology (see for example, Banathy et al. 1985, volumes 1 and 2).

One prominent recent theorist is Banathy (1988). His work on systems inquiry in education and on other topics has spawned a viable subfield. His concept of systems inquiry combines three interrelated components: systems philosophy, systems theory, and systems methodology. Systems philosophy stresses holistic thinking, systems methodology provides paradigms, and systems theory has resulted in a general theory of dynamic systems. Banathy shows how these tools can be used to improve the overly specialized and fragmented system of education.

Another important contemporary systems theorist is Troncale (1978, 1985). His work on hierarchy theory and isomorphies has helped unify a diversifying field. He has developed a number of linkage propositions to link various systems concepts. He presents a classification of fifty-seven systems concepts organized into eleven categories.

Cavalo (1979; Cavalo and Klir 1978) has surveyed the systems field, showing work from systems philosophy to fuzzy sets. He and Klir present the "general systems problem solver" (GSPS) based on four basic classes: systems, problems, requirements, and methodological tools.

Another recent important theory is the work on "soft systems" by Checkland (1981). Checkland's work focuses on organizations, and utilizes soft systems methodology (SSM) to analyze organizations from an hermeneutic perspective in the interpretive tradition of Berger and Luckman (1967; see also Davies 1988). Checkland summarizes the SSM as a seven-stage model of organizational activities. The model is deemed appropriate for situations dealing with human action and interpretation where a linear causal analysis might not suffice. It is less of a "positivistic" approach than the general stereotype of systems theory.

Yet another significant contemporary general systems theorist is Klir (1969, 1978). In addition to his work on the GSPS, Klir has noted the generality and thus neutrality of the general systems model, saying that the model is really quite neutral until specific data is loaded into it. In addition, he has developed a hierarchical classification of systems types by epistemological levels: presystemic; dataless systems; data systems; behavioral systems; structural systems; and metasystems at various levels.

Still another important book for sociological systems theorists is Bates and Harvey (1975). They use a systems perspective to discuss the micro structure of social systems, the macro structure, the actor, and dimensional change and analysis. They emphasize time and space as Giddens (1979) suggests. Their model is dynamic rather than static.

Another sociological general systems theorist is Luhmann (1982). Although I discuss Luhmann's autopoietic systems theory at some length in chapter 8, this is perhaps a good time to note some of his other systems work. Luhmann studied with Parsons, and some of his early work was on functionalism, such as on the function of religion. His later systems work centers around autopoiesis and complexity. Luhmann emphasizes the system as a means of dealing with increasing complexity, as in the case of complex society.

Luhmann (1982) stresses the notion of self-regulating, self-referential, or self-implicating relations and concepts in his systems theory. He says:

> Traditional theory conceived complex systems as 'wholes' made out of 'parts'. The basic idea was that the *order* of the whole accounts for qualities the isolated parts could never possess on their own. Recent systems theory, as I see it, has abandoned this traditional approach by introducing an explicit reference to the environment. The concept of an environment does not merely imply that something else exists outside the system being studied.

It is not a question of distinguishing between the 'here' and the 'elsewhere'. The novel thesis rather, runs as follows: the structures and processes of a system are only possible in relation to an environment, and they can only be understood if considered in this relationship. . . . These formulations make clear that systems are objects that generate and regulate self-implicating relations . . . An adequate social theory must therefore be able to exhibit and cope with the self-referential structures of its object . . . The advance from a traditional to a modern theory of systems (that is, the critique of the old idea of a self-sufficient whole that confers 'perfection' on its parts) hinges precisely on the development of self-referential concepts. (Luhmann 1982, pp. 257–58)

For further discussion of Luhmann's systems theory see chapter 8. I should also like to note in passing that systems ideas are making a recent impact on sociological thinking, even though the "old" systems theory is some two decades old, and the "new" systems theory is just now becoming accessible to sociologists. For examples, see the systems chapters in Collins (1988) and Turner (1991), as well as the discussion in Ritzer (1988), and the discussion of Bailey's work in Ritzer (1990b). See also the impact of the work of Luhmann and Miller in Turner's (1991, pp. 631–34) discussion of mesotheorizing.

For an analysis of GST from a sociological perspective, see Ball (1978). Ball does a good job of describing the systems emphasis on relationships.

NONEQUILIBRIUM THERMODYNAMICS

Prigogine

There have been a number of thermodynamic researchers whose work has been central to the development of GST. We have already presented Prigogine's formula for entropy in open systems. His work is very important and deserves further comment. In a sense, Prigogine is the latest of four great contributors to the thermodynamic tradition in GST. The first development, of course, is the development of the work on entropy, equilibrium, and the second law. Although developed by a number of people, the name of Clausius (1879) is dominant.

After this fundamental development, the next major development was the extension of equilibrium from relatively homogeneous thermodynamic phenomena to nonhomogeneous chemical phenomena. The name of J. Willard Gibbs is prominent here, as already discussed. A third great development was the development of the statistical or probabilistic model of entropy by Boltzmann (Olsson 1967). In some ways information theory (Shannon and Weaver 1949) can be seen as an independent but closely related discovery.

Prigogine's work constitutes the fourth great thermodynamic development. His place in history as the founder of nonequilibrium systems theory is secure. While Gibbs's work was an important broadening of the application of equilibrium, it had little impact on fields such as social science which are open systems where the equilibrium model will not hold. Inasmuch as Pareto, Henderson, and others promulgated this closed-systems approach in sociology, it had negative effects and caused confusion in equilibrium theory.

While Gibbs's work was exceedingly important for nineteenth-century science, it did not liberate social science—Prigogine's work did. His single most important contribution was showing that entropy could in fact *decrease* in open systems, thus explaining how organizational complexity could increase without violating the second law. However, his work has much larger significance as it *legitimates nonequilibrium theory*. Before Prigogine's contributions, Parsons could indignantly say "in my not so humble opinion" that without equilibrium there could be no order in society and thus no social science (Parsons 1961b). Prigogine has laid to rest such a claim for all time. Nonequilibrium theory is now not only credible but required—a universal equilibrium position is obsolete. Now that Prigogine has developed a solid basis for nonequilibrium analysis of open systems, the foundation has been laid for myriad developments in this area. Most important is the general analysis of how organizations grow in complexity, but also important are burgeoning areas such as autopoiesis (Luhmann 1986), which is discussed in detail in chapter 8.

Prigogine's work takes on added significance for social systems theory because he has studied not only physical chemistry, but also social phenomena—principally traffic flows. His work has been a major force in stimulating and giving credibility to work in the study of social phenomena by physical scientists. Most of this work entails the application of entropy to social phenomena such as traffic flows or economic and demographic phenomena. The general pattern is for scientists to first apply their model in physics, then transfer its application to social science. According to Miller, Prigogine reversed this by applying a model developed for traffic flows to physical phenomena.

Prigogine's fame has expanded since, as noted above, he received the Nobel Prize in 1979. He is one of a number of Nobel Prize winners whose work is relevant to systems theory, although not all have principal identities as systems theorists. Others include Samuelson (1983) and Simon (1964).

The work of Brillouin (1956) is important for establishing the link between entropy and information. There is still a great deal of confusion in the literature about this and I discuss it in more detail in chapters 5 and 6, as it remains on the frontier of systems theory. Szilard (1929) showed that entropy could be interpreted as the reverse of information or order (negentropy is order). Szilard's article is very important but was little known until translated for *Behavioral Science* (1964). Building on this work, Brillouin shows that information is *not* H, but rather the difference between H values. Since H is synchronic and categorical, we cannot define $-H$ or dH, but can define $H_1 - H_2$. This difference is information gain (or loss). This is an invaluable contribution that is frequently neglected. It is sound, though, and is confirmed elsewhere (Bailey 1990; Theil 1967).

Chaos, Catastrophe, and Artificial Intelligence

Chaos theory, catastrophe theory, and artificial intelligence (AI) must all be classified as "related developments" that are somewhat tangential rather than central to social systems theory. They are not, strictly speaking, part of the systems paradigm, but are of interest to systems theorists. Chaos theory is a statistical development that has had some application in sociology (Beauchamp, 1989). Chaos theory is directly concerned with entropy, as chaos can be defined as maximum entropy (equilibrium). Although the general thrust in science has been (until recently) to study order, there has recently been a great deal of interest in physics, mathematics, and other sciences, in the study of nonlinear dynamics, including the study of chaos, and specifically, the phenomenon of "order out of chaos" (Prigogine and Stengers 1984). For further discussion of chaos theory see Gleick (1987), Schuster (1984), and

Prigogine and Stengers (1984). Another basic source on chaos is the work on synergetics by Haken (1983).

World Systems Theory

Although sharing the same term, there is little overlap between general systems theory or cybernetics in America and the adherents of world systems theory. The one most familiar to sociologists is based largely on the work of Wallerstein (1974). This approach is essentially Marxian and historical. One notable methodological feature is that it generally eschews focus on national boundaries, and so is basically a "boundaryless" approach. This form of world systems theory focuses on interactions in the world without reference to national boundaries, and emphasizes changes in world political systems, such as the transformation of capitalism. This is a marked difference from our approach in the present volume, which emphasizes boundary formation and maintenance.

The second variant of world systems theory is exemplified by the "Club of Rome" approach, and the work of Forrester and Meadows (Forrester 1973; Meadows et al. 1972). Interestingly enough, this approach *also* eschews emphasis on boundaries in its analysis, focusing on developments for larger regions or the whole world. It has a quite different focus from Wallerstein's approach, however, being essentially a statistical model or projection of basic world variables including economic and social indicators of various sorts.

The "New Cybernetics"

The work of Fararo (1989) demonstrates the efficacy of classic cybernetic tenets in mathematically formalizing Parsons's action theory. Fararo's work seems to be independent of recent work in sociocybernetics, but quite congruent with it. The new approach in sociocybernetics gets away from the machine and circuit imagery of early cybernetics, and adapts an explicit, actor-oriented, action theory approach. Chief architects of the new sociocybernetics include Geyer and van der Zouwen (1978, 1982, 1986), Aulin (1982, 1986), and Parra Luna (1990). Another approach to sociocybernetics has also been recently developed by Busch and Busch (1984, 1988). This form appears to have developed relatively independently of the work of Geyer and his associates.

Geyer and van der Zouwen (1978) offer perhaps the most accessible statement on the new sociocybernetics, terming it "an actor-oriented social systems approach." The salient concept is "steering," particularly the "self-steering" of the system. Geyer and van der Zouwen (1978) discuss a number of characteristics of the emerging "new cybernetics". One characteristic of the new cybernetics is that it views information as constructed and reconstructed by an individual interacting with the environment. This provides an epistemological foundation for science, by viewing it as observer-dependent. Another characteristic of the new cybernetics is its contribution towards bridging the "micro-macro gap". That is, it links the individual with the society.

Geyer and van der Zouwen (1978) also note that a transition from classical cybernetics to the new sociocybernetics involves a transition from classical problems to new problems. These shifts in thinking involve, among others: (a) a changes from emphasis on the system being steered to the system doing the steering, and to the factors which guide the steering decisions; and (b) a new emphasis on communication between several systems which are trying to steer each other.

In summary, the new sociocybernetics is a much more subjective and sociological approach than the classical cybernetics approach with its emphasis on control. The new approach has a distinct emphasis on steering decisions. Furthermore, it can be seen as constituting a reconceptualization of many concepts which are often routinely accepted without challenge.

Parra Luna (1990) has an approach to sociocybernetics stressing organization and political sociology. Parra Luna stresses the concepts of globality and complexity. He says:

> The theoretical/methodological approach which derives from this position is what I call *axiological/operational* because:
> (i)   it focuses on values pursued and/or attained through social action; and
> (ii)   it proposes formal procedures and operational definitions (through empirical indicators) which result in the numerical expression of a set of concepts (*efficiency, socialization, change, progress, regression, deviation,* etc.), rarely quantified in sociological literature. (Parra Luna 1990, p. 355 emphasis in original)

Perhaps the most explicit and comprehensive work in the new sociocybernetics is by Aulin (1982, 1986). Aulin developed a critique of Marxism, and developed cybernetic laws of development.

His highly mathematical approach relies on a foundation of classical cybernetics, particularly the work of Ashby (1956), including the law of requisite variety.

Included in Aulin's theory of the origin of human value is a discussion of partial steering from the outside (power), complete steering from the outside (totalitarianism), and human self-steering action (human freedom). He also discusses the cybernetic world order, causality and freedom, and the development of social systems. The relation to entropy is clear. Self-steering human systems are purposeful systems which work in the direction of organizing, thus combating entropy production (à la Prigogine). Aulin's formulation is reminiscent of the entropy theory of Galtung discussed elsewhere (Bailey 1983; 1990). For good examples of the new sociocybernetics see the work of Baumgartner (1986) on "actors, models, and limits to societal self-steering," and the cybernetic action theory of Burns (Burns and Buckley 1976).

Three other chief developments in social systems theory date from the 1970s and 1980s. These are living systems theory, social entropy theory, and autopoietic theory. All three of these, while informed by GST and falling loosely within its rubric, nevertheless have somewhat the status of independent systems paradigms, or at least subparadigms. Of the three, living systems theory is the most general, and has the greatest claim to being a true general systems theory (while explicitly limited to living systems). The other two are narrower. Social entropy theory applies to human systems only, while the range of applicability of autopoietic theory is the subject of intense debate, as seen in chapter 8. These three recent developments *all* date since the last major review of systems theory in sociology (Buckley 1967). Hence they deserve increased emphasis, unlike the developments of this chapter, most of which, with the exception of the new cybernetics and recent work on nonequilibrium thermodynamics, were discussed by Buckley. The next chapter (chapter 5) deals with living systems theory, chapter 6 with social entropy theory, chapter 7 with a synthesis of these two, and chapter 8 with autopoiesis.

COUNTERPOINT

This is the final chapter in my reconstruction of systems theory. This chapter is a bridge or transition chapter from the old systems theory to the new systems theory. The bulk of the material is two or three decades old, but is "new" in the sense of largely escaping the equilibrium trap.

The breakthrough presented in this chapter is the emphasis on entropy and nonequilibrium analysis, with the attendant emphasis on organization (as negentropy). The problem that functionalism had with "leaving time out" (Giddens 1979) is also rectified in this approach. Notice that entropy is generally presented in diachronic terms (change or dS) rather than in synchronic terms, as things were generally expressed in functionalism. Notice also that while Giddens (1979) criticizes functionalism for operating blindly and having a simple model, he is more receptive to the more sophisticated and detailed cybernetic feedback model displayed here (Giddens 1979, pp. 78–80). Notice still further that all of the new sociocybernetics, with its emphasis on actor-oriented self-steering is new and was not subjected to critique by Collins, Giddens, and others. In fact, as far as I know, no mainstream theorist has yet critiqued the new systems theory of Geyer and van der Zouwen, Aulin, Burns, Busch, and others. Notice too, that the bulk of developments in this chapter are derivative from the "systems technology" movement approved of by Giddens (1979).

What can be said about metatheory in this chapter? Again, as in the last chapter, the most prevalent forms are the first variant of M_U in the form of the analysis of entropy, and the third variant of M_U in the multidisciplinary sense. Also, the attempt by GST to formulate an overarching theory might be labeled M_O in Ritzer's (1990a) terms.

What independent contributions were made in this chapter that have *not* been made in mainstream theory, and thus lend added breadth and richness to the mainstream? The emphasis on entropy is a chief addition. Sociology, after so warmly embracing equilibrium, has been woefully slow in analyzing entropy, even though it is one of the chief intellectual tools of the twentieth century. The related notion of nonequilibrium systems theory is another contribution of this chapter. Another is the concept of morphogenesis. Still another is the presentation of the new sociocybernetics of Aulin, Geyer, van der Zouwen, and others. All of these contributions are neglected in mainstream sociological theory.

CHAPTER 5

Living Systems Theory

Living systems theory is the life's work of James Grier Miller. It is the culmination of an interdisciplinary effort of some thirty years' duration. Although a number of scholars from several disciplines were members of regular meetings during which the theory was developed, Miller led the group and took the major responsibility. His name appears as single author on the major work, *Living Systems* (Miller 1978).

James Grier Miller studied psychology as an undergraduate at the University of Michigan. He then transferred to Harvard where he became a Harvard Fellow and protégé of Whitehead. As psychology was then still in the philosophy department, he studied philosophy, and his first publication (1937) was on the work of Whitehead. Although Whitehead did not use the term "system," he had an influence on young Miller that was to mold his career as a systems theorist. Whitehead taught Miller the "fallacy of misplaced concreteness" which was to become part of the foundation and inspiration for his distinction between concrete systems and abstracted systems (see Miller 1978; Bailey 1981, 1983). He also guided Miller in the general direction of interdisciplinary efforts, which he considered a crucial intellectual endeavor.

Miller prepared for his task by becoming widely versed in a number of disciplines. As a Harvard Fellow he was insured tenure for life at Harvard. The program was designed to give superior scholars intellectual freedom and time to do research and writing without the burden of formal degrees. This is the major reason why his "fellow Fellow" George Homans never received a doctorate. Miller, however, earned both the Ph.D. in psychology and the M.D. degree at Harvard, the latter primarily so he could learn biology, as he apparently never really intended to practice medicine. Another Harvard Fellow there at about the same time was Paul Samuelson, whose work on equilibrium was discussed earlier.

Armed with degrees in both psychology and psychiatry, and

already a formidable degree of interdisciplinary knowledge, Miller accepted a position offered him by Talcott Parsons as the only clinician in the new Harvard Department of Social Relations. Soon, he had the opportunity to assume the chairmanship of the University of Chicago Department of Psychology (he had already served a stint as founder of the psychological unit for the Veterans Administration). He became full professor and chair at Chicago at age thirty, after the humanitarian President Bernard Hutchinson gave him a grueling interview that dealt not so much with psychology as with theological philosophy (a topic in which Miller was also well versed due to an earlier year spent in a theological school).

At Chicago he decided to pursue his mandate from Whitehead by forming an interdisciplinary scholarship group, which would meet twice weekly for two hours. Among those attending were Rappoport (mathematical biology), Gerard (psychology), Easton (political science), and a number of others from science and social science. Shils (Parsons's co-author) also attended some of the time, but he divided his time between Chicago and Cambridge and was not always present. Bertalanffy later sought entrance to the group, but his application was denied by the group, who apparently considered his work in systems unoriginal and uninspiring.

Miller's attempt at integrated research was encouraged by others, including Fermi, who thought integrative interdisciplinary research was very important. The group applied to President Hutchinson for support. His reply was essentially, "Go ahead, but you will not succeed." Before plans could proceed, the University of Michigan offered them funding and a building, and several moved en masse. There Miller headed the Institute of Mental Health for many years.

Living systems theory is a concrete systems approach. Miller opens the volume by discussing the value of the systems approach and some criticism. He then differentiates between concrete and abstracted systems. This distinction was presented in chapter 2. To briefly review, a concrete system is anchored in physical space-time, and is an interrelated (nonrandom) set of objects such as persons or other organisms. In contrast, an abstracted system has relationships or roles as the basic units of analysis rather than objects. This form reverses nouns and verbs. For example, a concrete system would say that Washington (object) was the president.

An abstract system would say that the presidency (role) was occupied by Washington. Miller says that:

> The units of *abstracted systems* are relationships abstracted or selected by an observer in light of his interests, theoretical viewpoint, or philosophical bias. Some relationships may be empirically determinable by some operation carried out by the observer, but others are not, being only his concepts. (Miller 1978, p. 19, italics in the original)

Concrete systems were probably the norm in sociology up until the time of Parsons, and were generally accepted by early systems theorists such as Spencer (1892 [1864]), Pareto (1935), Cannon (1929), and Henderson (1935). Parsons (Parsons and Shils 1951) departed from this position by championing the abstracted system, and supporting it strongly until his death (Parsons 1979). Miller opts for the concrete system, thus placing himself in a direct confrontation with Parsons. Miller prefers concrete systems for a number of reasons. He says that they are easier to operationalize, more familiar and easier to understand, and provide clear links from the social sciences to other disciplines such as natural sciences which use concrete systems. Miller distinctly espouses a physical space-time framework for systems analysis, while Parsons (1979) vehemently opposes it. For further discussion of abstracted versus concrete systems see Bailey (1981, 1990).

Miller perceives a living system as a system which takes in energy and information so as to maintain a state of negentropy. In his words:

> The *living systems* are a special subset of the set of all possible concrete systems. . . . They all have the following characteristics:
>
> (a)   They are open systems, with significant inputs, throughputs, and outputs of various sorts of matter-energy and information.
>
> (b)   They maintain a steady state of negentropy even though entropic changes occur in them as they do everywhere else. This they do by taking in inputs of foods or fuels, matter-energy higher in complexity or organization or negentropy, *i.e.,* lower in entropy, than their outputs. (Miller 1978, p. 18, italics in the original)

Miller says that every living system is homeostatic. Responding to the question of whether a "nonhomeostatic" system might exist, Miller says:

> My answer is that no such concrete system could exist. Every one
> must maintain at least one variable in steady state. Locating a
> concrete system which is not homeostatic would create serious
> doubts about my whole conceptual approach. (Miller 1978, p. 50)

This model is minimal and does not block the study of change as
early equilibrium models did. Further, while homeostasis is mini-
mally assumed, it is not central to the approach, and is rarely
mentioned in living systems theory.

Miller's approach is a fortuitous blend of the old and the new.
As a student at Harvard he studied under Henderson and Cannon
(including the study of Pareto), and was subjected to the same
unquestioned socialization into equilibrium theory as was Parsons.
However, his model escapes the Spencerian dilemma entirely. By
spending thirty years working on the model, Miller was able to
build a foundation in classical theory, and then add new developments as they occurred. This is clearly in the Comtean tradition,
and is an example of "positive positivism." Miller's open system
model is cognizant of homeostasis, but utilizes primarily the concepts of negentropy, information, and organizational complexity. It
thus exemplifies general systems theory as opposed to Parsonian
equilibrium theory. Furthermore, Miller took great pains to emphasize both *process and structure* throughout the volume, and to
provide hypotheses that facilitate operationalization. He was keenly aware of the criticisms of Parsonian theory (such as untestability
and blocking of change) and took steps to avoid these problems.

When Miller's group started in the 1940s they had not even
decided on a name for their endeavor—the term "system" came
later. Miller claims credit for coining the term "behavioral science"
in its modern sense, and has edited the journal *Behavioral Science*
since its inception. The group's chief purpose was to facilitate the
integration of interdisciplinary research. They faced a basic semantic procedural dilemma: which was most efficacious, to choose a
term already extant in a single field, and generalize it, or to choose
an entirely new term to apply to all fields? Each strategy had
benefits and drawbacks.

If they chose an extant term, its meaning would already be
familiar to some, and thus might facilitate integration. However,
the term might also carry unwanted connotations, and other disciplines might resist its use. As for the second strategy, the neutrality
of a new term would be welcomed, but this would mean that all

disciplines would have a new term to learn, which might impede progress. Ultimately they chose the latter strategy, and developed generally new and basically neutral terms, although some may appear familiar today. For example, even the important term "boundary" was not standardized at the time. While all systems in all disciplines have boundaries, they were known (and still are) by various terms such as "border," "permeable membrane," or "wall."

While such semantic standardization may seem folly to eth-nomethodologists who see it as an attempt to overcome index-icality, or may seem unimportant to disciplinary specialists content with their own specific terms, systems theorists see it as *crucial,* as only through terminological standardization can we integrate the disciplines, and thus rid science of some of its many redundancies and contradictions. One of the chief goals of the general systems science movement is to avoid the situation where researchers in different disciplines are researching the same areas using different terms without knowledge of the other's efforts, or where the same term is being used for different applications and in different ways in different fields. However, the results from terminological standard-ization take many years to be accepted in some cases, and some of the terms in living systems theory may still appear unfamiliar and perhaps strange to some readers. Just remember that they are de-signed to be universal and neutral.

The Basic Theory

The basic living systems theory (LST) was built upon a search for the common properties of *all* living systems. Although it is limited to living systems and thus will not apply wholly to machines or cybernetic systems, it is still a true *general* systems theory. Of course, many of the principles apply to nonliving systems as well. According to the theory, all living systems are composed of sub-systems, each processing matter-energy or information, with two subsystems (the reproducer and the boundary) processing *both* matter-energy and information.

It should be noted at this point that the basic theory is in flux, and may continue to be. Originally, *Living Systems* (Miller 1978) presented nineteen basic subsystems at seven levels, and this is the formulation that I used when I began writing this volume. Since then, James Grier Miller and his wife and co-author Jessie Miller, have added a twentieth subsystem, the timer, and an eighth level, the community (see Miller and Miller 1992). The community level was added between the organization and the society. The timer subsystem was added to the original list of nine information-processing subsystems, making a total of ten (not counting the boundary and reproducer, which process both information and matter-energy).*

The twenty subsystems are shown in figure 5.1. These twenty subsystems are responsible for the ongoing day-to-day operation of the living system. In other words, it is these twenty subsystems which keep the system alive. Although Miller does not say so, or use the terms "function" or "functionalism," there is a clear parallel with functionalism, especially earlier anthropological functionalism (see Turner and Maryanski 1979). Clearly, each of the twenty subsystems serves a "function" for the system as a whole. But Miller's approach does not emphasize part-whole relations as functionalism did.

Living systems theory departs from functionalism in three significant ways: (1) it does not use the term "function" or identify with the functionalist perspective (in fact, it distinguishes itself from Parsonian abstracted theory); (2) it is more detailed than the listings of functions sometimes provided by the functionalists; and (3) it emphasizes the unity of all twenty subsystems working together for the whole, rather than emphasizing the function that each component separately serves for the survival of the whole, as functionalism did (Turner and Maryanski 1979).

Living systems theory is much more comprehensive and detailed than past approaches. It also emphasizes concreteness and de-emphasizes analytical separation. The biological and natural sciences have tended to emphasize the analysis of energy at the expense of information, while the social sciences have tended to

---

*I find it somewhat difficult to properly cite the basic LST since it is in flux. The proper reference for Miller 1978 is to nineteen subsystems and seven levels, rather than to twenty subsystems and eight levels. However, these are in a sense obsolete. Thus, we are in the awkward position of having a basic reference work (Miller 1978) with the incorrect number of subsystems and levels. In this situation, I beg the reader's indulgence if I sometimes slip and write in terms of the original formulation of nineteen and seven (which is of course correct when the reference is to Miller 1978). In general, however, I will attempt to conform to the current totals of twenty subsystems and eight levels.

FIGURE 5.1.

The Twenty Critical Subsystems of a Living System

SUBSYSTEMS WHICH PROCESS BOTH MATTER-ENERGY AND INFORMATION

1. *Reproducer*, the subsystem which carries out the instructions in the genetic information or charter of a system and mobilizes matter, energy, and information to produce one or more similar systems.

2. *Boundary*, the subsystem at the perimeter of a system that holds together the components which make up the system, protects them from environmental stresses, and excludes or permits entry to various sorts of matter-energy and information

SUBSYSTEMS WHICH PROCESS MATTER-ENERGY

3. *Ingestor*, the subsystem which brings matter-energy across the system boundary from the environment.

4. *Distributor*, the subsystem which carries inputs from outside the system or outputs from its subsystems around the system to each component.

SUBSYSTEMS WHICH PROCESS INFORMATION

11. *Input transducer*, the sensory subsystem which brings markers bearing information into the system, changing them to other matter-energy forms suitable for transmission within it.

12. *Internal transducer*, the sensory subsystem which receives, from subsystems or components within the system, markers bearing information about significant alterations in those subsystems or components, changing them to other matter-energy forms of a sort which can be transmitted within it.

13. *Channel and net*, the subsystem composed of a single route in physical space, or multiple interconnected routes, over which markers bearing information are transmitted to all parts of the system.

(continued)

FIGURE 5.1 (*Continued*)

SUBSYSTEMS WHICH PROCESS MATTER-ENERGY

5. Converter, the subsystem which changes certain inputs to the system into forms more useful for the special processes of that particular system.

6. Producer, the subsystem which forms stable associations that endure for significant periods among matter-energy inputs to the system or outputs from its converter, the materials synthesized being for growth, damage repair, or replacement of components of the system, or for providing energy for moving or constituting the system's outputs of products or information markers to its suprasystem.

7. Matter-energy storage, the subsystem which places matter or energy at some location in the system, retains it over time, and retrieves it.

SUBSYSTEMS WHICH PROCESS INFORMATION

14. Timer, the subsystem which transmits to the decider information about time-related states of the environment or of components of the system. This information signals the decider of the system or deciders of subsystems to start, stop, alter the rate, or advance or delay the phase of one or more of the system's processes, thus coordinating them in time.

15. Decoder, the subsystem which alters the code of information input to it through the input transducer or internal transducer into a "private" code that can be used internally by the system.

16. Associator, the subsystem which carries out the first stage of the learning process, forming enduring associations among items of information in the system.

17. Memory, the subsystem which carries out the second state of the learning process, storing information in the system for different periods of time, and then retrieving it.

18. Decider, the executive subsystem which receives information inputs from all other subsystems and transmits to them information outputs for guidance, coordination, and control of the system.

19. Encoder, the subsystem which alters the code of information input to it from other information processing subsystems, from a "private" code used internally by the system into a "public" code which can be interpreted by other systems in its environment.

20. Output transducer, the subsystem which puts out markers bearing information from the system, changing markers within the system into other matter-energy forms which can be transmitted over channels in the system's environment.

8. Extruder, the subsystem which transmits matter-energy out of the system in the forms of products or wastes.

9. Motor, the subsystem which moves the system or parts of it in relation to part or all of its environment or moves components of its environment in relation to each other.

10. Supporter, the subsystem which maintains the proper spatial relationships among components of the system, so that they can interact without weighting each other down or crowding each other.

Source: Miller and Miller 1992, p. 4. Reprinted by permission of the authors, Behavioral Science, and McGraw-Hill, Incorporated. This figure is an adaptation of Table 1.1, page 3 of James Grier Miller, *Living Systems*. Copyright © 1978 by McGraw-Hill, Incorporated.

emphasize information flows (as in communications) and to de-emphasize energy, except in a few specialties such as human ecology (Micklin and Choldin 1984). Parsons analytically separated energy and information in his cybernetic hierarchy of control (Turner and Maryanski 1979) so that information appeared to be at a "higher level" of analysis, with the implication that its analysis was more compatible with the analysis of cultural and social factors, and thus a more suitable topic for sociology than energy analysis. Parsons saw the study of matter-energy as squarely within the physical space-time framework which he rejected as the proper arena of sociology, preferring instead the analytical purity of the abstracted system (Parsons 1979).

In eschewing primary emphasis on the abstracted system, Miller is also eschewing the analytical separation of information and matter-energy flows. This approach seeks to integrate the analysis of energy and information, which Miller sees as analytically separated so that, in past approaches, one component (either energy or information) is emphasized at the expense of the other. While physics may emphasize energy and sociology may emphasize information, in reality they are seen as parallel processes in every living system. Further, information *cannot be transmitted* without a physical carrier of some sort. This physical carrier of information is known as a marker (Miller 1978; Bailey 1983).

Energy and information are symmetrically interrelated in a complex fashion in modern society. Efficacious usage of energy depends upon information, while transmission of information in turn is only possible through the use of matter to carry the message, and the expenditure of energy to move the message from its origin to its destination. The relationship between matter-energy and information is one of the most crucial issues in systems science, and is in a sense the key to understanding how societies operate. Unfortunately, this nexus has been neglected, and remains a frontier of systems science. I will consider it in more detail at the end of this chapter. It can be noted for now that the history of information transmission involves a clear trend from heavier to lighter markers or information carriers. While the original markers were the stone tablets of Hammurabi's time, modern markers are air waves, and film, with storage on tapes and computer diskettes. Not only are the markers becoming smaller and more efficient, but information transmission is taking place over longer distances (for example, with satellite transmission).

It should also be noted that the relationship between the amount of energy expended and the amount of information transmitted is imperfect and nonlinear. For example, if I pass you a note with random markings (no information) I expend almost as much energy as if I pass you a note filled with a high degree of information (maximum departure from randomness). Further, the "value" of the information is not perfectly correlated with the energy expended in its transmission. It takes me just as much energy to make and transmit a one dollar bill as a thousand dollar bill (see Miller 1978).

Although Miller does not use functionalist terminology, I shall, simply because it is efficacious in explaining the operation of the twenty subsystems. The twenty subsystems form a *concrete system*, having a *concrete boundary*, and *concrete substructures* (subsystems) which carry out twenty *processes*. Thus the subsystems can in a sense be viewed as *processes* rather than *structures*. In a sense this brings us back to the concrete/abstracted problem, especially in the case of multiple roles, as it is possible for one subsystem's *structure* to carry out more than one of the twenty vital processes. Conversely, it is possible for more than one substructure to carry out a single one of the twenty processes. That is, it is possible for a single concrete substructure to carry out multiple processes or for multiple concrete substructures to carry out a single process. In general, though, the assumption is that one structure carries out one process. However, the names tend to be *structure* names, rather than *process* names (for example, designation as "the decider" rather than "deciding," or the "distributor" rather than "distributing").

I can summarize the vital processes ("functions") of *any* living system quite succinctly. Matter-energy and information *must* be efficaciously brought into the (open) system from the environment. This dual function (both matter-energy and information transmission) is the burden of the boundary. The boundary really has multiple roles. It must admit (input) the *correct type* of matter-energy (including humans) and information, and in the *correct amounts*. It also must allow for proper exit (output) of the correct type and amount of matter-energy and information. The functions of the boundary thus include:

1. *Input* the correct type of matter-energy.
2. *Input* the correct amount of matter-energy (protect against underload (scarcity) or overload (surplus).
3. *Input* the correct type of information.
4. *Input* the correct amount of information (protect against overload or underload).
5. Protect the system against all other intrusion (provide security).
6. *Output* the correct type of matter-energy.
7. *Output* the correct amount of matter-energy.
8. *Output* the correct type of information.
9. *Output* the correct amount of information.

Besides the boundary, the other subsystem that processes both matter-energy and information is the reproducer (whose function is evident). As Miller (1978, p. 3) says, the reproducer is "the subsystem which is capable of giving rise to other systems similar to the one it is in."

After the reproducer and boundary serve their input functions, the other eighteen systems are seen to operate basically in parallel series (with some exceptions). The basic functions they fulfill (one for information and one for matter-energy) are decoding and incoding into a form suitable for processing by the system, distributing to the parts of the system where the information or matter-energy is needed, making decisions about how it is to be used, using the energy or information, and outputting (discarding waste and transmitting finished products).

The ten subsystems that perform these functions for information are: the input transducer, the internal transducer, the channel and net, the timer, the decoder, the associator, the memory, the decider, the encoder, and the output transducer. The eight parallel counterparts for matter-energy are: the ingestor, the distributor, the converter, the producer, the matter-energy storage subsystem, the extruder, the motor, and the supporter.

If all twenty vital processes are fulfilled, then the living system will function properly. In the ideal situation, not only its survival but its "good health" or optimal functioning will result. This will come about in the case where all of the necessary types and amounts of energy and information are satisfactorily input, decoded, processed, distributed, utilized, and output. In this case, external entropy flows or dS_e in Prigogine's (1955) equation will be negative (negentropy) and greater in absolute value than internal entropy increase (dS_i) which will always be positive according to the second law of thermodynamics.

This successful functioning of the system's twenty vital processes is contingent upon a number of factors. To begin with, the proper amounts of matter-energy and information must be available in the environment, so that they can be input into the system. Obviously any number of factors can hinder this, such as war or trade competition, scarcity, or receipt of unusable or counterfeit products or information.

The hallmark of Miller's scheme in a real sense is its comprehensiveness. Most physical, social, or biological models incorporate *some*, but generally not *all*, of his twenty processes. One of the chief benefits of his systems model is that it can guide the construction of other models, including models of nonliving systems as well as living systems (such as space stations, for example). Further, inasmuch as the living systems model was derived independently of earlier functional models, it serves as a replication of them. Consider the work on "functional prerequisites" by Aberle et al. (1950). All of the functions listed as being necessary for any society are clearly presented in living systems theory (for example, the reproducer). Notice, however, that their scheme is clearly incomplete. Notice also, that according to living systems theory, these functions are necessary not only for societies as Aberle et al. imply, but for all living systems.

The twenty subsystems obviously provide a wide array of "functions," from security, input of materials and information, and internal distribution of materials and information, to decoding and encoding, decision making, reproduction, and extrusion (output). Are some of the twenty subsystems more important than others to the operation of the living system? Should they be rank-ordered? Further, are some of the twenty more crucial to the definition and existence of a system?

There is no rank-ordering of systems (Miller 1978). *All* are simultaneously or sequentially in operation. All are *necessary*, and together the *set is sufficient*. While the variety of the twenty subsystems' structure (and process) may appear quite diverse within a single system, all twenty are compatible and work together harmoniously (in most cases). Thus, they are interrelated. Also, each *part* (subsystem) clearly has a *function* for the *whole*. Miller (1978, pp. 32–33) calls the subsystems "critical", meaning that they are all necessary for a system to survive. However, in some cases the function of a given subsystem may be carried out by some other system, as in the case of living systems which are parasitic on or symbiotic with some other living system. The only subsystem that is essential and cannot be substituted through being parasitic or symbiotic, is the decider (Miller 1978, p. 32). For further discussion of the functions of the twenty subsystems see Bailey (1993).

Notice that with the exception of the reproducer and boundary, all of the other eighteen subsystems are listed as processing either matter-energy or information, but not both. Although I said above that the matter-energy and information systems run in parallel, this is only partially true, and deserves discussion. Of the eight subsystems which process matter-energy, five are said by Miller to have direct one-to-one counterparts which process information (although obviously with some differences due to the differing natures of their task). These five processing pairs (with the matter-energy subsystem listed first) are: the ingestor/input transducer, the distributor/channel and net; the converter/decoder; the producer/associator; and the matter-energy storage subsystem/-memory. In addition to these five one-to-one parallel pairs, there is another parallel set where two matter-energy subsystems are paired with one information subsystem. This set consists of the extruder and motor (matter-energy) paired with the output transducer (information). This leaves five remaining subsystems which are "singles" and do not have parallel counterparts on the other side of the chart in figure 5.1. One of these is a matter-energy subsystem (the supporter). It maintains correct spatial relationships among components of the system. Since information does not have spatial relationships in the same sense as matter-energy, there is no information-processing counterpart to the supporter. There are also four information-processing subsystems which have no matter-energy parallels. These are the internal transducer, the timer, the decider, and the encoder.

The parallel nature of these processes was discovered inductively in the course of the thirty-year development of living systems theory. Recently, parallel processing has been seen as very significant in computer science. While complex mathematical problems (such as engineering applications in fluids) often tax sequential-processing machines, computers which can process in parallel are much more suited to the solution of such problems. Note, though, that in living systems theory, the parallel material and information-processing acts will not necessarily occur simultaneously. Due to the complex interrelationship of energy and information (to be discussed later), it may be necessary for matter to be processed first to obtain the energy to process information. Conversely, it may be necessary to process information in order to learn how to correctly carry out some matter-energy processing operations. In general though, the tasks of the system (for example, distributing) can be seen as tasks which need to be carried out in parallel for *both* matter-energy and information. Further, the distinction between matter-energy and information is not always as empirically clear as it seems. It has been noted that two of the twenty critical subsystems (the boundary and the reproducer) process both matter-energy and information. Besides these two, there are other subsystems such as the input transducer and the internal transducer which also process matter-energy (by changing the matter-energy form of the marker) in the course of facilitating the transmission of information within the system.

While the twenty subsystems are not "officially" rank-ordered in living systems theory (although adherents may of course have their own "favorites" which they consider of primary interest or importance), ranking does occur when the eight levels of living systems are analyzed. According to living systems theory (Miller 1978), all twenty subsystems operate in all living systems. These living systems constitute eight hierarchical "levels"—cell, organ, organism, group, organization, community, society, and supranational. The twenty subsystems at each of the eight levels are shown in figure 5.2.

The eight levels are hierarchical or "nested" in the sense that each higher level contains the next lower level in a nested fashion. Thus, the cell is the bottom level. The second level, the organ, contains cells as subsystems. The third level, the organism, contains organs as subsystems and thus cells as "second-order" subsystems. The fourth level, the group, contains organisms as subsystems, organs as second-order subsystems, and cells as third-order subsystems. The fifth level, the organization, contains groups as subsystems, organisms as second-order subsystems, organs as third-

FIGURE 5.2.
The 160-Cell Matrix Showing all Twenty Subsystems for All Eight Levels of Living Systems Theory

Subsystem Level	Reproducer	Boundary	Ingestor	Distributor
Cell	DNA and RNA molecules	*Matter-energy and information:* Outer membrane	Transport molecules	Endoplasmic reticulum
Organ	Upwardly dispersed to organism	*Matter-energy and information:* Capsule or outer layer	Input artery	Intercellular fluid
Organism	Testes, ovaries, uterus, genitalia	*Matter-energy and information:* Skin or other outer covering	Mouth, nose, skin in some species	Vascular system of higher animals
Group	Parents who create new family	*Matter-energy:* Inspect soldiers; *Information:* Television rules in family	Refreshment chairman of social club	Father who serves dinner
Organization	Chartering group	*Matter-energy:* Guards at entrance to plant; *Information:* Librarian	Receiving department	Assembly line
Community	National legislature that grants state status to territory	*Matter-energy:* Agricultural inspection officers; *Information:* Movie Censors	Airport authority of city	County school bus drivers
Society	Constitutional convention that writes national constitution	*Matter-energy:* Customs service; *Information:-* Security agency	Immigration service	Operators of national railroads
Supranational System	United Nations when it creates new supranational agency	*Matter-energy:* Troops at Berlin Wall; *Information:* NATO security personnel	Legislative body that admits nations	Personnel who operate supranational power grids

Part 1. Source: Miller and Miller 1992 p. 6.

Converter	Producer	Matter-Energy Storage	Extruder	Motor	Supporter
Enzyme in mito-chondrion	Chloroplast in green plant	Adenosine tri-phosphate	Contractile vacuoles	Cilia, flagel-lae, pseu-dopodia	Cytoskeleton
Gastric mu-cosa cell	Islets of Lan-gerhans of pancreas	Central lumen of glands	Output vein	Smooth mus-cle, cardiac muscle	Stroma
Upper gas-trointestinal tract	Organs that synthesize materials for metabolism and repair	Fatty tissues	Sweat glands of animal skin	Skeletal mus-cle of higher animals	Skeleton
Work group member who cuts cloth	Family mem-ber who cooks	Family mem-ber who puts away groc-eries	Mother who puts out trash	Driver of family car	Birds that build nests
Operators of oil refinery	Factory pro-duction unit	Stockroom personnel	Janitorial staff	Crew of company jet	Building re-pair and maintenance personnel
City stock-yard organi-zation	Bakery	County jail officials	City sanita-tion depart-ment	City transit authority	Maintenance crew at capi-tal building
Nuclear in-dustry	All farmers and factory workers of a country	Guards at na-tional armory	Export orga-nizations of a country	Trucking in-dustry	Officials who operate na-tional public buildings and lands
EURATOM, CERN, IAEA	World health organization	International storage dams and reservoirs	Downwardly dispersed to societies	Operators of United Na-tions motor pool	People who maintain in-ternational headquarters buildings

FIGURE 5.2.

The 160-Cell Matrix Showing All Twenty Subsystems for All
Eight Levels of Living Systems Theory (cont.)

Subsystem Level	Input Transducer	Internal Transducer	Channel and Net	Timer
Cell	Receptor sites on membrane for activation of cyclic AMP	Repressor molecules	Pathways of mRNA, second messengers	Fluctuating ATP and NADP
Organ	Receptor cell of sense organ	Specialized cell of sinoatrial node of heart	Nerve net of organ	Heart pacemaker
Organism	Sense organs	Proprioceptors	Hormonal pathways, central and peripheral nerve nets	Suprachiasmatic nuclei of hypothalamus
Group	Lookout of gang of thieves	Group member who reports members' attitudes to group decider	Person-to-person communication channels among group members	Mother who wakens other family members on time
Organization	Secretaries who take incoming calls	Factory quality control unit	All users of corporate phone network	People who operate factory whistle
Community	Representatives who report from state capital to local voters	Neighborhood watch groups	Telephone linesmen in city	Caretakers of clock on city hall tower
Society	Foreign news services	Public opinion polling organizations; voters	Telephone and communications organizations	Legislators who decide on time and zone changes
Supranational System	UN Assembly hearing speaker from nonmember territory	Speaker from member country to supranational meeting	Universal Postal Union (UPU)	Personnel of Greenwich observatory

Part 2. Source: Miller and Miller 1992 p. 7. Reprinted by permission of
the authors, Behavioral Science, and McGraw-Hill, Incorporated.
Adapted from material in James Grier Miller, Living Systems.
Copyright © 1978 by McGraw-Hill, Incorporated.

Decoder	Associator	Memory	Decider	Encoder	Output Transducer
Molecular binding sites	Unknown	Unknown	Regulator genes	Structure that synthesizes hormones	Presynaptic membrane of neuron
Second echelon cell of sense organ	None found; upwardly dispersed to organism	None found; upwardly dispersed to organism	Sympathetic fibers of sinoatrial node of heart	Presynaptic region of output neuron	Presynaptic region of output neuron
Sensory nuclei	Unknown neural components	Unknown neural components	Components at several echelons of nervous system	Temporoparietal area of dominant hemisphere of human cortex	Larynx
Member who explains rules to team	Parents who teach good behavior	Father who stores family records	Parents, family council	Writer of group communication	Jury foreman
Foreign language translation group	People who train new employees	Filing department	Top executives, department heads, middle managers	Annual report writers	Public relations department
Attorney general of state who interprets law	City school teachers	Operators of central police computer	Governor, legislators, judges of state	Writers of city ordinances	Representatives from state to rational legislature
Cryptographers	All teaching institutions of a country	Keepers of national archives	Voters and officials of national government	Drafters of treaties	National representatives to international meetings
Simultaneous translation staff of supranational organization	FAO units that teach farming methods in third world nations	Librarians of UN libraries	Council of Ministers of the European Communities	UN Office of Public Information	Top official who announces decisions of suprarational body

order subsystems, and cells as fourth-order subsystems. The sixth level, the community, contains organizations as subsystems, groups as second-order subsystems, organisms as third-order subsystems, organs as fourth-order subsystems, and cells as fifth-order subsystems. The seventh level, the society, contains communities as subsystems, organizations as second-order subsystems, groups as third-order subsystems, organisms as fourth-order subsystems, organs as fifth-order subsystems, and cells as sixth-order subsystems. The eighth and last level, the supranational system, contains societies as subsystems, communities as second-order subsystems, organizations as third-order subsystems, groups as fourth-order subsystems, organisms as fifth-order subsystems, organs as sixth-order subsystems, and cells as seventh-order subsystems. Thus, each K-th level has K-1 respective subsystem levels within it.

These levels require some clarification. Remember that the system represented at each of the eight levels is specifically designated as a *concrete* system by Miller as opposed to a conceptual or abstracted system (Miller 1978). As such, it is amenable to empirical observation and induction. The twenty subsystems were derived inductively ("grounded theory" in the terms of Glaser and Strauss 1966) by examining actual cases, or by secondary analysis of research reports.

The lowest level is the cell. Going below this level would require observation of subcellular objects that function as unitary concrete systems. It is possible to identify *organelles* as components of cells (Miller 1978, p. 205), but Miller decided on the cell as the lower limit. For further discussion see Miller 1978. The upper level is the supranational system. This is a system composed of interrelated societies. Examples are the United Nations, the European Economic Community (EEC), the Organization of Oil Producing Countries (OPEC), the North Atlantic Treaty Organization (NATO), and the Warsaw Pact. Notice that this level is *not* a "world system" in the sense defined by either Wallerstein (1974) or Forrester (1973). Though quite disparate, the two basic world system approaches both eschew analysis of national boundaries, and look at relationships for the entire world. Miller's supranational system does *not* include the whole world, but only certain member nations, and so it is called a supranational system rather than a world system.

The cell, organ, and organism (individual) levels are definitionally and empirically quite clear. The group level is defined as a dyad or larger, as long as no internal hierarchy is present. That is, a group is a nonhierarchical system of size N, where $1 < N \leq K$.

All groups are of size 2 or more. A "group" of one individual would be treated as an organism. The upper bound K is not mathematically set. Rather, a "group" becomes an "organization" when there is evidence of an internal "echelon," or internal hierarchical authority structure of chain of command (president, vice-president, and so forth). While Miller does not set a numerical maximum size for a group, he does discuss a large amount of literature on internal communication structures within groups (Miller 1978, pp. 515–93).

Once a system exhibits an internal rank-ordered authority structure or chain of command, Miller calls the system an "organization" instead of a "group." He uses the term "echelon" instead of "hierarchy" for this internal ordering so as to distinguish it from the "hierarchy" constituted by the eight levels. The nature of the echelon, including the number of levels, the number of people in the echelon, and so on, depends upon a number of sociological variables as just discussed, including the size of the organization, its goals, the ascribed and achieved characteristics of its members, and other factors, including the type of technology utilized.

Miller makes two main points about the eight levels: they are formed through evolutionary processes, and they are characterized by "emergence." Both of these points require discussion. Miller takes it as axiomatic that the higher levels evolve from the lower ones. Thus, organs evolve from cells, organisms from organs, groups from organisms, organizations from groups, communities from organizations, societies from communities, and supranational systems from societies. While he says that all sociologists he has talked to agree with him that social systems have evolved from the biological level (Miller 1987), sociologists seem to devote very little if any attention to this matter, judging from the literature. Only rarely do sociologists discuss the "evolution" of organizations (see Stinchcombe 1965). There is a potential link between Miller's evolutionary development of hierarchical levels and the evolutionary explanations of functionalism. Classical functionalists sometimes fell back on "evolution" to explain how particular institutions (parts) were selected to have a particular function for the society as a whole (Merton 1949). However, the evolutionary explication of a function in part/whole analysis is quite a different application than Miller's explication of the *entire* system (not just a part) on all eight levels.

Miller's evolutionary schema is much more familiar to biologists. Most biologists accept the evolutionary development described by Miller. However, there is a controversy over evolution in biological clustering, or numerical taxonomy, a field which has a number of distinct parallels to general systems theory (see Bailey 1982, 1985, 1987). Numerical taxonomists construct classification schemes mathematically, generally through Q-analysis (see Sokal and Sneath 1963; Sneath and Sokal 1973). While accepting the cladistic (diachronic) evolutionary theory, their analyses are often necessarily cross-sectional or synchronic (phenetic), thus arousing criticism and controversy from evolutionary purists.

Miller describes the evolution of the eight levels in terms of the principle of "shred-out" (Miller 1978, p. 1). More recently this same principle has been referred to as "fray-out" (Miller and Miller 1992, p. 2). This is an analogy to a ship's hawser, or a large rope or cable in general. This is illustrated in figure 5.3. While the hawser appears unitary, upon dissection it is found to contain a number of smaller hawsers, each representing the larger, but in miniature. Imagine that the cable had a plastic cover (boundary). When this is removed, the cable can be "shredded out" to reveal a lower level of smaller cables, all combined to form the whole. These in turn can each be shredded to reveal another lower level of still smaller cables, which in turn can be shredded, and so forth. The analogy is a little bit confused by whether one is going "up" the chain from the cellular level, or "down," from the supranational level. Thus, figure 5.3 can be read in either direction (i.e., the illustration can be turned upside down).

Miller's evolutionary schema is quite compatible with Spencer's evolutionary framework in *First Principles* (Spencer 1892 [1864]). Miller avoids the Spencerian dilemma by eschewing reliance on equilibrium in this schema. Contemporary sociologists have had little experience with evolution except in a few specialties such as social change, and with the exception of a brief resurgence of interest in evolutionary theory in the 1960s by Parsons (1966) and others (see Ritzer 1988). Contemporary sociologists seem to have no trouble accepting the similarity of the top five levels (group, organization, community, society, and supranational). They can

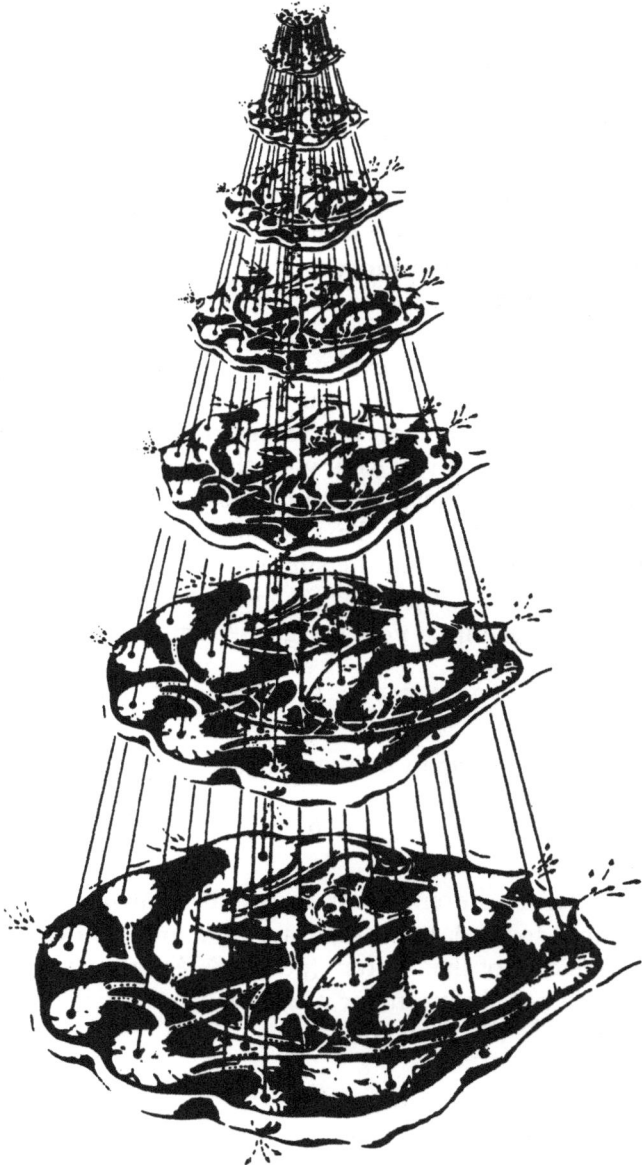

FIGURE 5.3.
An Illustration of the Eight Levels of Living Systems Theory and the
Principle of Fray-out

Level

Cell
Organ

Organism

Group

Organization

Community

Society

Supranational
System

Source: Miller and Miller 1992, p. 2. Reprinted by permission of the authors,
Behavioral Science, and McGraw-Hill, Incorporated. This figure is an
adaptation of Figure 1.2 on page 4 of James Grier Miller, *Living Systems*.
Copyright © 1978 by McGraw-Hill, Incorporated.

see the system's counterparts at all levels. Some can accept the general schema for all six top levels (individual, group, organization, community, society, and supranational). However, below the level of the individual (cell and organ), many social scientists seem to see (or "feel") a discontinuity. They have difficulty accepting the cell and organ as living systems "in the same way" that the individual, group, organization, society, and supranational are. Thus, even though Miller (1978; Miller and Miller 1992) sees the eight levels as "continuous" and all the same in that they are all living systems and all have the same twenty subsystems, some observers, both social scientists and biological scientists, see the organismic level as a point of disjuncture. If not a "boundary" or border it is at least discretely different, or an intellectual dividing line. Many sociologists probably are not ready to accept a supranational system as a cell—seven evolutionary steps removed.

A second main point to be made about the eight levels deals with the phenomenon of "emergence." Emergence can generally be defined as the process whereby a phenomenon that does not exist at lower levels of an analytical scheme "emerges" at higher levels of the scheme. This is the classic systems notion that "a whole is greater than the sum of its parts." In statistical terms, if five persons $(P_1, P_2, P_3, P_4, P_5)$ were "summed" or placed together to form a group, a nonemergent (additive) equation would be:

$$Y = \Sigma P_i = P_1 + P_2 + P_3 + P_4 + P_5. \qquad (5.1)$$

An emergent equation would be:

$$Y = \Sigma P_i + C_E = P_1 + P_2 + P_3 + P_4 + P_5 + C_E. \qquad (5.2)$$

Here C_E is the emergent constant. It always appears when the group is formed, but is separate from and "above and beyond" the summation of the "parts" (persons). Another analytical or mathematical possibility (never mentioned) is that the whole is *less* than the sum of the parts, or

$$Y = \Sigma P_i - C_E = P_1 + P_2 + P_3 + P_4 + P_5 - C_E. \qquad (5.3)$$

Something interesting is clear here. While "the whole is greater than the sum of its parts" is usually stated verbally, when we attempt to express it mathematically, we see some clear confusion (an "anomaly"). Specifically, what are the "parts"? Are they objects (for example, humans), or "variables"? In other words, is the "system" a Q-system or R-system?

If the system is a Q-system, then the "parts" are objects (let us say humans, for example). This is the case symbolized in equations 5.1, 5.2, and 5.3. However, the question arises, what is the "emergent property," symbolized in these equations as C_E? Is it a sixth person? No, because there is no way for a sixth person to "emerge" through mere formation of systemic relationships. Then, if it is not a person (Q), it must be a variable, or a constant, or some other conceptual symbol. If that is the case, are not equations 5.1, 5.2, and 5.3 incorrect or inaccurate? Should not they be written in R-terms (variables) instead of Q-terms (objects) (see table 2.1) as

$$Y = \Sigma \, X_i \qquad\qquad (5.4)$$

$$Y = \Sigma \, X_i + C_E \qquad\qquad (5.5)$$

$$Y = \Sigma \, X_i - C_E. \qquad\qquad (5.6)$$

Or still further, is the emergent a constant, or a variable? If the latter, the correct equations would be

$$Y = \Sigma \, X_i \qquad\qquad (5.7)$$

$$Y = \Sigma \, X_i + Z \qquad\qquad (5.8)$$

$$Y = \Sigma \, X_i - Z. \qquad\qquad (5.9)$$

The question still remains, what is the nature of this "emergent" Z? Is it multiplicative, so that the equation becomes nonlinear, as in:

$$Y = \Sigma \, X_i \qquad\qquad (5.10)$$

$$Y = \Sigma \, X_i + X^2 \qquad\qquad (5.11)$$

$$Y = \Sigma \, X_i - X^2 \qquad\qquad (5.12)$$

or, for example,

$$Y = \Sigma\, X_i + X_1 X_2 \qquad (5.13)$$

or

$$Y = \Sigma\, X_i - X_1 X_2. \qquad (5.14)$$

Although the verbal statement that "a whole is greater than the sum of its parts" will always be imprecise, the most faithful mathematical rendering out of all the possibilities would seem to be either equation 5.2 or a new equation, equation 5.15:

$$Y = \Sigma\, P_i + Z. \qquad (5.15)$$

The only difference is that in equation 5.2, the "emergent" phenomenon is a constant (C), and in equation 5.15 it is a variable (Z). Both fit the verbal statement.

It is significant that in *both* equations 5.2 and 5.15, while P is an object (Q-analysis), C or E are symbols representing concepts (R-analysis). Thus, the equations represent a combination of Q- and R-analysis. This is mathematically clear, though computationally difficult as only a few mathematical or statistical techniques yield simultaneous Q- and R-interpretations (Lingoes 1968). Most techniques such as cluster or factor analysis allow *either* Q- or R-interpretations, but only one at a time (see Sneath and Sokal 1973).

Equations 5.2 and 5.15 are very significant epistemologically for systems theory because they show that the *concrete* or *abstracted* system debate is largely illusory. There is no need to choose one or the other—it is not an "either-or" situation. In fact, *both* Q and R exist simultaneously in systems and need not be analytically separated. Concrete systems exist as groups of persons (P_i) but a number of variables (X_i) must be analyzed, as they order the actions of the persons. The congruence of Q and R (persons and variables) is an action/structure (process/structure) relationship which sometimes appears dialectical. The understanding of this relationship is fundamental to the understanding of social order. It was discussed in chapter 2, and will be examined in detail in the next chapter (chapter 6) which deals with social entropy theory (SET).

Miller (1978) says that living systems theory is an "emergent" theory, yet he stipulates the identical twenty subsystems at all eight levels. How can something (for example, the "memory") "emerge," for example, at the societal level, when it already existed at the level of the cell, organ, organism, group, and organization? It appears that confusion surrounding the issue of "emergence" will never cease.

What Miller means by "emergence" is *not* that a new variable or concept will "emerge" when individuals are grouped, or will "emerge" between one level and the next, but simply that the twenty subsystems are "emergent" in that they take on a new character or manifestation at each higher level. For example, the "memory" at the organism level is the brain, while at the societal level it is a library, museum, centralized data bank, or some similar repository for storing information, according to Miller (1978, p. 767). The memory thus "emerges" in Miller's schema from a brain at the organismic level to a library at the societal level.

For want of a better term I will call this phenomenon "transformational emergence," to describe a phenomenon that exists at *all* levels, but is transformed or "emerges" into newer forms at higher levels. Notice that this is *not* morphogenesis or growth, as here the units of analysis are different at each level, while morphogenesis implies transformation of the same unit. Transformational emergence can be contrasted with *new-variable emergence*. In the latter form a variable emerges at a given level that was not present at former levels.

Transformational emergence is probably not what most sociologists mean by the term "emergence" (they probably mean new-variable emergence). In Miller's scheme, emergence does not mean that new variables emerge, or even that the whole becomes progressively more than the sum of its parts as one proceeds up the levels. The "whole" is composed of the same twenty subsystems whether at the level of the cell or society.

However, it is interesting to note that transformational emergence plays somewhat the same role with regard to reductionism as does new-variable emergence. New-variable emergence is clearly a barrier to reductionism, because the reduction must necessarily eliminate the emergent variable *whenever reduction proceeds below the level where the variable emerges*. This would be "reverse emergence." Thus, with new-variable emergence, one can never fully have reductionism, as the reduced form can *never* fully represent the emergent form.

It would seem at first glance that Miller's scheme would be vulnerable to reductionism if it does not exhibit new-variable emergence (it does—as will be seen shortly). There would seem to be little reason to analyze all eight levels when all twenty subsystems are represented on one level—and with exactly the same names (input transducer, decider, extruder, and so forth). If this were the case, figure 5.2, which is now a 160-cell matrix (twenty subsystems × eight levels) would reduce to a 20-cell matrix, and the other 140 cells would be unnecessary. Under Miller's schema of transformational emergence, however, this cannot be done. In fact, the bulk of some thirty-years of research has been the task of discovering empirical examples of all 160 cells. In other words, it is taken as axiomatic that *at least* 160 cells exist empirically.

Since the chief task at present is to find examples for all 160 cells, reduction is in a sense undefined. It could be empirically defined if the same example were found for two or more cells of the classification, and given the same name. Then the cells could be merged, thus reducing the typology.

For the present, no reduction has occurred, either empirically or theoretically (analytically). On the contrary, the number of cells has been expanded from the original 133, formed from seven levels and nineteen subsystems (Miller 1978), to the present 160, formed from eight levels and twenty subsystems (see Miller and Miller 1992). Adding the community level and the timer subsystem has expanded, rather than reduced, the number of cells. When work began, numerous cells were unfilled in figure 5.2. Now only 6 cells remain unfilled. These are for the associator and memory (information-processing subsystems) at the three lowest levels—the cell, organ, and organism.

Notice that while Miller does not describe his schema as "holistic," the proper interpretation of figure 5.2 assumes a holistic or at least a nonreductionist view. If any cell at the higher levels is analyzed piecemeal or in and of itself ("in a vacuum") it often appears truistic, commonsense, and generally very intuitive. For example, at the societal level, the "memory" is the library (National Archives). Critics can judge this as obvious and unimportant. However, when viewed collectively and holistically, in the context of all 160 cells, it is important and nontrivial. It demonstrates (*a*) that the memory exists on this level as well as the others and (*b*) that an empirical example can be identified. Thus, the viability of Miller's scheme depends upon the identification and analysis of empirical examples of all 160 cells, no matter how familiar or unfamiliar each may be to a particular observer. Although a library (memory) or a garbage truck (extruder) may seem more familiar and mundane than their counterparts of the cellular level to a sociologist, they are no less important theoretically and empirically. Further, they are emergent, for while they fill the same role for their level as their counterparts on the lower level, they are not the same phenomena.

New-Variable Emergence

Although Miller (1978) says early in *Living Systems* that living systems theory is emergent, he does not pursue this. He seems to be referring chiefly to the transformational emergence in which, through evolution, the same twenty subsystems are replicated on each new level, but "emerge" on each "higher" level in a new form.

In addition to transformational emergence, however, one can identify at least two "transition points" at which new-variable emergence takes place in the eight levels. One of these emergent points is at the group level. The other is at the organizational level. We have already mentioned that the individual (organism) level is a disciplinary focal point. Biologists may be able to accept the analysis through the cell, organ, and organism levels, but feel uncomfortable with the analysis beyond that. Similarly, organization theorists may accept the living systems approach for the organization, group and individual levels, but resist if the organ and cell levels are brought into the same analysis (much of Miller's emphasis is on analyzing several levels simultaneously). However, social scientists who feel uncomfortable with the cell and organ levels must remember the *extreme generality* of the living systems approach. This schema is formulated for *all living systems,* including plants, and thus must be much broader than a purely social systems framework. This does not mean that the analysis is superficial, as the wealth of detail in *Living Systems* will attest.

Quite apart from the discomfort that may result around the pivotal individual level from disciplinary ethnocentrism, the group and organization levels constitute two *real* emergent points. The group level—as opposed to the individual level—is the first level at which collective perceptions (X) can occur in the three-level model (see figure 2.2). At the group level perceptions (X) such as "patriotism," which are collective conscious representations that individuals cannot conceive of because they require a group basis, can occur. There is no way that patriotism or treason can be defined at the individual level. This also holds for "hierarchy." A hierarchy can be conceived of for two people, and thus the individual/dyad transition is the "real" point of new-variable emergence in the eight levels. At the individual level or below, none of the collective consciousness representations can occur. Thus, new-variable emergence occurs at the group level. Furthermore, perceptions (X) or the whole perception level of the three-level model, cannot occur at the cell and organ levels, which is one reason social scientists object to these levels. For example, much of French structuralism (Althusser 1970), symbolic interactionism (Blumer 1969), ethnomethodology (Garfinkel 1967), and other "subjective" or interpretive approaches, to say nothing of psychology, are hindered or precluded at the cell and organ levels, thus hinting at the blocking phenomena we have attempted to avoid throughout this volume.

The second point of new-variable emergence is the organizational level, where internal hierarchies (echelons) develop that are not found at the group level (by definition). While hierarchies could occur at the group level, Miller defines hierarchy as an organizational property. This is somewhat definitional and arbitrary, but is in line with bureaucratic theory à la Weber.

The eight levels appear quite arbitrary. This contention is reinforced by the fact that the number of levels has already been increased from seven to eight (Miller and Miller 1992) with the addition of the community level between the organization and society, and could be increased further in the future (although it is unlikely that the number of levels will ever be decreased). While the distinctions between levels are based at one point on the new-variable emergence of the echelon (to separate the group and organization), in general the possible number of new levels above the individual is almost infinite, as one could identify a new "level" for each larger system when one person is added (for example, a system of size N is one "level," a system of size $N + 1$ is another "level," and so forth). This of course has ramifications not only for the number of cells in figure 5.2, but also for the theoretical validity of the analysis (and labels) of these cells.

As it now stands, distinctions above the individual level seem somewhat arbitrary. While the group/organization distinction in terms of an internal hierarchy has a solid basis in organizational theory, distinctions above this (community, society, supranational) seem based on a combination of system size in terms of numbers of individuals, spatial area, and the nature of boundaries. The practice followed in SET (chapter 6) is to recognize the societal (macro) and individual (micro) levels, and then to consider all others as intermediate (eschewing world systems for now). See also Ritzer's (1990b) division into micro, meso, and macro.

Let us return briefly to the question of whether the twenty subsystems should be ranked in any way, perhaps in "importance." Although Miller eschews such ranking, there is a hierarchical nature in the eight levels. A common view, clearly embodied in Parsons's (1966) "cybernetic hierarchy of control" is that the environment and energy flows are to be emphasized at the lower levels (e.g., cell, organ, organism) while the higher levels emphasize information processing. Is not the decider the most important subsystem for NATO or the EEC, while the distributor or even extruder is more important for the lower levels? Do not the higher levels emphasize information-processing systems such as the decider and channel and net? Are not these much more important in the computer age, with material processing assuming less importance for the higher levels (organization, society, and supranational)?

This would seem to be Parsons's view, but Miller resists it. He feels strongly that natural scientists have overemphasized energy to the neglect of information, and social scientists have overemphasized information to the neglect of matter-energy. Since integration is one of the main goals of systems science, and since *all* living systems process *both* matter-energy and information, all of these flows must be studied, and analytical emphasis on one more than others is misleading and counterproductive.

The tendency of Parsons (probably shared by many sociologists) is to focus on the higher levels so as to better understand social and cultural factors (the true subject of sociology) and not be subject to biological reductionism by slipping to the lower levels. However, from the standpoint of living systems theory, such a view is misleading. The sociologist who wants to study only bureaucracy (the organizational level) in living systems in *forgetting that each level contains all others below it*. Thus, even if it is true that decision making is of paramount importance in the bureaucracy, we cannot study this at the exclusion of matter-energy, because the *bureaucracy also contains the four lower levels—group, organism, organ, and cell*. This point has ramifications for the current controversy over whether autopoiesis can be applied to social groups, as discussed in chapter 8. If matter-energy is an important topic for these lower levels, then by deduction it is an important topic for bureaucracy, as the lower levels form the foundation for the upper levels, and all four lower levels are present in the bureaucracy. To emphasize only decision making or other "social or cultural factors" is to go against the living systems goal of integration and indulge in a form of piecemeal research. It is a form of dissective reification, when a particular level is seen as separate, when really it is interrelated or intertwined with the seven other levels. This is somewhat like separating the sections of an orange to study each individually. Then there is danger of being unable to study the whole, and the problem is reintegration. Note too that the inclusion of the lower levels (cell, organ, organism) is *not* reductionist. Miller is not saying that the study of the higher levels can be reduced to these. On the contrary, *all eight levels* must be studied, and so this is an essentially antireductionist position. The higher levels contain the lower, but all must be studied, and none can be omitted or reduced.

This point can also be made from the perspective of what earlier I have called "positive positivism." Positive positivism requires that the study of "higher level" social and cultural factors such as information processing, networks, authority structures, decision making, ethical and value systems, and so forth, must not be studied in isolation, as they depend upon the material base studied by the natural sciences. A social system can successfully deal with these "higher" or "sociocultural" issues only if its matter-energy processing functions are satisfactorily carried out. Thus, I agree with Miller that an integrative approach—as opposed to a piecemeal approach—requires emphasis on all twenty subsystems, regardless of the level of analysis.

Perhaps the clearest way to state the systems position is to say that as parallel processes, matter-energy and information flows are inextricably linked in each system. Even if one or the other were more important at some level, this is immaterial from the holistic standpoint, since all levels are inextricably linked, and in fact are nested rather than being independent.

Historical and urban sociologists have long recognized the interplay between material flows and information flows. It is clear that the formation of early cities required not only a material food supply, but also organizational means for coordinating this surplus. These means were "social organizations," generally taking the form of either the state (political organizations) or religious bodies, or both. This is an example of the dual "material factors—social factors" integration or nexus emphasized by Miller. However, this example also introduces the role of values (religion) or political ideology, a topic which Miller generally excludes analytically as "nonscientific." I will discuss this further below.

The links between the eight levels—specifically the way the higher require the lower—can be shown by briefly examining the world situation. Rhetorically, could the entire world function as a single system in Miller's terms? Probably not, because the development of a successful world system depends evolutionarily upon the maturity of the lower systems which support it. While many societies may be said to be "mature," the infrastructure needed for all the twenty subsystems to operate successfully on a worldwide level has not developed ("evolved"), as evidenced in a set of partial world systems or "supranational" systems in their infancy.

According to the evolutionary perspective inherent in living systems theory, at any given point in time there is always some level which exhibits relative autonomy, while the lower levels function as subsystems. For quite some time, the society (state) has been the dominant level, with it. Before that it was sometimes the organization (e.g., the church) that was dominant. Only recently have supranational links begun to take some autonomy over states. These include not only associations of states such as the United Nations or NATO, but also alliances of businesses (or multinational corporations), world churches, and international financial markets. I will argue in SET (chapter 6) that to fully understand this situation one cannot rely upon the concrete systems model to the degree that Miller does, but must analyze macro variables—not only values and information, but also such things as population size, spatial area, technology, organization, and level of living. This is essentially the Q-R conjunction. Only a broad systems model can even recognize this crucial problem, and this is the only perspective that can hope to solve it. The Q-R problem is fundamental to the understanding of social order, but most sociological models are either so narrow, or so analytically specialized (á la Parsons and his abstracted systems and anti-empiricism), that they miss this crucial issue of Q-R relationships.

Miller (1978) begins *Living Systems* by discussing the merits and criticisms of the systems approach, by defining types of systems, including abstracted and concrete, and by a general discussion of basic theory such as open and closed systems, feedback, entropy, negentropy and information, and so forth. After presenting the twenty subsystems and eight levels, he then proceeds to a detailed analysis of each level. The chapters are parallel, with all twenty subsystems being discussed at each level. There is also a discussion of both structure and process in each chapter, and the book discusses *dissimilarities* in systems as well as similarities. While all eight levels are similar in exhibiting all twenty subsystems, the examples of each subsystem are different at each level ("transformational emergence"). Also remember that new-variable emergence occurs at the group level with the possibility of "collective consciousness" or "collective conscience" (Durkheim's meaning can be translated both ways into English) representation (our perceptual or X level in figure 2.2), and at the organizational level with the emergence of links or networks—specifically authority structure (networks were possible at the group level). Also note that Miller's eight levels are true hierarchies and are nested, and so should not be confused with the three "levels" of the three-level model (figure 2.2), which are independent dimensions.

A frequently heard criticisms of integrative systems theory from disciplinary specialists is that systems display a great deal of diversity as well as similarity (for example, political systems, educational systems, and religious systems are different). Miller shows a sensitivity to this issue in several ways. He agrees that many differences exist, and discusses this. He also says that many people (in various disciplines) are studying the differences, but few are studying the similarities. He suggests that perhaps 5 percent of scholars should study systems integration or interdisciplinary endeavors. He shows further sensitivity to the disciplines by insuring that each of his individual chapters meet the criteria and standards of its respective discipline, in addition to embodying the principles of systems theory. Thus, the early chapters were published in biology journals, the later ones in social science journals (see Miller 1978). Miller's earliest work on the general systems approach appeared in 1955 (Miller 1955).

Most of the chapters also exhibit certain other similarities besides discussion of the twenty subsystems and of process and structure. Miller also shows sensitivity to the issue of external pressures on the system and internal pressures within the system. He adopts the term "stress" to refer to external pressures, and "strain" to refer to internal pressures. While consistent with engineering usage, this external/internal distinction is generally not made by sociologists, with most sociologists writing generically of "stress."

Another persistent theme is operationalization and testing. Miller believes interdisciplinary integration between natural science and social science is being hindered by the social scientist's failure to establish basic mensurable dimensions (such as centimeter/gram/second) which engineers and scientists interested in interdisciplinary work can understand and work with. Miller believes that precise quantification of living systems theory is necessary, not only for integration of disciplines (particularly the integration of social science and physical science) but also for hypothesis testing. General systems theory still devotes a lot of attention to systems methods and systems philosophy. Miller sees attention to philosophical matters as appropriate for early development of the paradigm, but feels that it is now time for general systems theory (and living systems theory in particular) to move from the stage of philosophy to mathematics.

As was noted in chapter 3, Parsons was criticized for his lack of operationalization and his general failure to subject his theory to empirical testing. He was also criticized for an obtuse writing style which tended to obfuscate his ideas, and further contributed to the lack of testability. I can list at least five factors which probably contributed in some way to the absence of testing in Parsonian theory:

1. An anti-empiricist stance
2. A complex writing style
3. Failure to formulate hypotheses

4. Failure to measure variables
5. Use of abstracted systems which seem at times to lack empirical links or to be difficult to test.

Miller makes a concerted effort to avoid these problems. He chooses a concrete systems model which is easier to measure empirically than Parsons's abstracted model. He seeks parsimony and clarity in his writing style (he has been an editor for many years, and is keenly conscious of writing style). He espouses empirical, positivistic, mathematical science (in direct contrast to Parsons). He presents many hypotheses suitable for testing, in direct contrast to Parsons. These include "cross-level hypotheses" which pertain to the same phenomena (for example, decision making) at two or more levels (for example, the individual, group, and organization).

In spite of all these contrasts to Parsons, Miller generally stops short of actual testing of his hypotheses. He does "test" a great many of them by presenting secondary data, as his book is replete with the results of experimental and other studies. He also reports his own hypothesis testing in a number of cases, especially with regard to information-input overload (Miller 1978). But many hypotheses remain untested. Thus, *Living Systems* (Miller 1978), like *Social Entropy Theory* (Bailey 1990) and *Inequality and Heterogeneity: A Primitive Theory of Social Structure* (Blau 1977), does not test all its hypotheses, but it does at least show they *can* be tested by presenting them in a form suitable for testing.

While Parsons disregarded testing, contemporary work is sympathetic to the aims of testing (Miller 1978; Bailey 1990; Blau 1977). However, the authors cited share with Parsons the basic problem that they are physically unable to both develop a broad approach (which takes many years of work) and simultaneously test all of the myriad hypotheses (in some cases hundreds) that the broad approach generates. The difference between Parsons's *The Social System* (1951) and contemporary work is that the contemporary ones (Miller 1978; Blau 1977; Bailey 1990) have the potential for testing, while Parsons does not. But while Miller, Bailey, and Blau present testable hypotheses while Parsons does not (and probably cannot), nevertheless, all four can generally be classified as verbal theories rather than mathematical models (although Bailey does present mathematical entropy models). For further discussion see Miller 1986.

Miller is quite sensitive to this, and sees operationalization as the next step. The question now is how to proceed. I will discuss this further in chapter 9. A related question is how to quantify basic variables for congruence with physical science. Should social scientists choose the centimeter/gram/second measurement of physical science, or would some other scales be more efficacious (or is the idea even feasible or beneficial)? I will deal with this question further in chapter 9.

Information-Input Overload

It is clear from the presentation of Miller's twenty subsystems that the overall system represents a complex state of functioning, requiring much coordination among the twenty subsystems. Until Miller's work there was little emphasis on sensory overload. Yet it is clear from perusal of Miller's model that system dysfunction can stem *not only* from scarcities of energy and information (which directly fail to input sufficient negentropy to offset internal entropy buildup), but also from input overloads. If too much matter-energy and/or information is brought into the system all at once, the system simply cannot deal with it all. The system faces not only problems of processing, but simple problems of how to store the surplus in a secure way until it can be processed. Most systems can accommodate a certain maximum amount of information or matter/energy (which may vary with form and quality) and beyond that problems arise.

Miller (1978) performed classic studies of information-input overload. This research is quite fully tested and replicated, and is perhaps the best-tested portion of living systems theory. The amount of information is quite quantifiable at most of the eight levels of Miller's model. He and his colleagues measured information flows and processing through the system for different system levels—the cell, organism, group, and organization. They found parallel (similar) curves at all levels, but found that processing times slowed as the level increased.

Notice that when too much information (or matter-energy) arrives too quickly, this can be "dysfunctional" for the system, or have negative consequences such as impairing the function not only of the subsystem dealing with it at the time, but of the nineteen related subsystems as well. This type of phenomenon is what some people call "negative feedback" because of its negative consequences. However, in cybernetic terminology this is *not* negative feedback. Negative feedback is a negative relationship which balances a positive relationship to establish a steady state. In the original cybernetics terminology, "negative" refers only to the sign of the relationship, not to value judgments about whether the effects will be "positive" (functional) or "negative" (dysfunctional) for the system.

Cross-Level Hypotheses

In addition to advocating the operationalization of living systems theory, Miller also advocates what he terms "cross-level" research. This is any research on the same phenomenon studied simultaneously for two or more of the eight levels. An early example is a course that Miller and Kenneth Boulding co-taught at the University of Michigan on decision making at the individual, group, organization, societal, and supranational levels (Miller 1984).

In addition to listing single hypotheses, Miller (1978) also presents a number of cross-level hypotheses. He also presents scenarios in the form of research suggestions for cross-level hypotheses. Here he not only presents the cross-level hypothesis, but also tells how it can be measured and tested on various levels. An example of across-level hypothesis has just been presented in the case of information-input overload. Another example involves cross-level research on entropy.

One preliminary research project had a number of researchers analyze all of the cross-level hypotheses in *Living Systems* with a view toward finding the ones best suited for a study of human/machine interfaces, most specifically as applied to information processing at the individual, group, and organizational levels. A clear consensus emerged among the cross-level hypotheses. This was hypothesis 3.3.3.2-3 (Miller 1978, pp. 96, 110). I have selected it here for citation. The following is the illustration of proposed cross-level research on hypothesis 3.3.3.2-3 as presented by Miller.*

---

*From James Grier Miller, *Living Systems*. Copyright © 1978 by McGraw-Hill, Incorporated. Reprinted with the permission of the author and the publisher.

Suggested approaches to cross-level researches on selected hypotheses

HYPOTHESIS 3.3.3.2–3: *In a channel there is always a progressive degradation of information and decrease in negative entropy or increase in noise or entropy. The output information per unit time is always less than it was at the input.* (H)

This information processing hypothesis is in some ways like matter-energy processing Hypothesis 3.2.2.2–2 (see page 94). It is also related to Hypotheses 3.3.3.2–2, 3.3.3.2–7, 3.3.3.2–11, and 5.4.3–6 (see pages 96, 97, and 109). It is explained, fundamentally, by the second law of thermodynamics.

Cell. Neural signals are progressively altered by random noise as they cross synapses and go through neurons. Increasing noise in such transmissions can be demonstrated in electrical recordings of input and output neural pulse sequences.

Organ. Sound frequencies pass through the cochlea of the ear and go through several neurons before reaching the auditory cortex. Electrical recordings at each echelon in the pathway can demonstrate the increasing amount of noise in the signal.

Organism. A subject carrying out a repetitive task, such as typing from copy, makes errors. These are more frequent if the task is carried out at a forced pace or if the subject is tired.

Group. The old parlor game sometimes called "telephone" illustrates how noise can enter into interpersonal communications. In this game one person whispers a story to his neighbor on the right, and he to his neighbor on the right, and so on around a circle. The message is progressively altered, and when the message gets back to the originator, it may be unrecognizable.

Organization. Noise may affect the accuracy of messages as they pass from one group to another. When an employee brings a verbal report from a field office in a corporation to the home office, and it is then relayed from one executive unit to another, errors in the transmission increase progressively.

Society. In ancient times messages coming to the king of a large country from the provinces were notoriously full of error, making governing difficult and often hazardous. Historical research could show whether the messages from the most distant provinces were the least accurate.

Supranational system. In general the channels in supranational systems are longer than in nations. A great deal of noise or error has always been present in supranational communications because of linguistic and cultural differences among component countries. Studies could indicate whether the amount of this noise in general is greater in long than in short supranational channels. (1978, pp. 110–11, emphasis in original)

Miller feels strongly that the same processes occur simultaneously at different systems levels (though, as has been seen, in different "emergent" forms). He thinks that cross-level research is the frontier in systems science (and in interdisciplinary research in general) and sees some urgency in conducting it. He feels that in spite of its obvious importance, cross-level research is neglected. There are a number of reasons for this. One is that some researchers simply do not recognize its significance. This is not surprising, as researchers working with a narrow approach within a single discipline often will not have a theoretical framework large enough to encompass two or more levels, and thus cross-level research will be precluded (or "blocked" in our earlier terminology).

Other researchers (such as systems researchers) may have a model sufficiently broad to recognize cross-level hypotheses, but may still eschew cross-level research. One reason is that it is complex and difficult. It is tantamount to studying one phenomenon, but at several different levels, possibly with different measurement and verification problems at each level. After all this "extra" work is done, the researcher may only achieve recognition for one research project. Thus, the cost/benefit ratio is not good. The researcher may understandably prefer to expend precious research time and resources on new single-level projects instead of one cross-level project, or may prefer to replicate a single-level project instead of conducting a cross-level project. Other systems theorists who recognize the issue of cross-level research may prefer theoretical analysis of hierarchies or of "isomorphisms" among levels instead of the sort of empirical research that Miller suggests. In spite of the obstacles and the general lack of precedent for cross-level studies, their importance seems obvious. Their time will come—and may already be here. For a discussion of the rise of cross-level research and examples of such research see Miller 1986.

Researchers in various areas may perform a variety of cross-level research (for example, simultaneous studies at the individual and organizational level) without labeling them as cross level or even being aware of the concept. Miller's research agenda for cross-level research serves as a guide to help such researchers conduct more complete studies, and to conduct studies more systematically, rather than in the common ad hoc manner, with no linkage between the respective cross-level researchers, or knowledge that other such researches exist. It should be noted in conclusion that Miller's work in cross-level research has parallels in sociology, although often under the name of multilevel research (see Mason, Wong, and Entwisle 1983; Markovsky 1987).

Applications

While Miller sees particular merit in cross-level hypothesis testing, this is part of a larger zeal for testing and empirical application of all systems theory, but especially living systems theory. Miller essentially feels that the task of science is to establish general principles through theory, hypothesis formulation, measurement, and verification (often in the experimental laboratory). His approach is a rather classical "positivist" approach which recognizes values and subjective phenomena as legitimate aspects of the world, but as *not* falling within the purview of science. Thus, values and subjective phenomena in general are essentially excluded (by design) from systematic treatment in living systems theory.

While there have been some cross-level applications of living systems theory, most applications have been single level, or at the most two level or three level. A number of applications and theoretical extensions have appeared since the publication of *Living Systems* in 1978, some carried out by Miller and his students and colleagues, and some conducted by researchers not personally acquainted with Miller who have read his theory.

The largest application has been carried out by Miller and his co-workers (Miller 1985). It is a study of forty-one United States Army battalions. It is one of the largest known studies of organizations. Most research at the organization level analyzes only one or a few organizations, while most studies with really large sample sizes use the individual as the unit of analysis. This massive study is both single level and cross level. This study concluded (among other things) that living systems concepts are understandable to Army personnel, that Army units can be described as living systems, and that LST can help not only in describing phenomena, but also in identifying sources of problems. Besides the army, LST has been applied to a number of different areas, including the family (Miller and Miller 1980), and small groups (Miller and Miller 1983).

Critiques and Reviews

Living Systems (Miller 1978) received a large number of reviews, most of them positive. It was the subject of a review symposium in *Contemporary Sociology* (1979). The symposium included long reviews by prominent systems theorists, including Boulding, Kuhn, and Parsons. These reviews were reprinted in *Behavioral Science* (1980) with an additional review by Rapoport. While the reviews were generally positive, a number of issues were raised. Boulding referred to the book as "universal physiology," implying a biological emphasis. Parsons and Kuhn both focused on the abstracted/concrete distinction. Parsons, in one of the last publications before his death, argued eloquently in favor of the abstracted systems approach. He criticized living systems theory for being static like a "snapshot" (shades of déjà vu—Parsons was often the recipient of a similar criticism), and criticized the physical space-time framework included in the definition of a concrete system. Parsons preferred a more social and cultural framework, without the possibility of physical reduction. all in all, Parsons admired the scope of the work on concrete theory, and lamented that no one had done a parallel treatment of the abstracted approach. He concluded his essay (and apparently his publishing career, as Miller [1986] says this was the last paper written before his death) as follows:

> The path that leads through a series of abstracted systems seems to be much more fruitful. I only hope that there will be another Miller who will explore this alternative path as thoroughly as Miller has explored his. (Parsons 1979, p. 705)

COMMENTARY

Contributions

Clearly living systems theory is a very valuable contribution not only to social systems theory in particular, but to social theory in general. It deserves careful reading by sociologists. Among Miller's distinct contributions to social systems theory are:

1. His detailed analysis of types of systems and their roles in social systems theory, including abstracted, concrete, conceptual, and totipotential systems.

2. His equation of negentropy with information and order, and his clear explication of the role of entropy in social research.

3. His presentation and exhaustive analysis of the twenty subsystems and their interrelations. This is by far the most comprehensive formulation of system properties to date, going far beyond any previous explication of the interrelationships between matter-energy and information-processing subsystems.

4. His careful analysis of the eight levels and their interrelationships. This analysis not only links the social levels to biological levels in a new and modern way, but also contributes to understanding of hierarchy, emergence, and reductionism.

5. His analysis of various "organizational pathologies" or irregularities of systems functioning. This includes not only information-input overload, but other pathologies such as communication gaps within organizations, and his general analysis of feedback irregularities as well as analysis of system stress (external) and strain (internal).

6. His utilization of "positive positivism" (my term—not his) in a number of ways such as basing analysis of social phenomena on their biological foundation through his nested hierarchy of the eight levels, and his contemporary application of evolution in the eight levels in a manner completely devoid of the social Darwinism that plagued nineteenth-century evolutionary theory in social science.

7. His presentation of a relatively exhaustive framework to guide integrative efforts of the future. Living systems theory is unquestionably the "most integrative" social systems theory to date. It not only integrates information processing and matter-energy processing in a very detailed manner, but it simultaneously integrates all eight levels as well. There is no other framework comparable to Miller's 160 cells (twenty subsystems by eight levels). He has made a start on the Comtean philosophy through a long overdue integration of the biological and social sciences, not through a "sociobiology" ap-

proach which links biological causes to social effects, but through a careful explication of how biological and social systems are distinct but interrelated. Many narrower frameworks can be seen to include a number of Miller's twenty subsystems (for example, see Aberle et al. 1950) but virtually none include all of them. Instead of proceeding on an ad hoc basis in which they rediscover the wheel in some cases, and omit important subsystems in other cases, future researchers can follow Miller's guidelines to identify all twenty subsystems in widespread applications, and perhaps even to extend his 160 cells, either through new subsystems or new levels (or both).

8. His formulation of cross-level research. As with the twenty subsystems, there are also many examples of cross-level research (and they seem to be increasing). However, most (that are not informed by Miller's approach) are ad hoc and unsystematic, and formulated without recognition of cross-level research in general.

9. His legacy of numerous hypotheses, both single level and cross level, as a guideline for empirical researchers to mine for years and years.

10. His emphasis on both structure and process, and their interrelationship, at each of the eight levels of analysis.

11. His emphasis on the recognition and discussion of dissimilarity as well as similarity, both within and among systems. This is generally not seen in systems theory. Most authors stress similarities among systems, but a comprehensive approach acknowledges dissimilarities as well.

Clearly these eleven are not all the contributions that this exhaustive work has made, but are perhaps some of the most visible. There are also a few limitations from the standpoint of sociology.

Limitations

Miller's decision to eschew study of values and subjective phenomena as outside the realm of science is unfortunate. This is perhaps the Achilles' heel of living systems theory in relation to sociology. Subjective phenomena and values play such an important role in complex society that their omission is a significant theoretical loss. The basic question is whether living systems theory merely *omits*

FIGURE 5.4.

A Comparison of the Systems Hierarchies of Miller and Parsons

Miller's Eight Levels	*Parson's Cybernetic Hierarchy of Control*	
Supranational	Cultural System	Information Flows
Society	Social System	
Community	Personality System	
Organization	Organismic System	Energy Flows
Group		
Individual (Organism)		
Organ		
Cell		

(See Miller and Miller [1992], Parsons [1966], and Turner [1991, p. 66].)

analysis of subjective phenomena, or *blocks* (precludes) their analysis. In other words, is living systems theory unsuited to the study of values in the manner that equilibrium theory was unsuited to the study of social change? Fortunately, the answer appears to be no. Living systems theory does not block the analysis of values, but merely omits them. Thus, future theorists can add the analysis of subjective phenomena in order to make the integrative potential of living systems theory even more powerful.

This point may be clearer if I briefly compare the hierarchies of Miller and Parsons, as in figure 5.4. Parsons's hierarchy is such an extreme analytical abstraction that it loses all of its concrete grounding. There is no analytical mechanism for moving from the perceptual or conceptual level (X) of the analysis to the empirical level (X') (see figure 2.2). This is because even though Parsons (belatedly) included the biological level, he consciously excludes the individual actor who serves (either individually or in groups) as the "carrier" or agent of transmission for the personality, social factors, and culture. In contrast, Miller includes this agent. Since the agent is present, values or other excluded variables can be added to the analysis.

A related weakness in living systems is the overemphasis on concrete analysis. In SET (chapter 6), concrete analysis is seen to be a viable starting point, but the full analysis must include both concrete and abstracted systems, as they are two sides of the same coin (see chapter 2). I have hinted that Miller's concrete system and Parsons's abstracted system are two polar extremes. While Parsons's model is so abstract to be debilitating in terms of such matters as boundary determination, measurement, and hypothesis testing, Miller's scheme is too concrete. The history of classical social science has been the history of analyzing important macro-sociological variables (R-analysis). Each of the classic theorists earned his fame by analysis of one or more of these variables. For example, Weber analyzed values and organization (bureaucracy), Malthus analyzed population size, Durkheim analyzed occupational division of labor. The chief problem with the classics is their piecemeal nature—each theorist only analyzed part of the whole picture. We need to analyze the *relationships* among these major macrovariables, and we cannot do it merely by summing the classics.

Unfortunately, Miller emphasized *concrete objects* (Q-analysis) to the virtual exclusion of variables (R-analysis). Thus, we have little explicit theoretical treatment, not only of values and subjective variables (which were purposefully excluded), but also of crucial sociological variables such as norm, role, status, race, gender, age, population size, technology, spatial area, and level of living. While obviously there is some analysis of variables (R-analysis) in living systems theory, it is minimal, and "comes in the back door" (see chapter 7). An explicit integration of concrete and abstracted systems (R- and Q-analysis) is overdue. If social systems theory claims to be an integrative approach it cannot neglect this crucial nexus.

There are other problems in living systems theory, but most of these are simply unresolved issues in the future of general systems theory. Examples include the interpretation of H (as entropy or information), and perhaps a lingering and slightly exaggerated acceptance of equilibrium (but balanced by the inclusion of entropy). In general, Miller shows a keen awareness of unresolved problems in general systems theory, such as the relation of energy to information, and of entropy to information. Where one ends is partly a function of where one starts, and by any criteria, living systems theory has come a long way on the integrative path of social systems theory. For an approach that starts at a different place and heads in generally the same direction, I now turn to the examination of social entropy theory (SET) in chapter 6. But before that, it is time for counterpoint.

COUNTERPOINT

While the prior chapters dealt largely with reconstruction, this chapter marks in earnest the presentation of the new systems theory. It is clear that while Miller's approach may bear some resemblance to functionalism, while one may speak of the twenty subsystems as serving "functions" (although Miller does not), and while some of the twenty subsystems have the same names as "functional prerequisites" (e.g., "reproducer"), this is an entirely different approach, and the old criticisms about functionalism do not apply. Notice that while Miller does retain the notion of homeostasis, it is a very mild version (requiring only a single variable in the system to remain in homeostatis), and is written in terms of negentropy and process. It may be instructive to simply go through my list of counterpoint terms (from chapter 1) to see how this new systems approach compares with mainstream sociological theory.

Swift perusal suffices to show that there is not much congruence between Miller's approach and some key terms of functionalism and neofunctionalism, such as action, order, equilibrium, and idealism/ideational (Alexander). Miller (1984) does not rely on such terms, and indeed seems to have no clear conception of what constitutes of what constitutes sociological functionalism. This is not to imply that Miller has no mention of these terms (see for example his discussion of order and disorder, 1978, pp. 43–44), but only that he does not devote substantial attention to them in a manner parallel to neofunctionalism or other sociological approaches (see Bailey 1993).

There is much more congruence between living systems theory and the counterpart terms derived from Giddens's structuration theory: agency, structure, system, structuration, system integration, time, and space. In general, there is no discussion in living systems theory of the terms "agency" and "structuration," as these are specific to sociological theory. However, the other terms are relevant for an analysis of living systems theory.

As for time and space, Giddens (1979, p. 54, italics in the original), says, "social theory *must acknowledge, as it has not done previously, time-space intersections as essentially involved in all social existence.*" Thus, although his approach prompted an argument with Parsons (see Parsons 1979), Miller's emphasis on time

and space, and his inclusion of these concepts in his basic definition of concrete systems, can be seen to be basically in tune with
Giddens's admonitions. That is, Miller's concrete system in anchored in physical space-time, and so meets Giddens's requirements for emphasis on space-time intersections.

System integration is also a point of congruence between Giddens's theory and living systems theory. System integration is discussed by Miller throughout the analysis of the various levels (see
Miller 1978, pp. 80–88 and 576–77, for example), and hypotheses concerning system integration are presented (see p. 109).

As for structure, Giddens says that structure is "Rules and
resources, organized as properties of social systems. Structure exists only as 'structural properties'" (1979, p. 66). Miller, on the
other hand, takes Giddens's own admonition to incorporate time
and space into theory seemingly much more literally than Giddens
himself does, as Miller incorporates both into his definition of
structure. Miller original) says,

> The *structure* of a system is the arrangement of its subsystems
> and components in three-dimensional space at a given moment
> of time. This changes over time. It may remain relatively fixed for
> a long period or it may change from moment to moment, de-
> pending upon the characteristics of the process in the system.

He says further that "all change over time of matter-energy or
information in a system is process" (Miller 1978, p. 83).

These two definitions (by Giddens and Miller) are quite different, but seem parallel and compatible rather than contradictory.
Giddens would probably applaud Miller's explicit inclusion of
time and space in the definition of structure, but would perhaps be
dissatisfied with the actual juxtaposition with process, so that
rather than achieving the relation between synchronicity and diachronicity that Giddens seeks, Miller's definition is more reminiscent of the old "snapshot" synchronic structure of functionalism
(see Miller 1978, p. 22).

I will turn now to the variables derived from Collins's theory—
conservatism, conflict, age, sex, and hypostatization. Miller discusses conflict in some detail (see, for example, Miller 1978, pp.
962–64 and 995–1000). Most of this analysis is in the context of
models of conflict and conflict resolution among nations as part of
Miller's analysis of supranational systems. However, he also dis-

cusses conflict as resulting from conflicting commands (Miller 1978, p. 39) and conflict among values (p. 804). The point is that the new systems theory is not limited to the study of consensus or equilibrium, but also analyzes conflict and change. Miller (1978) also has formulated hypotheses dealing with conflict.

Age and sex are other variables used by Collins which may serve as points of comparison. Miller deals with aging of humans, groups, and cities in some detail, and again presents hypotheses (see Miller 1978, pp. 481–82, 576–77, 726–31). Miller also deals with sex throughout, but in a rather sporadic and uncodified manner. There can be no claim that sex is a major variable of emphasis for him.

The next question from Collins is, "Is living systems theory conservative?" Answering the question first with regard to the theory and not the person, living systems theory *is not* conservative as compared to classical functionalism. Conservatism was endemic in classical functionalism in at least three ways. First, the equilibrium model insured perpetual return to the status quo power structure. This was politically and ideologically conservative. Second, the equilibrium model fostered *methodological conservatism,* in the form of teleology, tautology, and absence of traditional causal analysis. Third, functionalism was conservative in the sense of being positivistic or science based.

Miller's living systems theory is not conservative in the first two ways. It emphasizes the processing of information and energy flows over time, and thus escapes the return to the status quo and the teleology and tautology that plagues functionalism. Cursory perusal will suffice to show that it is a very different model. In the third sense of conservatism, it is true that Miller advocates a quite rigid scientific model, which might be termed "conservative" in that regard, but his theory clearly escapes the unacceptable conservatism that plagued functionalist theory.

The last issue from Collins deals with hypostatization. I would like to combine this discussion with discussion of the micro-macro linkage issue which is of interest to all three authors—Alexander, Giddens, and Collins. In criticizing functionalism, Collins (1975, p. 21) said that "system is hypostatized" and " 'system' is usually a myth," while the conflict perspective, in contrast, "grounds explanations in real people pursuing real interests."

I would submit that while this criticism may have been true for some functionalist theories, and for Parsons's abstracted systems model, *it does not apply to Miller's concrete system as defined in this volume and in Miller 1978*. Miller's analysis is every bit as concrete as is Collins, and uses "real people pursuing real interests." As random examples (admittedly taken out of context), consider Miller's discussion of the reproducer subsystem for the organization. Miller says, "Members of an industrial organization delegated to set up a subsidiary are examples of such components, as are groups of citizens of a community who separate off to form a new political unit" (1978, p. 605). Contrast this with Collins's discussion of membership-controlled organizations. He says, "The typical membership-controlled organization is also a pyramid, except that it is upside down" (1975, p. 329). I leave it to the reader to judge which discussion is more susceptible to hypostatization, the new systems theory's discussion of citizens of a community ("real people pursuing real interests"), or the "upside-down pyramids" of conflict sociology.

The reason that Miller's model is *not* hypostatized or reified is that he takes pains to define his model in terms of relations between concrete individuals, specified in physical space-time. This is about as "real" as one can make it. Miller does make some claims to "emergence" in his theory, but this is another issue and has been discussed in detail.

In terms of the micro-macro link, most mainstream theories start at the micro level, and are extended to the macro level, or start at the macro level (the minority), and are extended to the micro (see Ritzer 1990b). Miller's approach is somewhat novel as he analyzes *all eight levels simultaneously,* and using the same twenty categories. Thus, it is difficult to tell whether he starts at the "top" or "bottom," but his contribution is novel, and certainly adds depth and richness as a supplement to the mainstream approaches.

What types of metatheory are seen in living systems theory? The most prevalent are M_P (prelude to theory development), and the first variant of M_U (internal-intellectual). Also present is M_O, the attempt to build overarching theory.

What independent achievements are made in living systems theory that have not been made in mainstream sociological theory, and thus add breadth and richness to the mainstream? The contributions of LST have already been summarized and need not be repeated. Chief among them are the simultaneous analyses on eight levels which add a new dimension to micro-macro analysis, specification of the twenty subsystems, the emphasis on energy and information (relatively neglected in the mainstream—especially the former), and myriad other contributions such as the notions of information-input overload and organizational pathology, and numerous empirical applications (not to mention the array of testable hypotheses).

CHAPTER 6

Social Entropy Theory

Social entropy theory (Bailey 1990) was basically inspired by intense study of functionalism (see, for example, Turner and Maryanski 1979) and a desire to escape some of the quagmire of problems that functionalists encountered, as discussed in detail in earlier chapters. Some of these problems were direct reflections of the intellectual currents of the time (e.g., equilibrium), while others were reactions to intellectual currents (e.g., Parsons's emphasis on abstracted systems and German idealism as a response to the evolutionary theory of Spencer's generation). Social entropy theory (SET) is based on two critiques of classical functionalism plus a number of principles of its own (discussed below). The two critiques are:

1. Functionalism suffered from overreliance on outmoded concepts such as equilibrium.
2. Functionalism was not sufficiently broad to achieve an adequate analysis of complex society.

We have shown in earlier chapters that Parsons's concept of equilibrium encompassed a wide variety of sometimes contradictory or disparate connotations, and was based primarily on nineteenth-century or early-twentieth-century equilibrium analysis in physics. Indeed, there is close similarity between the equilibrium of Parsons and the nineteenth-century physical equilibrium of Pareto and his protege Henderson (who taught Parsons). When we peruse the influence of modern physics on systems theory, we see quite another emphasis as we approach the twenty-first century—an emphasis on nonequilibrium analysis as exemplified by the work of Prigogine (1962). This work uses entropy rather than equilibrium as its cornerstone.

The first critique of functionalism is consistent with (and exemplifies) the second. Functionalism, partly because Parsons reacted polemically to oppose Spencer's evolutionary theory, was overdependent on equilibrium, and thus overly narrow. By expanding our model to *include* nonequilibrium analysis as well as equilibrium analysis, we are less likely to preclude important phenomena that we must study in order to truly understand the social world. Specifically, broadening the model to incorporate both equilibrium and nonequilibrium analysis means not only that we have removed the basic classical complaint about functionalism (that it does not facilitate study of social change—see the critique of functionalism in chapter 2) but also that we have updated the model, and allowed it to incorporate modern physical approaches to systems theory such as the work of Prigogine.

But critique #2 goes beyond simply expanding the model to include instances of nonequilibrium. Writ large, critique #2 says that the functional model precluded analysis of certain phenomena, not just social change. In correcting the ills of functionalism, it is not sufficient to simply update the model through incorporation of the work of Prigogine and other recent writers. Rather, one must be sure that the model is sufficiently holistic that crucial elements of the society are not neglected.

At this point there is a dilemma. How can we be sure that crucial elements are not neglected? The problem is that until one has a complete model of society, one is not even aware of all the phenomena it encompasses, and so cannot insure that all of them are included. To state the dilemma another way: By definition we know our model is complete *only* when it includes all societal phenomena we wish to study, but we cannot recognize all the societal phenomena we need to study *until* we have a complete model.

This contradiction places constraints upon the strategy we can use to construct the systems model. One obvious strategy would simply be an additive one in which we add to the existing functional model. The problem here, as our dilemma illustrates, is that we do not know what to add. Is it sufficient to simply expand the model to include a class of nonequilibrium phenomena? We have already answered this to the contrary. If not, how can we identify phenomena to be added? One possible procedure is to simply encounter them through analysis, and then add them as they are encountered. That is, when we find phenomena that the model does not include, we simply add them.

There are a number of problems here. One is that since the model does not include these phenomena, it does not guide us to identify them. Another is that the history of science shows that phenomena that do not fit a model are often explained away or ignored rather than incorporated through revision of the model (Kuhn 1962). Perhaps the best reason for not adding to the original functional model is simply that models which are isomorphic with one perception of society and which are constructed in a specific language, are not easily expandable to domains outside of, and perhaps contradictory to, their original purview. A prime example of this is Parsons's definition of a system as being in equilibrium (see chapter 3). As long as a system is *by definition* in equilibrium, then it cannot easily be expanded to incorporate that which it is designed to exclude: nonequilibrium phenomena.

It seems then that our dilemma is best avoided by pursuing what we have elsewhere labeled the direct approach (Bailey 1990). In contrast to an additive strategy in which we sum narrower models, or a divisive strategy in which we divide broader models, the direct strategy identifies the phenomena to be modeled, and then constructs a model which fits it as exactly as possible. That is, there should be the potential for point-by-point congruence between the model and the empirical phenomena being modeled. Such point-by-point congruence is termed isomorphism (Miller 1978; Bailey 1990). The entire procedure of model construction is best illustrated via the three-level model of figure 6.1.

FIGURE 6.1.
The Three-Level Model as Applied to the Task of Systems Model Construction

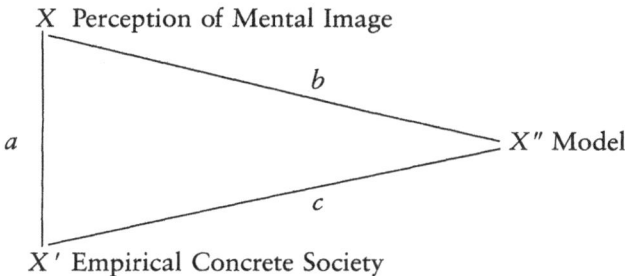

X Perception of Mental Image

a b X'' Model

          c

X' Empirical Concrete Society

Figure 6.1 can be seen as a somewhat different application of the same basic three-level model presented in figure 2.2. Level X is one's mental image of the society (concept or perception). Level X' is the actual empirical concrete society (the phenomena that is being modeled). Level X'' is the model, which is formed as a combination of both X and X'. For example, one could form a model X'' that was purely hypothetical. This could be done from mental analysis, and spawned only along path b, without direct observation of the empirical society X'. Conversely, one could conceivably form a purely descriptive model X'' by direct observation of the totality of the empirical society X' (path c). Direct observation of the totality of society by any one scholar is clearly impossible, and so the descriptive road (along path c) to the isomorphic model (X'') is generally precluded (it simply is not feasible). Rather, one must combine direct observation of X' along with our perception of society (X) to arrive at X'' from both paths b and c. This provides probably the strongest model, as it is the result of both "ivory-tower theorizing" (X) and "real-world observation" (X'). The question remains, how can one isomorphically model the entire society (X'')? Can we even perceive it mentally (X level)?

There was much hesitation and uncertainty during the initial formulation of SET about how to proceed (see Bailey 1990). This centered around definition of a system, or the conceptual-level perception (X). Since we probably can never simultaneously observe the empirical system X' in its entirety, our perception of it (X) is crucial to our final model (X''). Hesitation centered around the question of whether the system is initially conceived as concrete or abstracted. This distinction has already been discussed in some detail in chapters 2 and 5, and further discussion can be found in Miller (1978) and Bailey (1981, 1983, 1990).

The basic issue concerns what is the best initial concept of a system for analytical purposes. While Parsons (1951) championed the social role as the basic systems unit (abstracted system) and Miller (1978) championed the individual as the unit (concrete system), SET is based on the premise that *both* roles (relationships) and persons must be analyzed. There is no choice, and no conflict. The only question is whether one can best formulate an isomorphic three-level model with an abstracted system or a concrete system. After much analysis I concluded that I must begin with a concrete system, consisting of a spatial area (S), with definite boundaries. This area contains a population (*P*) of individual persons.

Why the Entropy System Is Concrete

I am not saying that one *cannot* begin analysis with an abstracted system and proceed to the analysis of a concrete system. It may be possible. However, I do not at this time see how to do it. If one begins with an abstracted system in which the basic units are roles or relations, then one is essentially limited to this level of analysis. This is what Parsons described, but the cost was a limited model. This was perhaps the optimal model for its time, but I am committed to the formulation of a model which can generate testable hypotheses and facilitate operationalization (unlike Parsons's model), and a model which *includes* rather than precludes. While Parsons's model precluded nonequilibrium analysis and concrete systems analysis, my model *must include all of these—equilibrium, nonequilibrium, abstracted, and concrete.* The only way I see to do this is to begin with a concrete system, and then I can easily proceed subsequently to analysis of abstracted systems.

Why does this basic asymmetry exist? Why can one proceed from concrete to abstracted (from persons to variables) but not vice versa (from variables to persons)? The answer is simple enough—because the basis of the system is a set of *persons* acting and interacting in physical space-time to process matter-energy and information. This is the basis of the human system. This is true for the case of a single individual as well, and thus the human system is presocial. As one adds other humans, they all process matter-energy and information, and as a system interrelate so that entropy levels are kept below the maximum. At this point the abstracted system becomes predominant. Parsons (1951) recognized this. He simply wanted to limit sociology to the study of the abstracted relationships among humans ("social and cultural variables") and to leave the analysis of the material basis to other disciplines.

I am saying that a satisfactory model must not preclude this basic material analysis lest it become a house built upon sand. When this happens, the model does not include but precludes, and this violates the dictates of SET. If one recognizes human systems as concrete systems which process matter-energy and information in order to reduce entropy, then one is facing reality, and has a firm foundation upon which to build an analysis of social and cultural phenomena. If one attempts study of social and cultural phenomena without sufficient understanding of their material foundations, one has most likely undertaken a task that cannot be successfully completed. Such a model will likely be plagued by myriad nagging classical problems, from reification to reductionism to simple lack of testability to tautology and teleology, and all of these are essentially due to faulty epistemology in the lack of a proper foundation. We do ourselves a disservice by placing the sociocultural cart before the entropic horse. By building our model on a natural and material basis we are not only being realistic by building a model that is isomorphic with reality, but we are also following the classical dictates of Comte, Marx, and others.

This point is easily illustrated. Imagine that I decide, in spite of this argumentation, to proceed with an abstracted system, à la Parsons, in which the social role is the basic unit of analysis. I recognize that individuals exist empirically, process matter-energy (eat, drive cars, etc.) and process information (watch television, use computers, etc.), but decide to preclude or at least de-emphasize these factors from my initial model. Thus, the empirical system (X') is reduced to a set of roles—the politician, police officer, fire fighter, judge, etc. How do I get from this set of roles to the whole of human society—the energy crisis, the economic crunch, the birth rate, etc? If I limit the social systems model (X'') to that consistent with German idealism, it *cannot* be isomorphic with the empirical world (X'). Thus, the choice is simple.

Our model (X'') consists of the spatial area, the individuals, and their interrelationships. The population (P) of N individuals has a set of M characteristics or variables that are relevant to its existence. These are shown in the score matrix, or S matrix, previously presented as table 2.1. The relationships among the N individuals are Q-relationships. The relationships among the M variables are R-relationships (see Bailey 1990).

Verbal theorists (excluding action theorists such as Parsons) generally analyze relationships among individuals, and so are essentially doing Q-analysis (although nonstatistically). Clinicians also do this type of analysis. For example, they may study how Robert relates to Charlene. Often this is expressed in terms of *roles*, such as how the husband relates to the wife. This is essentially an abstracted system, in which roles are units. Quantitative social scientists (statisticians) generally analyze relationships among variables, and so are doing R-analysis. For example, they relate age to income, or race to income.

A salient point here is that if we champion either form of analysis (Q or R) we are precluding or at least de-emphasizing the other, because Q uses the N persons and R uses the M variables. Our basic concrete model uses the *full matrix,* both N and M. While we will not always analyze both N and M simultaneously due to complexity, we recognize the salience of each and can rely upon it when needed.

The combined Q-R analysis represents both sides of the systems' coin when viewed synchronically, or at one point in time. This is *structure.* However, it has been emphasized that concrete systems operate in space-*time* to process matter-energy and information to keep entropy below the maximum. Thus, one must also take care not to preclude time or diachronic analysis, classically referred to as "process."

The human actors are the agents which, over time, propel the system. They were insufficiently emphasized in functionalism, leading to tautology and teleology, as structural relations just seemed to happen or be deterministically arranged. This led many sociologists to oppose systems theory, as it seemed that humans were not in control.

There has been lengthy discussion in chapter 4 and elsewhere about the importance of control in human-systems analysis. If humans do not control systems, then systems seem to be regulated by economic determinism or other forces, perhaps vague and mysterious. Further, without any clear human agents of action and control, functional systems always returned to equilibrium (apparently whether this was what their human members wished or not) and thus were always essentially *synchronic* over the long run. By this I mean that although time passed, analysis at time 1, time 2, or time t would always reveal the same equilibrium condition, except in cases such as "moving equilibrium" (see Bailey 1984a, 1990 for a critique of "moving equilibrium" and other equilibrium concepts).

In SET, social change is *not* blocked as in functionalism. There is no automatic "return to equilibrium" (although return to equilibrium is not blocked either, and the model can accommodate it if it is found to occur empirically). Rather, in SET, human actors set system goals and work to maintain them over time. Thus, *process is emphasized along with structure,* and neither are blocked or precluded by the model.

While structure and process are often discussed, their relationship is seldom analyzed. In SET I show that structure and process are highly complementary, and indeed each has little meaning without the other. The relationship between structure and process is analyzed in SET through two tools or submodels which have already been introduced in this volume and examined in some detail—the three-level model (figure 2.2) and the Q-R distinction (see chapter 2).

Examining Q first, it is clear that the set of objects (persons) relate to each other, and this is Q-analysis. This takes place *over time,* and is process, or *diachronic.* For example, two people who date each other (or "have a relationship") are relating over time. They interact with each other physically and spatially, and over time. Thus, their action in physical time-space fits Miller's (1978) definition of a concrete system. In addition to this process, the social system also is composed of *structure.* One form of structure is the relationship between variables (R-analysis) which persists over time. For example, the relationship between education and occupation. The crux of the relationship between structure and process (so defined) is between R- and Q-analysis. Q-analysis is the action of people over time and the interrelationship of individuals. The regularities in Q-analysis (process) lead to regularities in R-analysis (structure). Subsequently, this structure (R) guides future action (Q). Thus, Q→R→Q→R . . . is a never-ending cycle of Process→Structure→Process→Structure . . . .

As an illustration, people go through the *process* of seeking jobs—learning of an opening, being interviewed by the employer, etc. This is Q-analysis, as the individuals meet face to face. This is also *process.* If women obtain certain jobs during this employment process and men obtain other types of jobs, *the result from the Q-process is R-structure, or a statistical relationship between the two variables of sex and occupation.* Thus, regularities in diachronic processes (Q) lead to regularities in synchronic structure (R) in one time period. In the next time period, job seekers and employers know that the R-correlation exists, and that *certain jobs are women's jobs.* This knowledge *structure* informs their *process,* and provides the type of job they seek. Thus, in the first time period, Q leads to R. In the second, R leads to Q. This is a never-ending cycle of replication. Compare this discussion of the structure-process relationships with Giddens's theory of structuration (see Giddens 1984), and see also the discussion in chapter 9.

The structure-process endless cycle can also be easily illustrated in the three-level model. Let X be our perception of the situation, X'' be the written record ("structure") and X' be the empirical world. Imagine that we had a formal norm in the form of a written law saying that women could not be automobile mechanics (X''). This "structure" in the form of a constant written record appears synchronic as it does not change over time. But it constrains and regulates action (process) by precluding women from going through the process of obtaining automobile mechanics jobs. Thus the law (X'') says that women cannot be automobile mechanics, they do not apply for or receive such jobs in the "real world" (X') and this precedent in the form of process bolsters the continuity of the law. However, if a perception (X) arises in the minds of some that this is an "unjust law" (X''), then it can be challenged by women working as automobile mechanics (X'). Such action may simply lead to punishment of law breakers, or over time, could put pressure on the legal structure, leading to revision of the law (X''). This is another instance of the fact that while structure guides process, process can also guide (or change) structure. More specifically, process either replicates structure or puts pressure on it to change.

THE CHALLENGES

If SET wishes to avoid blocking, to include and not preclude, and to simultaneously study both order and action (which Alexander 1987, says theory does not do), it obviously faces many challenges. A successful social systems theory faces at least thirteen basic problems:

1. An adequate definition of system.
2. Adequate specification of boundaries for the system as a whole, for system components, and for subsystems (if any) and their components.

3. An adequate measure of system state, and adequate operationalization of such a measure.

4. The attainment of isomorphism between the theoretical systems model and the actual operating, empirical, complex system.

5. The selection of a suitable set of explanatory variables out of the almost infinite number that could be identified in a complex social system.

6. An adequate understanding of the relationship among component parts of the system, and between each part and the whole, so that the problem of unwitting displacement of scope can be satisfactorily overcome.

7. An adequate analysis of both micro and macro levels and their interrelationships so that problems such as reductionism and emergence can either be solved or avoided.

8. A recognition of the needs of the individuals and subgroups within the system, and of the system as a whole.

9. An adequate defense against the almost certain criticism that the systems analysis is an inappropriate organic or mechanical analogy.

10. The recognition of individual, subgroup, and system goals, and understanding of how they are attained.

11. An understanding of the role of matter-energy and information in ongoing system functioning.

12. An adequate diachronic analysis of the system, so that change over time can be understood.

13. The attainment of adequate explanation and prediction (including verification) of salient aspects of the complex system via utilization of the social systems model.

SUMMARIZING SET

Social entropy theory uses the structure-process analysis throughout via the three-level model and the Q-R distinction. The combination of structure-process, three-level model, and Q-R is difficult and complicated, but also necessary and efficacious in understanding how the social system works on a day-to-day basis. Social entropy theory postulates a set of structural variables with the social system as the unit of analysis. Recall that earlier I said that when Parsons used a plurality of individuals as the system unit, he really was defining the *individual* as the unit, not the society. Social entropy theory uses the *society* (in its entirety), or *sui generis* as the basic unit. It is *not* viewed as a "set of individuals" but as a concrete system or population of individuals interacting over physical space-time within boundaries. This leads to a whole set of supra-individual variables not easily definable under Parsons's definition. These are macrovariables, or *social facts* in Durkheim's terms. They are properties (R-variables) of the society as the object (Q). These are *global variables* in Lazarsfeld's (1937, 1958) terms, meaning that they cannot be defined in terms of individual properties, but only as units of a society. In addition to global macrovariables, SET also uses the distinctions of mutable distributions (macrovariables) and immutable variables (micro).

I will first derive the global macros. Remember that the structural model (X'') must be isomorphic with the empirical world (X') and must be large and flexible enough so as not to block or preclude elements needed for analysis. What then, are the key structural variables that we must include in our model? Remember that these are the minimum, but that the model must not preclude the later inclusion of other necessary globals if they are found.

The first several of these globals are relatively evident from the definition of a concrete system as already discussed. I said that a concrete system operates in space-time. Thus spatial area within boundaries is one key macrovariable (S). Another of course is the population (P) of humans required to form a social system. I said further that these humans must process matter-energy and information in order to maintain entropy levels below the maximum. If this is not accomplished, the society cannot survive. While one could list "matter-energy" as a global variable, it must also be considered as included within the boundaries of the spatial area, and thus already included as part of S. If not included as "natural resources" which are indigenous, the needed matter-energy must be imported. Thus, space (S) in the SET model includes all area (land, water, air) within the boundaries and thus comprises what is often called the "environment." This environment contains matter-energy as raw materials (oil, coal, etc.). Also, food supplies come from the environment.

However, the available matter-energy cannot be properly processed without technology, which I define simply as tools. These tools are not limited to primary extraction of matter-energy (as in mining or farming), of course, but can also be used for secondary or tertiary manufacturing or distribution, and of course technology is also used for processing information as well as matter-energy (although information transmission generally involves a matter-energy carrier or "marker" of some sort, such as this sheet of paper, film, computer diskettes, etc.—see Miller 1978).

Thus, population (P), space (S), and technology (T) are all major macrostructural components of the system. The fourth major global (already mentioned) is information level (I). For simplification, I can define information operationally as any perception, conception, or knowledge $(X$ level) which the society utilizes in making decisions. Thus, information guides decision making. This definition is obviously very broad (and stated differently than in Bailey 1990), and includes technical knowledge (physics, chemistry, medicine, etc.), but also includes statements of fact of all sorts, as well as ideology, beliefs, values, norms, prejudices, etc. Any cognitive content which guides societal functions will be included in this broad definition of information, including not only norms, values, and ideology, but also myths and erroneous information ("misinformation").

In addition to population, space, information, and technology, what other global variables are necessary for a society to operate? Obviously, all of these entities have to be coordinated. A wealth of resources and a large spatial area are insufficient to maintain a population, particularly a large one, unless they can be properly coordinated. Thus, organization (O) is a crucial variable. A study of the development of urbanization (Light 1983) shows clearly the role that social organization played in coordinating surplus in such a manner that early cities could develop.

The last global variable is level of living (L). This can vary from zero (the society disappears) to some designated maximum. There are many ways to measure level of living, or standard of living, or quality of life. The most common is income or wealth. However measured, this is a crucial variable, and is a minimum (zero) when physical entropy is maximized.

Together, these six factors form *PISTOL* (or *PILOTS*), the complete set of globals. A number of points about these six globals must be made clear:

a. By definition, as globals they cannot be defined in terms of characteristics of individuals. They are unitary macroproperties, and are *not* formed by aggregating or analyzing properties of individuals. There is one partial exception in the set. This is population (*P*) which has elsewhere (Bailey 1990) been referred to as a "semiglobal." This is because *P* requires knowledge only of whether a person exists or not (dummy variable coding, "1" if exists, "0" if does not), but requires no information on variables such as age, sex, race, etc.

The nature of *P* is more clearly seen in the *Q-R* distinction, as illustrated in the score matrix presented in chapter 2 (table 2.1). The marginal column along the edge of table 2.1 shows a population of *N* objects (but no variables). Thus, the population by itself does not represent a matrix, but only a column vector, and so no real decision-making information in the form of variables known about each person is represented. Thus, while *P* may not be a "strict" or pure global, it is still essentially a global, or at least a "semiglobal" (see chapter 7). The other five factors (*ISTOL*) are full globals and require no information about *P* or about individuals in the system for their definition.

b. These six globals are strictly speaking not single variables, but factors or components. That is, each can be split into sets of variables and can be operationalized in myriad alternative ways. For example, level of living (*L*) might be operationalized in terms of number of food calories in one society and by the amount of cash wealth in another society.

c. As globals, these are not distributions, "relational variables," sociometric or network variables, etc. Rather, they are sum totals of the amount of the entity in each case. They are ratio variables, with a minimum of zero in each case, and a maximum to be empirically determined, depending upon the particular operationalization and application. For example, *P* is the sum total of all individuals (e.g., one million persons); *S* is the sum total of all spatial area (e.g., one million acres); *O* is the sum total of all occupational positions (e.g., one million jobs); *T* is the sum total of all tools, from spoons to lasers; *L* is the sum total of all optimal operational units measuring level of living (calories, dollars, etc.); and *I* is the sum total of all information units (bits, books, etc.).

d. The model is so general as to apply to all societies in the world regardless of level of development. Thus, all societies have some levels of *P, I, S, T, O,* and *L,* although obviously these will be operationalized much differently for a small undeveloped society than for a large industrial society.

e. These six global macroproperties are all interrelated properties of the society as the basic unit of analysis. Although *L* may be sometimes conceived as a "dependent" variable and the others as independent, in fact, in systems terms, all are interdependently interrelated

$$P = f(I,S,T,O,L) \tag{6.1}$$

$$I = f(P,S,T,O,L) \tag{6.2}$$

$$S = f(P,I,T,O,L) \tag{6.3}$$

$$T = f(P,I,S,O,L) \tag{6.4}$$

$$O = f(P,I,S,T,L) \tag{6.5}$$

$$L = f(P,I,S,T,O) \tag{6.6}$$

Thus, all six can alternatively be seen as "dependent variables" (in five of the equations) and an "independent variable" in one equation. For example, while level of living (L) is certainly dependent upon a proper population size (P), spatial area (S), information level (I), technology (T), and organization (O), it in turn is necessary for the others. That is, population size (P) *cannot* be increased if the level of living is insufficient, for example, if L is at the subsistence level and there is simply no food for additional people. Similarly, technology (T) cannot be expanded if the level of living (L) is so low that resources for research or purchase of technology are simply precluded.

f. The globals can also be seen in terms of process. The six globals can all be written as six interrelated synchronic variables (R-analysis). They provide a structure for the social system. However, they of course can be analyzed diachronically (in terms of process) as well. Human decision makers assess and change the levels of the various six variables. For example, population policy makers assess (P), politicians seek to maximize (L), and (I), etc. However, since the six are interrelated the level of each imposes constraints upon the others, so social planning cannot be done piecemeal, but ideally should take all six variables into account simultaneously.

g. Just as the six globals apply to *all societies* at all stages of development, they also apply to different levels within the society and to supranational systems as well. The discussion of chapter 5 showed that Miller (1978; Miller and Miller 1992) analyzes eight levels, supranational, society, community, organization, group, organism, organ, and cell. Choosing three of these for illustration—society, organization, and group—it is clear that the six globals exist at all three levels. Actually, the organization is a subsystem of the society, and the group is a subsystem of the organization (and thus of the society). Thus, a *subset* of P, I, S, T, O, L exist for the organization and for the group. Another way to say it is that a portion of the P, I, S, T, O, L for the whole society is possessed by these smaller systems.

h. The *PISTOL* or *PILOTS* formulation can be seen as similar to the *POET* model (Duncan and Schnore 1959). Holistically, it is much different, as it includes level of living and information—both crucial to the set, but missing in *POET*. Also keep in mind that as systems theorists we are not analyzing six individual variables piecemeal, but are studying the *relationships* among the six. Thus, while the *POET* model has only six relationships (*PO, PE, ET, OE, OT, PT*), the *PISTOL* model has fifteen—all crucial to the model.

IMMUTABLES

We now turn to the individual in society. The individual exists within the structure of the six globals, plus a host of other social and cultural phenomena. The six globals obviously serve as constraints for the individual. His or her life and life choices are very different in a small, poor, technologically underdeveloped society than in a large, bureaucratic society with the latest technology, from computers to lasers. Within a given society, the life chances of an individual also vary with the relations between the structural globals. For example, if $P < O$, there are fewer people than organizational positions, and vacant jobs exist. But if $P > O$, there are more people than jobs and competition for available positions may be fierce.

Within the context formed by the global structure, the individual has personal characteristics or variables (R-analysis) which greatly affect his or her life chances. Many of these personal characteristics are *immutables,* meaning that they cannot be changed. These are *micro* properties with the individual as the unit of analysis. Among notable immutable micro properties are race, ethnicity, gender, and age (birthdate).

Many immutable variables are commonly known as *ascribed characteristics* (see Davis 1949) because the individual is born with them (they are ascribed at birth). For our purposes here it is of secondary interest whether or not a characteristic is ascribed. The salient issue is whether or not it can be changed, particularly if it is affecting the individual adversely (e.g., if it causes him or her to be discriminated against).

Most variables are not *totally* immutable, but are generally so, being immutable in the vast majority of cases in a given society. While birthdate is totally immutable and *cannot* ever be changed, age changes constantly. Skin color, as an empirical entity (X') can be changed, but this is very difficult and very rare. It may be even more difficult to alter racial identity (X). Similarly, while both sex and gender are immutables (and are likely consistent) for the vast majority of people, it is possible to change empirical sex (X'), but again changing gender identity (X) may be more difficult, especially after a certain age. In fact, one rationale for changing sex (X') is to make it consistent with gender identity (X) for a given individual. Both gender and sex are considered immutable here. Three very salient immutables in many societies are gender, race, and age (birthdate), or *GRA*. Changes to these variables, while possible, are so comparatively rare as to be the exceptions that prove the rule. This topic deserves more analysis than can be devoted to it here.

I have said that the six globals are sums of units. As such, they form six dimensions along which individuals can be distributed. That is, each individual, at a given point in time, occupies a position not only as a member of the population (P) but also in the information structure (I), spatial area (S), technology (T), organization (O), and level of living (L). One of the most crucial questions in SET is the issue of how individuals are allocated into this multidimensional structure. Allocation cannot be random, or there would be no social order and no predictability. In this case multidimensional entropy (H) would be maximum. It is a postulate of

SET that the entropy level is always below the maximum. More likely, there is an orderly process by which individuals are allocated. This orderly process is regulated by norms of various types, from federal or international laws, to local regulations, to folkways, mores, customs, and precedents.

One of the chief ways that individuals are selected for allocation is through the immutable variables. Characteristics such as race, age, and sex are ideal for use by those in power who make societal decisions regarding allocation. Such immutables are immediately visible and cannot be changed. Thus, decisions as to allocation can be made quickly without extensive prior research, and require only perusal of the person's salient immutables. After the decision is made, power is not threatened by alteration of characteristics, as the individual cannot do this (cannot change immutable factors such as skin color or sex). If an individual is dissatisfied with his or her allocation (e.g., a particular job), he or she has little recourse through change of immutables. For example, a woman stuck in a "woman's job" cannot change her sex. She must either work for "social change" in the form of change in the norms, or for change of designation of the job as "women's work," or change her position through some other variable such as residence. This brings us to the third class of variables in SET (after globals and immutables)—the mutables.

Globals are purely *macro* variables. They are defined only for societies, and do not require information about individual characteristics such as race or sex (the special case of population has been discussed). Immutables, on the other hand, are purely *micro* variables, and are definable only for individuals (societies do not have race or gender). Mutables are intermediate. They have both macro and micro qualities, and so in a real sense are the true micro-macro link.

Mutable variables take the form of *distributions*. There are five basic mutable distributions in SET, and they are formed by allocating or distributing the population of individuals, a semiglobal, into the other five globals. That is, each individual has a position not only on the level of living variable in the form of income or wealth, but also has a position in the spatial areas (residence), in the information dimension (e.g., educational level), in the technological realm (tools that can be operated, such as flying airplanes or programming computers) and in the occupational structure (a job).

Thus, each individual has a set of mutable characteristics in addition to his or her immutables. By definition, the mutables can be changed (at least potentially or theoretically, although it is sometimes difficult). Mutables are essentially the variables often described as *achieved characteristics*. For our purposes the salient point is that they can be changed, while immutables cannot be.

I should first discuss the *micro* aspect of mutable characteristics. Each person in the social system has a relatively large set of characteristics which can be changed (at least potentially) and thus are mutable. For each person generally has a residence, an education, income, occupation, etc. These are mutable characteristics and can be changed over time. They are also clearly micro characteristics, as they are properties of the individual, with the individual as the basic unit of analysis.

Thus, each individual possesses a set of micro *mutable* characteristics as well as a set of micro *immutable* characteristics. The immutables and mutables are clearly interrelated. For example, individuals may be barred from certain mutable positions (or given entrée to others) solely on the basis of one or more of their immutable characteristics. Thus, females may be barred from certain occupations or black persons from certain neighborhoods.

On the other hand, since the mutable characteristics are changeable, they do sometimes offer recourse to a person blocked from his or her goals because of his or her immutable characteristics. For example, a person desiring a certain occupation who is barred from it because of race or sex (immutables) might be able to gain entry to it by changing a mutable position (e.g., by moving to another state where the immutable restrictions do not apply). It is thus possible for persons whose immutables restrain their life chances to remove or at least lessen these restrictions through the use of their mutable characteristics. However, it also works the other way around. Rather than mutables lessening restrictions on immutables, immutables can also work to restrict mutables, sometimes removing much of their mutable quality.

For example, if correlations exist between immutable and mutable characteristics, then the former can affect the latter (and vice versa). For example, if an immutable (gender) affects education (young women are discouraged from taking mathematics courses) (I) and technology (T) (women are discouraged from using mechanic's tools), then mutables become less mutable, and tend more to immutability (if information $[I]$, technology $[T]$, and occupational [O] choices for women are fixed [by norms, precedent, etc.]).

I said that immutable characteristics are useful for those in power because they are instantly visible (their determination requires no background research or documentation) and they cannot be changed. Conversely, mutables are not quite as convenient for those in power because mutable levels are more difficult to determine, and they can be changed over time. For example, one cannot instantly used education (I) or occupation (O) or technology (T) to reward or punish a stranger, as these cannot be instantly determined as can gender or race. Rather, the mutables are generally determined through documents such as diplomas, licenses, etc.

We are now ready to discuss the macro aspects of mutables. While each individual in the system has a set of mutable characteristics, the system as a whole also possesses a *mutable distribution*. This mutable distribution is a macro property of the society, but it is not a global. It cannot be defined without information on characteristics of individuals. Rather, it is an "analytical property" in Lazarsfeld's (1958) terms, as it is formed from aggregating individual mutable properties such as occupation into a distribution. While the mutable micro characteristic is a property of an individual (e.g., his or her occupation), the mutable macro distribution is a property of the society (e.g., the occupational division of labor). While a distribution such as the division of labor cannot be defined for an individual (by definition), it can be defined for subsystems of the society such as an organization or group.

Thus, there are three main types of variables in SET, globals (macro only), immutables (micro only), and mutables (taking the dual form of both *macro* mutable distributions and *micro* mutable individual properties). Since mutables have dual micro-macro properties they serve as the link between micro and macro, but are often confused as either micro or macro, rather than being seen as dual properties.

Allocation Theory

I have indicated that the allocation of individuals into structural positions is the most crucial process in SET. The allocation of individuals entails social interaction processes, or Q-relations among individuals. However, allocation decisions are based upon

R-variables (mutable and immutable micro characteristics of individuals). Thus, Q and R are closely related.

Where an individual is located in the mutable structure at any point in time depends, among other things, upon where he or she began. The minimal set of dimensions for analysis of allocation is the set of five mutable distributions (information, space, technology, organization, and level of living—*ISTOL*) *plus* three immutables—race, sex, and age (birthdate). This individual is at some position on each of these eight dimensions at any given point in time. His or her position on one (e.g., occupation) depends upon his or her prior position on that same dimension (prior occupation) as well as on the other mutables (I, S, T, L) and on his or her immutables. The context for allocation depends not only upon mutables and immutables, but also (as has been shown) upon levels of the globals.

I can now illustrate the allocation process via the three-level model. Suppose a business consists of an employer (person in power or decision maker) and eight employees. The empirical system (X') consists of occupational positions, all with incumbents. Suppose that the employer has a perception (X) of an additional position (O). The boss may first hire a person (X') and then create a position (X), but generally he or she will first perceive of or envision a position (X), create it on paper (X'') and then seek to fill it (X'). The position may be created on the indicator level (X'') through a written description of duties, qualifications, and pay in the company personnel manual. The paper description in the manual (X'') may describe the mutable characteristics desired in the incumbent, such as education (I), residence in the city (S), prior experience (O), a driver's license (T), etc. It may also include a description of immutable characteristics such as age and sex.

The next step may be for the employer to advertise the position in a newspaper classified advertisement (X''). Then applicants will peruse the advertisement (X'') and decide whether to apply. The next step is for the applicant (X') to go for an interview.

There are at least two applications of the three-level model at work here. The employer has a perception (X) of the perfect incumbent (X') for the position as advertised (X''). The applicant also has a perception (X) of the advertised position (X'') and of himself or herself as an applicant (X'). What is needed is not only congruence isomorphism between the position (X'') and the incumbent who fills it (X') but also between the employer's perception (X) of the position (X'') and of the applicant (X').

There are a whole set of possible congruences that must be isomorphic in order for hiring to take place. From the applicant's standpoint, his or her perception (X) of the position must be congruent with the actual position (X'). If he or she finds the empirical situation (X') much different than expected from the advertisement (X''), then there will be dissatisfaction. Congruences should be attained if the applicant's perception (X) is congruent with the position as described on paper (X'') and if the paper description is representative of the actual position (X') rather than misrepresenting it. If it is misrepresented, the applicant will be dissatisfied (e.g., if job duties are different than described or the pay is less). From the applicant's standpoint there should also be congruence between his or her perception (X) of his or her empirical qualifications (X') and actual empirical qualifications (X') needed for the job.

From the employer's standpoint there are also a number of necessary isomorphisms. The employer's perception (X) of the applicant's qualifications (X') must be isomorphic with the qualifications (X') actually required to adequately perform the job duties. There may well be discrepancies between the employer's perception (X) of the applicant's qualifications, and the applicant's own perception (X) of the same qualifications. These examples do not exhaust the points of isomorphism in this hiring process. If isomorphism is relatively good, hiring is successful. If isomorphism is lacking between any of these sets of perceptions of reality (X), paper descriptions of reality (X'') and empirical reality (X'), for either the employer or the applicant, then hiring will be unsuccessful.

Both the mutables and immutables figure prominently in the hiring process. When the applicant sends his or her resume to the employer, it is scanned for information on the five basic goals— residence $(S;$ and how long at that residence), information level $(I;$ education), technology $(T;$ e.g., computer skills, typing skills), occupational history or experience (O), and past salary (L). But the employer also likely attempts to discern information on the immutables (age, sex, race) which may or may not be evident. This initial reading of the resume is necessary to determine information the employer needs about the mutables, as these are often not directly observable but must be documented at the indicator level (X'') (e.g., on a resume). Resumes may be bolstered by additional documents providing verification such as university transcripts, driver's licenses, birth certificates, etc.

After a perusal of these indicators (X'') the employer can reach an initial perception (X) of the applicant (X'). However, hiring generally is not done solely on the basis of this perception, but includes a face-to-face interview. Here the employer is able to supplement his or her initial perceptions, both by verifying these initial perceptions regarding the mutables (originally obtained from the paper resume) and by now securing needed additional information or points of clarification. In addition the employer is of course forming a perception (X) of the applicant's visible (X') immutables (age, sex, race) which were lacking or insufficient at the indicator (X'') level. The employer can use the interview to alter his or her initial perceptions (X) on the basis of the observed immutables or further evidence on the mutables, while the applicant can use the interview to attempt to repair negative impressions or inconclusive information present on the resume. In other words, the employer may feel that there is a negative lack of congruence between the skills needed by the actual job (X') and the skills actually possessed by the applicant (X'), while the applicant may try to convince the employer that the needed congruence exists.

This hiring process is an example not only of the use of the three-level model, but also of the structure/process interrelationship that was mentioned earlier. The paper structure (X'') provided by the applicant in the form of a resume or vita represents synchronic structure at a given point in time. Some information on this vita will not ever change (including the immutables such as race or sex, and perhaps some mutables, such as education or residence) while some will change, but perhaps not for a long time, or only very slowly. However, the information selected to be included on the vita (X'') can change, or the way it is presented can change. Thus, structure (X'') guides process (the interview) which can lead to changes in structure (X'') through revision or clarification of the document (resume or X'').

ORGANIZATION

According to SET, there are two principal forms of organizational formation. These are *agglomerative* and *divisive*. Agglomerative organizations are formed by beginning with a nucleus of one or more persons, and adding members over time. Divisive organiza-

tions are formed by the branching or splitting of pre-existing organizations. Branching may occur because of internecine fighting, or
may simply occur because the organization has become too large
to be efficiently administered.

Agglomerative groups are generally known as organizations or
associations. They are usually goal oriented, and possess a communications or interaction network. Some divisive groups do not engage in face-to-face interaction among all members. They may
indeed have a group existence only as a class or category for statistical purposes (e.g., black males). Divisive subgroups are homogeneous or monothetic (Bailey 1975) on the defining characteristics,
but are generally polythetic (heterogeneous) on remaining mutables and immutables. Agglomerative groups are generally polythetic in terms of immutables (ascribed characteristics). Examples of
divisive and agglomerative groups are shown in figure 6.2 (from
Bailey 1990, p. 156).

Social entropy theory recognizes at least three types of organizational boundaries:

1. The actual boundaries of the physical plant occupied by the
   organization.
2. The bounding characteristics of the members of the organization.
3. The territorial limits or "outreach" of organizational operations (e.g., the limits of a salesperson's "territory").

External boundaries are entropy breaks. There must always be
different entropy levels on different sides of the organizational
boundary. If not, the organization cannot be distinguished from its
environment, and is best viewed not as a different system, but only
as a subsystem of some larger encompassing system.

FIGURE 6.2.
Divisive and Agglomerative Groups

	Network	No Network
Divisive	NOW (subgroup of all women)	All women
Agglomerative	A business	Undefined

Source: Bailey (1990, p. 156).

Boundaries become structures which not only define spatial area (S) but also constrain social interaction. Thus, boundaries are in a dialectical relation to process (social interaction) just as all structure is. For example, one pertinent "chicken or egg" question is: Do boundaries determine social interaction or does social interaction determine boundaries? The obvious answer is both. When a boundary is formed (for a society, organization, or system of any level) it becomes a structural constraint which social interaction must deal with. If the new boundary includes more territory than before, technology (T), information (I), organization (O), and population (P) all must be adapted so that changes in space (S) will not result in a lower level of living (higher entropy). Thus, the boundary structure affects process. On the other hand, process affects the boundary, and may even change its placement. Thus, $S{\rightarrow}P{\rightarrow}S{\rightarrow}P$, etc. Organizations must protect and maintain their boundaries. This includes not only preservation of the physical walls of the building, but also the screening of members so that nonmembers are excluded.

Each organization has a set of globals, just as each society does. Each organization has its own population (P), information (I), space (S), technology (T), set of positions or organizational structure (O), and level of living (L). Each individual within the organization is thus constrained not only by the six globals and five mutables operating at the society level (and above that at the world level) but also by the six globals and five mutables possessed by his or her own organization and by the work group (department) within it. For example, it is possible for the society as a whole to have insufficient income, technology, and information, and to have overpopulation and lack of space, and for the individual's organization to also be so affected, but for his or her department to be well paid, provided with the latest technology and up-to-date information, and to be understaffed. In general, though, since globals and mutables at lower levels (group and organization) are a subset of those globals and mutables for the whole society, they will often mirror the conditions of the larger system.

Entropy Management in Organizations

Organizational administrators at all levels are responsible for maintaining proper levels of the six globals and five mutables (although they may not realize it, and certainly do not use these terms). If personnel levels (P) fall too low, they must hire; if they are too high, they must fire. They also decide to buy computers (T), send employees for further training (I), expand plant capacity (S), etc. They generally manage these levels with a view towards maximizing profits (L) either in the short run or the long run.

Although such managers generally do not describe themselves as managing entropy levels, this is in fact what they are doing. They are continually balancing the constant increase of internal entropy through decay of physical plant ("depreciation" in tax terms), use of materials, obsolescence of information, etc. This internal entropy increase is offset through inputs of new raw materials (matter-energy), new information, new technology, etc.

The job of entropy management thus consists of regulating inflows, outflows, and throughflows. Inflows must bring in negentropy. This means proper food, raw materials, personnel, etc., must be in order. Outflows must not deplete negentropy. This means that security must be regulated, so that information or technology is not stolen, etc. Security must insure not only that supplies and computer information are not stolen, but also that the proper personnel are admitted, and improper personnel are excluded. This is mainly a boundary maintenance task.

THE CENTRAL PROBLEM OF SOCIAL ORDER

Proper regulation of globals and mutables leads to maintenance of social order (as well as adequate entropy levels). But this can only be done if individual action is also maintained through the use of immutables. I have discussed two types of order—diachronic Q-order among individuals (social interaction or process) and synchronic R-order among variables (structure). I have indicated that there is a continuous and ongoing dialectical relationship between the actions of individuals and the structure of variables. Structure informs action, which then reshapes structure, etc. This is accomplished through use of the three-level model (X, X'', X') to allocate individuals (on the basis of immutables and mutables) into the mutable structure within the context of the globals. When this is done in an orderly fashion, social order results.

How is order achieved? What constitutes orderly allocation? I have defined order as departure from randomness (from maximum entropy). I have also said that the human interactions which allocate people into the mutable distributions will generally show some degree of order. This order is empirically variable, but generally stays below maximum entropy (randomness) and above minimum entropy (zero).

What constitutes nonrandom or orderly behavior (process)? Orderly action is replicated action. All that is necessary for departure from randomness is replication or regulation ("reproduction") of behavior. This is true regardless of the cause of the replication—i.e., regardless of whether the regulation results from norms, customs, habit, precedent, etc.

This leads us to the Central Axiom of Social Order:

> Orderly process relationships between human actors (diachronic Q-relationships), when based on specific variables, will result in orderly relationships (correlations) between these variables (synchronic R-relationships).

It is not the content of the variables that decides their degree of correlation, but only the degree to which they were used in replicated interaction.

Another salient conclusion can be reached from all this. The relationships are the substance of abstracted systems. The concrete actions by human individuals are the substance of concrete systems. We have just seen that *abstracted and concrete systems are inherently related.* That is, *concrete action* results in *abstracted relationships.* Thus, we see that the relationships between variables (R-relationships) that are the focus of the abstracted system are generated by interaction among concrete actors (the focus of the concrete system). Thus, our holistic model deals with both concrete and abstracted systems, and shows their nexus. The point is that an adequate model cannot exclude *either* abstracted or concrete systems, but must include both. For further discussion see Bailey (1990).

Markers and Information

Information is carried by *markers.* These are physical matter-energy substances which can serve as carriers. Common markers include paper, film, computer tapes, computer diskettes, etc. Humans are also markers. They carry information in the form of immediately visible immutables (e.g., age, sex, race). Mutable information is less visible, and is generally coded upon paper indicators (X'') such as licenses, diplomas, resumes, vitae, letters, bank statements, letters of credit, etc. All of this information is used in allocating persons into the mutable structure. Unfortunately, decisions that preclude the attainment of an individual's goals may be reached prematurely by someone in power (a decision maker) on the basis of immutables before information about mutables is even processed. This is because information concerning mutables may take longer to process and may be attained later in the interaction process.

Statistical Entropy

But whether decisions about allocation (e.g., hiring) are based on mutable or immutable information, as long as they are replicated, the resulting statistical order (*R*-order) or structure will be nonrandom to some degree. This degree of statistical order can be measured through various correlation coefficients. The most generic measure and the one which has both statistical and theoretical interpretations is entropy (*H*). The various statistical entropy measures, including univariate entropy, multivariate entropy, and conditional entropy, have been discussed at some length in chapter 4 (see equations 4.1 through 4.15). Some of these equations are repeated here for ease of application in our discussion of SET. One of these is the basic *H* measure:

$$H = \sum_{i=1}^{K} p_i \log 1/p_i. \tag{6.7}$$

If allocation were random, *H* would be maximum (log *K* where *K* is the number of categories). If allocation were perfectly replicated, *H* would be zero.

I have noted previously that the objection to past systems theories (and other models as well) is that they sometimes appeared to critics to be deterministic, leaving little or no role for human action. For example, economic phenomena are determined by the "market," and the functionalist society returns to equilibrium "automatically" after being disturbed. In SET, equilibrium can occur, but there is no automatic return to equilibrium. Ironically, equilibrium in entropy terms is maximum entropy. This represents stability, but only in the negative sense of systems death (the system has degraded as much as it can). In SET, human actors regulate matter-energy and information flows so that maximum entropy is avoided.

Thus, the Q-R connection is seen to be much more than a statistical phenomenon. It has great theoretical relevance, as it explains not only how action (process) (Q) leads to structure (R), but also how human action rather than economic determinism, physical determinism, or some other form of determinism, leads to the state of the system.

At this point some measure of system state is required. Past models have used various measures of system state. Ecologists have used the "subsistence level" as a benchmark for progress in adaptation. Other models use the "poverty level." Functionalists used "equilibrium." Equilibrium is an indicator of system state, but an inadequate one. It is very simplified. It is a single point. If the system is not in equilibrium it will return to it, or achieve a new equilibrium (as in "moving equilibrium"). Methodologically and theoretically there are a number of problems with this, and these are discussed in more detail elsewhere (Bailey 1984a) and in chapter 7. One major problem is that equilibrium is really system death (maximum entropy) in physics. If functionalists do not follow this usage, then equilibrium loses its theoretical foundation, and becomes an arbitrary point. Thus, "equilibrium" can occur at any level of societal development, and we have no way to determine this level. This is theoretically inadequate. Further, equilibrium is a dichotomy—the society is either in equilibrium or is out of equilibrium. If in, the state is arbitrary. If out, what is the level? It could be anything. Obviously, a measure is needed which varies from zero societal integration to maximum societal integration. Entropy is such a measure. There is still some confusion regarding the interpretation of entropy (Bailey 1983, 1990). While recognizing this, we need not become mired in it. We can define entropy both in terms of matter-energy and information. The equation for information entropy is presented in equation 6.7. The equation for matter-energy (for isolated systems) is presented in equation 6.8:

$$q/T_{max} < dS, \tag{6.8}$$

where q = the heat increase of the system; T_{max} = the maximum temperature of the system; and dS = the change (increase) in entropy of the system.

If all human action were random, with no replication, then *H* would be maximum. If all human actions were identical, then *H* would be the minimum, or zero. In reality, *H* is generally somewhere in between. Purposive action guides systems—not return to equilibrium. This purposive action is guided by customs, laws, five-year plans, short-term goals, etc. It thus is to some degree "rational" and replicated, thus keeping *H* below the maximum. It is also imperfect and subject to error, thus keeping *H* above the minimum.

The statistic *H* suits our purposes as a measure of system state. It varies between zero and a (variable) maximum. It can be interpreted both substantively and theoretically (as we have done throughout this chapter) as well as statistically.

Table 6.1 shows univariate entropy, as applied to one of the six globals, level of living (*L*) as operationalized in terms of income.

If true thermodynamic equilibrium were present for income, then allocation of persons into income categories would reveal *no discrimination* by race, gender, age, residence, etc., but would be totally random. In this case, each person in the population (*P*) would be as likely to be placed in the highest income category as the lowest, or in any of the three intermediate categories. According to entropy theory, maximum entropy is statistically the most probable. Substantively, we are saying that a truly egalitarian, non-discriminatory society would be statistically most probable. Why then does it not occur? Perusal of Table 6.1 reveals that actual entropy levels are far from maximum (from the perfectly non-discriminatory state).

Table 6.1.
Percent Distribution of Aggregate Income, by Fifths of the Population, for Families and Unrelated Individuals, United States, 1978

| | Percent Distribution of Aggregate Income | | | | |
	Lowest Fifth	Second Fifth	Middle Fifth	Fourth Fifth	Highest Fifth
1978	3.8	9.7	16.4	24.8	45.2

Source: United States Department of Commerce, *Money Income of Families and Persons in the United States: 1978*. Washington, D.C.: U.S. Government Printing Office, 1980.   From Bailey, 1990, p. 217.

The answer has already been discussed in some detail. Persons are allocated according to immutables and mutables. The correlation between income and both immutables (such as race, age, and sex) and mutables (such as residence [S], technological skills [T], education [I], and occupation [O]) lead to regularities far from the maximum (log K). This can easily be analyzed through bivariate or multivariate entropy:

$$H(XY) = -\sum_{i=1}^{K}\sum_{j=1}^{L} p_{ij} \ln p_{ij}. \tag{6.9}$$

At this point some readers might be confused about which of the six globals or five mutables serve as indicators of system state. Be reminded that all of the globals are closely interrelated, so when you are measuring one you are measuring them all to some extent. The mutables are formed from the globals, and are in a proper categorical format for measurement by entropy (H). Thus, while we can measure system state by applying H to a single variable such as level of living (L), a stronger approach is to apply H to *all five mutables,* or ideally to all five together (multiple or multivariate H). Multiple H (equation 6.9) is theoretically clear for *any number of dimensions,* but of course computation becomes more difficult as variables are added. H can also be used to measure distributions of immutables (see Bailey 1990).

Another form of H is conditional entropy:

$$H(Y|X) = \sum_{j=1}^{L} p_{j|i} \ln p_{j|i}. \tag{6.10}$$

This measures the entropy level for one variable when the value of another variable is known. Although not optimal as a measure of system state, conditional entropy may be valuable in assessing relations between parts of the system, for example relationships between an immutable and a mutable. Conditional entropy can tell us, for example, the entropy level of occupation when race is known, or when gender is known.

Statistically, H has analogues to correlation (multiple H), and regression (conditional H). Also, as a measure of dispersion it is analogous to variance. For further comparison of H with other common sociological statistics see Krippendorff (1986).

COMMENTARY

Contributions

Like LST, social entropy theory has made a number of distinct contributions. Among these are:

1. A methodological critique of equilibrium theory (most past critiques had been theoretical or ideological). See also Bailey 1984b.
2. A thorough discussion of the virtues of both concrete and abstracted systems (see also Bailey 1981).
3. A demonstration (via the *Q-R* distinction) that concrete and abstracted systems are two sides of the same coin, and are thus complementary and should both be used simultaneously.
4. An explication of the link between process and structure.
5. Development of the three-level model.
6. Development of the global-mutable-immutable distinction.
7. The explication of social entropy, both qualitatively and quantitatively.
8. The development of allocation theory.
9. A novel explication of social order.
10. A novel explication of power.
11. A utilization of the macro, micro, and organizational levels of analysis (compare this with the macro-meso-micro formulation of Ritzer 1990b).

Limitations

Social entropy theory also has some salient limitations.

1. From the standpoint of this book, SET has the limitation of not being a true general systems theory. Rather, it is an application of some systems principles to the study of society. It neglects discussion of a number of salient systems issues such as feedback loops, self-steering, and autopoiesis (leaving these instead for the present volume).
2. Social entropy theory is not as detailed as LST, and does not simultaneously discuss all twenty subsystems and all eight levels.

3. Social entropy theory does not test hypotheses (but does present a set of testable hypotheses).

These of course do not exhaust either the contributions or limitations of SET, but are just some of the principal ones which come to mind. While this short synopsis cannot do justice to social entropy theory, it does provide an introduction. Further details are given in Bailey 1990. I shall also discuss SET in more depth in the next chapter. In chapter 7, I compare and contrast social entropy theory and living systems theory, and seek a congruence between the two approaches.

COUNTERPOINT

As I said in chapter 1, social entropy theory (SET) has clear parallels to neofunctionalism, to structuration theory, and even, to some extent, to conflict theory, as seen respectively in the work of Alexander, Giddens, and Collins. As such, SET is a third-phase theory (in Alexander's terms) which has goals of synthesis and interaction. At the end of chapter 1 I posed seventeen concepts as a checklist for comparing the new systems theory with mainstream sociological theory. I will consider each of these in turn.

Let us first consider the group of variables derived from Alexander's neofunctionalism: action, order, equilibrium, and idealism/ideational. Social entropy theory deals extensively with these. The central axiom of social order from SET states that order is derived from replicated action, without specification of how this replication is obtained (internalization of norms, ritual, coercion, custom, etc.). Thus, order is a central concern of SET, as is action. Action, of course, is the motor through which order is derived. Action is central to SET, but is analyzed after first specifying macro variables as a context for it. Action is never analyzed in a vacuum, nor outside of the macro context of the *PILOTS* variables.

Equilibrium has been analyzed a great deal in this volume, and little more need be said about it. Suffice it to say that SET completes the reconstruction of equilibrium by adopting a nonequilibrium entropy model. As for the idealism/ideational component, it is included, but relatively lacking in SET as compared to Parsonian functionalism. However, SET does not preclude it, but has simply not stressed it to the degree that Parsons did.

The next set of comparison variables is derived from Giddens's structuration theory: agency, structure, system, structuration, system integration, time, and space. As mentioned earlier, the "synchronic structure/diachronic process" interaction chain described in SET is a never-ending dialectic. Structure is encoded on markers such as the Constitution, sets of statutes, the municipal code, dictionaries, etiquette books, rule books for sports, etc. This synchronic structures guides action (agency) in an orderly fashion, but the result of the action can in turn be a change in the synchronic structure.

This is quite parallel to Giddens's formulation, but with a different definition of structure. It is very compatible with and complementary to his work, and uses time in the fashion that he advocates. Also, space is a central variable in both the set of *PILOTS* globals and the set of *LOTIS* mutable distributions. No concept comparable to structuration is found in SET, and SET does not discuss system integration to the degree that Giddens does.

In SET, the parallel to systems integration is system state. The discussion of system state in SET subsumes the notion of system integration, but in a more methodological way. Again, the two approaches are compatible. The differences in the definitions of system for these two approaches have already been discussed in the counterpoint for chapter 1, and need not be repeated here.

The third set of counterpoint variables is derived from Collins's conflict theory, and includes: conservatism, conflict, age, sex, and hypostatization. As said earlier, without the concept of equilibrium and the return to the status quo (which excludes the possibility of revolution) the critique of systems theory as conservative becomes much less valid. Social entropy theory is science based, and so might be called "conservative" by some in that respect. But it is not conservative in the other ways previously discussed. It does not use equilibrium, is not teleological or tautological nor deterministic, emphasizes individual action (in a macro context), emphasizes the study of social change, and includes factors such as values and norms. With its frequent examples of race and gender it is by most standards quite a liberal theory for 1990s sociology.

While not its main thrust, SET also contributes to conflict theory and the study of power. Among the contributions here is the notion of immutable and mutable variables and their role in conflict and exploitation. Specifically, those in power can discriminate against persons with immutable characteristics, and there is little that these persons can do except attempt to alter their mutable characteristics. Age and sex are relevant because both are immutables (if age is operationalized as date of birth) and both are emphasized throughout SET.

This brings us once again to hypostatization, and to the study of micro-macro links (from Alexander, Giddens, and Collins). I took great care in SET to explicate the notion of the need for isomorphism between the model of complex society (X'') and the society itself (X'), with the proviso that for explanation to be adequate, the model needs to be as complex as the social phenomenon it is modeling. I see that as a powerful methodological bar to easy charges of reification or hypostatization. This is because isomorphism requires a point-by-point linkage between the model and the empirical phenomenon. When this is accomplished, hypostatization is difficult. The SET model is *not* hypostatized. It is a concrete system model that is *subsequently* used to develop analyses of ideational and abstract phenomena. As a concrete model it deals with "real people pursuing real interests" just as Collins's theory does.

I hope that the innovative contribution made by SET to the micro-macro linkage problem can be recognized (Ritzer 1990b has already commented favorably on the methodological approach). Social entropy theory does not proceed from micro to macro as do most formulations, or even from macro to micro, as do most others. Rather, it develops the *PISTOL* macro variables as a macro context (both material and ideational) for the analysis of individual action. This nonhypostatized system provides a grand forum for viewing action. The macro variables or globals (*PISTOL*) and the micro variables (immutables) are connected by the true linking or mediating variables (mutables) which can be used to formulate both macro variables (the mutable distributions or *LOTIS*) or micro variables (mutable characteristics of individuals, such as the individual's residence, occupation, or educational level).

Returning to the structure/process relationship, we see that individual action results in change in synchronic structure (at time 1) which results in change in regulation of individual action at time 2, etc. For example, individual demographic behavior such as birth, deaths, and migration during the period 1990 to 2000, when viewed collectively, leads to a set of synchronic statistics (as in the

2000 United States census figures) which are then used by policy makers to regulate future individual behavior and set goals (employment, agricultural production, etc.). This is just one example, as the model is very broad. It utilizes the Q (person) and R (variable) distinction *in conjunction with the three-level model* (X, X'', and X') to analyze behavior *over time*. The result is a "rolling model" of synchronic-diachronic overplay ($Q{\rightarrow}R{\rightarrow}Q{\rightarrow}R{\rightarrow}$etc.), while ($X{\rightarrow}X'{\rightarrow}X''{\rightarrow}X{\rightarrow}X'{\rightarrow}X''{\rightarrow}$etc.).

The marker level or operational level or indicator level (X'') is instrumental in storing most of the synchronic variable data which is used to guide future behavior. The model is one of the most complete (and certainly the most methodologically complex) micro-macro models that I have seen to date. It parallels Giddens's model and is quite complementary to it, as Giddens's structuration model includes a number of the subjective elements not stressed in the SET model.

What forms of metatheory are evident in SET? Again, as in the preceding chapter, the first variety of M_U (the internal-intellectual) is prominent in the analysis of entropy. Also evident are M_P (prelude to the development of theory), and M_O (metatheorizing to create an overarching perspective).

What independent contributions have been made in SET that have not been made in mainstream sociological theory, and thus might add breadth and richness to the mainstream? Again, the contributions have just been summarized and need not all be repeated. The contributions that stand out as especially complementary to mainstream theory and which add breadth and richness to it are the methodological contribution to the analysis of society, the operationalization of order (in terms of entropy), the presentation of the notion of entropy, the three-level model, the contribution to the analysis of the micro-macro link in the form of the global-mutable-immutable distinction, the process-structure interaction model, the inclusion of time and space (à la Giddens), and the theory of allocation.

CHAPTER 7

Living Systems Theory and Social Entropy Theory: A Congruence

I have presented summaries of both living systems theory (chapter 5) and social entropy theory (chapter 6). Both of these are detailed and relatively recent systems approaches to the study of society. Living systems theory and SET share a number of common goals and features. Both are relatively recent (LST, 1978; SET, 1990). Both utilize modern systems concepts, with an emphasis on the study of entropy as well as on the processing of matter-energy and information. Both recognize and analyze several hierarchical systems levels (e.g., individual, organization, society). Both distinguish between concrete and abstracted systems, and begin analysis with the former.

DIVERGENCE

But besides these similarities and others, LST and SET display parallel or even seemingly divergent features which make it difficult for some readers to mentally merge the two approaches. A few such divergent features are prominent.

1. Living systems theory is more "biological," beginning its analysis of the eight levels with the cell (Miller 1978, pp. 273–314) and the organ (Miller 1978, pp. 315–360) while many sociologists seem to encounter difficulty in conceptualizing below the level of the organism (Miller 1978, pp. 361–514).

2. Living systems theory is truly a *general* systems theory encompassing all living systems, and generalizing about their subsystems. In contrast, SET is essentially a sociological theory, with the society as the basic unit of analysis, but utilizing systems concepts.

3. Living systems theory studies all twenty subsystems for all

eight levels (cell, organ, organism, group, organization, community, society, and supranational system) while SET primarily focuses on the society, with considerable attention to the subsystems of individual, group, and organization.

4. By focusing on concrete systems, LST is guilty of some neglect of not only abstracted systems, but of the relationships between concrete and abstracted systems. Social entropy theory begins analysis with concrete (object) systems, but is careful to analyze the manner in which abstracted (variable) systems develop and the role they play in society. Also, the link between concrete (Q) and abstracted (R) systems is carefully analyzed in SET.

5. In contrast to SET, LST is more reliant upon troublesome classical concepts such as equilibrium and emergence which, if not carefully defined, can lead to problems. Social entropy theory does not preclude use of equilibrium or homeostasis, but eschews them as general properties of systems, preferring instead to analyze the state of a given system in terms of its entropy level.

6. While LST eschews analysis of values and subjective phenomena as integral parts of the social system, SET includes them and relies on them to some degree. Miller's reluctance to include subjective phenomena stems from his understanding that "science" does not include study of these subjective phenomena. In this sense, perhaps SET is less "positivistic" than LST.

7. Living systems theory relies on analysis of twenty subsystems for each of eight levels. In contrast SET relies on three categories of variables. Notice that while the subsystems of LST are concrete systems, the focus in SET is on abstracted or variable systems (that were derived after beginning from the standpoint of a concrete system). The three categories of variables in SET are as follows.

a. *Globals*. These are the *PISTOL* variables (population, information, space, technology, organization, and level of living). Globals were defined by Lazarsfeld (1958) as properties of societies which can be defined *without information* about individuals within the societies. As such, the globals are true *macro* properties, not requiring micro information on individuals for their formation.

There is some disagreement over whether population (P) is a true global, as the population size cannot be determined without information as to whether each individual in the population exists (e.g., coded 1) or does not exist (e.g., coded 0). The "1"s are summed to obtain the population size P. However, while one must know whether a person exists or not to compute P, one need not know any additional information regarding the individual, such as race, age, gender, income, or any of myriad other possible characteristics. Thus, P is essentially a global. However, if one wishes, P may be termed a "semiglobal" to distinguish it from the other five globals (*ISTOL*) which are "true" globals (in that their computation requires no knowledge of characteristics of individuals in the society).

   b. Mutables. Five *mutable distributions* are defined in SET by distributing the P individuals in the population into each of the other five global variables. The special status of P as a semiglobal is clear here, as it is the variable involved in creating mutable distributions from the five basic globals (*ISTOL*). Thus, there is in each society not only a "division of labor" or occupational mutable distribution (O_m) but also distributions of the population by residence (S_m), class or level of living (L_m), by the technology they utilize (T_m), and by the information they utilize (I_m). The mutables have the character of a true "micro-macro link" as both macro and micro forms of mutables are found. For example, each society possesses an occupational division of labor (O_m), and this mutable distribution is thus a *macro* (but not a global) property of that society.

   Each individual person also has a place in the particular mutable distribution possessed by his or her society. For example, if your society possesses a mutable level of living or class distribution (L_m) as a macro property and also an occupational division of labor (O_m) as another macro property, you as a member of that society have a *position* in each mutable distribution. To say it another way, you have a class position (L) as one of your individual characteristics, and an occupation (O) as another of your individual characteristics. Thus, L and O are micro properties which are possessed by you as an individual rather than by the society as the unit of analysis. Thus, the *total mutable distribution* (e.g., the society's occupational division of labor) is a *macro* property, while an *individual's mutable characteristic* (e.g., occupation) is a *micro* property, and one part of the overall aggregate mutable distribution. A macro property such as a mutable distribution, which can be seen as an aggregate of individual properties was termed an *analytical* property of society by Lazarsfeld (1958).

   c. Immutables. Properties or characteristics of individuals (as opposed to societies) are termed *micro* properties. Micro properties such as class, residence, and occupation have already been identified as *mutable* micro properties of the individual, as these identify the position of the individual within the mutable distribution of the entire society. Mutable micro properties are amenable, at least potentially, to change. Thus, one can potentially (if not actually) change his or her residence, class position, occupation, education (*I*), and so forth.

In contrast, *immutable* micro properties cannot be changed. These are properties such as birthdate (age), sex, and race. The mutables (both macro and micro) link the globals (purely macro) and the immutables (purely micro). An individual is allocated into a particular position in each macro distribution on the basis of his or her overall pattern of mutables and immutables.

After perusing chapters 5 and 6, the reader might have a relatively clear picture of the scope and goals of both LST and SET, but feel that the two approaches are generally divergent, or at best parallel. One reason for this apparent lack of congruence between LST and SET may be the reader's inability to discern clear analytical links between the twenty subsystems of LST and the six *PISTOL* variables of SET. That is, suppose that the societal level is chosen for analysis (the society as the unit of analysis). The two approaches, LST and SET, would proceed quite differently. Living systems theory would analyze the twenty subsystems of societies (see Miller 1978, pp. 747–902) while SET would begin with the *PISTOL* variables. For example, Miller (1978, pp. 766–68) identifies the decoder subsystem of a society as an organization such as the State Department (among others) and the decider as the society's voters.

Thus, the ultimate portrayal of a society is different for LST and SET. Social entropy theory provides a more holistic picture, showing how the society alters levels of the six key *PISTOL* variables and reacts to changes in these levels. In contrast, LST provides a more fragmented picture, showing separately the working of the twenty subsystems of the society. For example, Miller (1978) discusses separately all of the twenty subsystems, classifying them as both matter-energy and information (reproducer and boundary only) or matter-energy processing subsystems (for example, ingestor, distributor, converter, producer, matter-energy storage) or information processing (for example, input transducer, internal transducer, channel and net, decoder, associator, memory and decider). Each of these is discussed in turn, with the discussion divided into analyses of structure and process. For example, in discussing the decider, Miller (1978) notes under *structure* that "A society's decider consists of its central governing components" (p. 799) and under *process* that "A society's rights over people, physical resources, land, money, and credit supercede the rights of lower-level systems" (p. 800).

How can these two seemingly different pictures of the societal system (LST and SET) be reconnected? Or further, are the two portrayals too disparate for reconciliation? Closer perusal shows clear linkage between the two approaches. These are the topic of the next section.

TOWARDS CONGRUENCE

Congruence between LST and SET is best illustrated by the parallel comparison of their basic features. For SET these are six basic variables (*PISTOL*) at the level of society, and for LST they are the twenty subsystems. It seems most parsimonious to focus upon the six *PISTOL* variables of SET, and compare features of LST with each of them (although alternatively one could begin with the twenty subsystems of LST).

The best starting point is the definition of abstracted and concrete systems. While these terms have been discussed previously (see chapters 2 and 5), it seems wise to repeat their definitions due to their importance in this analysis. Both LST and SET begin with concrete systems, and indeed two of the basic *PISTOL* variables (population and space) are evident in the definition of a concrete system. As defined by Miller,

> The units of *abstracted systems* are relationships abstracted or selected by an observer in light of his interests, theoretical viewpoint, or philosophical bias. Some relationships may be empirically determinable by some operation carried out by the observer, but others are not, being only his concepts. . . .
>
> A *concrete, real,* or *veridical* system is a nonrandom accumulation of matter-energy, in a region in physical space-time, which is organized into interacting interrelated subsystems or components. . . . The units (subsystems, components, parts or

members) of these systems are also concrete systems. (1978, pp. 17–19)

Miller comments further:

> A theoretical statement oriented to concrete systems typically would say, "Lincoln was President," but one oriented to abstracted systems, concentrating on relationships or roles, would very likely be phrased, "The Presidency was occupied by Lincoln." (1978, p. 19)

Note that both LST and SET begin analysis with the concrete system. This means that each uses a set of *objects* as the basic units of analysis rather than relationships or variables. Confining analysis to the society, it is clear that SET uses the *individual* as the basic unit of the social system. Thus, a society as a concrete system is composed of individuals and their interrelations. Living systems theory shares this view, although analysis of the society is sometimes conducted in terms of groups or organizations rather than individuals (see Miller 1978, p. 767). Congruence between LST and SET can now be shown by analyzing each of the six *PISTOL* variables of SET in turn and seeing how it is utilized in LST.

Population

Population size is one of the fundamental macro variables of sociological systems analysis in the SET approach. Population size is seen to be crucial in SET as it determines in large part the five mutable distributions. That is, the information, space, technology, organization, and level of living mutable distributions are all directly affected—and indeed defined—by the population size. For example, remembering that global properties are defined in absolute values and not as distributions (see chapter 6), imagine that a society has five global levels of X_o jobs (O), X_s plots of land (S), X_l dollars (L), X_t computers or other technology (T), and X_i college vacancies (I). In some sense, the chief job of decision makers in the society ("decider" subsystems in Miller's terms) is to insure that the number of residences or plots of land (S), dollars (L), school slots (I), computers (T), and jobs (O) is consistent with the population size (P) (or vice versa). If this is not true, then the effects on the society can be disastrous. For example, if $P > S$, then there is a housing shortage; if $P > O$ there is unemployment; if $P > L$ there is poverty, and so forth. Thus, population size is a determinant of a variety of crucial conditions—from full employment to unemployment, from prosperity to poverty, and so forth.

Granted that population size (P) is a crucial variable in SET, does it have a role in LST? Can it provide a point of congruity between the two approaches? It is clear that population size, per se, is a variable rather than a subsystem. Thus, it does not have the prominence accorded to the twenty subsystems in the general schema of LST. However, population size is recognized in LST, if only in passing. In discussing the variables of a concrete system, Miller (1978, p. 17) lists "its size" as one example. In discussing the society, Miller says:

> The number of people in a society is of critical importance. If its population falls below the minimum needed to perform the society's essential subsystem processes, particularly the producer processes that provide the successive new generations upon which the system's continuity depends, its future is threatened. . . . Strains arising from excess population are being experienced in many contemporary societies . . . (1978, p. 840)

Miller (1978), pp. 842–43) also mentions the spatial distribution of the population. This is, in terms of SET, the spatial mutable formed by distributing the population size over the global amount of space available to the society.

The examination of population size shows that this key variable of SET is accorded only scant attention in LST. Nevertheless, it is recognized as important, and does form one small link or point of contiguity between LST and SET. Further, the comparison of how population (P) is viewed in LST and in SET provides not only a clue as to how the other key SET variables are analyzed in the two approaches, but also offers a general view of differences in the purviews of LST and SET. To generalize, SET takes a holistic approach, viewing the whole society as it operates on a day-to-day basis, and identifying the major variables involved in this operation. From this perspective, population size is extremely salient. In contrast, LST focuses upon structure and process in each of the twenty subsystems for eight labels. From this perspective, population size is less central, but it is recognized as a relevant variable for all concrete systems at all levels. When specifically analyzing the societal level, population size attains increased salience. This is illustrated by Miller's 1978 discussion on pages 839–42, even though it is clear that discussion of the twenty subsystems takes precedence in the LST schema.

In other words, SET shows how the actual social system works as a whole, while LST describes salient parts of the system and the structure and process of each. Note also that while population size is not explicitly included in the definition of concrete system, there is emphasis, in the notion of concrete system, on the object (person) as the basic unit of analysis. There is implicit understanding that the totality of persons, or population size, constitutes the parts of the concrete system.

Space

While population size is not explicitly included in the definition of a concrete system, space is. Miller (1978, p. 17) defines a concrete system as existing "in a region in physical space-time . . ." Commenting further on the use of space in LST, Miller says:

> My presentation of a general theory of living systems will employ two sorts of space in which they may exist, *physical* or *geographical* space and *conceptual* or *abstracted* space. . . .
>
> The characteristics and constraints of physical space affect the actions of all concrete systems, living and nonliving. . . .
>
> Physical space is a common space because it is the only space in which all concrete systems, living and nonliving, exist (although some may exist in other spaces simultaneously). Physical space is shared by all scientific observers, and all scientific data must be collected in it. This is equally true for natural science and behavioral science. (1978, pp. 9–10)

Miller provides several examples of the effect of physical space on living systems, stating that "on the average, people interact more with persons who live near to them in a housing project than with persons who live far away in the project" (Miller 1978, p. 9). Further, he (Miller 1978, p. 560) says that the distance between the spatial positions of group members is significant, noting that when the distance between members of a dyad is increased, costs increase, and thus rewards must also increase.

In addition to physical space, Miller also recognizes conceptual or abstracted space in LST, saying, "Scientific observers often view living systems as existing in spaces which they conceptualize or abstract from the phenomena with which they deal" (Miller 1978, p. 10). Examples he provides include social-class space in which social classes are located, social distance among racial or ethnic groups, political distance among political parties, sociometric space, semantic space, and intermarriage space, representing frequency of intermarriage among ethnic groups.

Miller's definition of concrete system is explicitly defined in physical space, as opposed to conceptual or abstracted space. Thus, while he recognizes that scientists often utilize conceptual or abstracted space, particularly in the social and biological sciences, he eschews it for LST. One reason is ease of measurement. Miller contends that measurement is more difficult in conceptual or abstracted spaces, saying:

> Scientists who make observations and measurements in any space other than physical space should attempt to indicate precisely what the transformations are from their space to physical space. Other spaces are definitely useful to science, but physical space is the only common space in which all concrete systems exist. (1978, p. 10)

He goes on to say that scientists working in spaces other than physical are "fractionalizing science" unless they can relate the space they are working to physical space. He says further:

> Any scientific observations about a designate space which cannot be transformed to other spaces concern a special theory. A general theory such as I shall develop here, however, requires that observations be made in a common space or in different spaces with known transformations. (Miller 1978, p. 11)

The concept of space is also crucial to Miller's concept of structure. He says,

> The *structure* of a system is the arrangement of its subsystems and components in three-dimensional space at a given moment of time. . . . This process halted at any given moment, as when motion is frozen by a high-speed photograph, reveals the three-dimensional spatial arrangement of the system's components as of that instant. (Miller 1978, p. 22)

Parsons (1979) disagrees with Miller's choice of physical space for his systems formulation. In his review of *Living Systems,* Parsons says:

> One of the fundamental postulates of his [Miller's] option in favor of concrete as distinguished from abstracted systems is the apparent belief that physical space is ontologically absolute and

not subject to any kind of relativizing interpretation. It is this postulate which I would like, in concluding, to question. . . . It would be my own view that even the categories for organic systems ought to be treated as differentiated from those of physical space. (1979, p. 704)

It is clear that both LST and SET have in common a basic reliance upon physical space for their concrete systems analysis. A basic difference is that while LST relies primarily upon concrete systems, SET uses them as a starting point rather than an ending point, and also heavily utilizes variable or abstracted analysis (R-analysis).

A basic difference between SET and LST is that while LST sees space (like population) as a "given," and indeed builds it into the definition of concrete system, it incorporates little systematic analysis of space after this initial definitional formulation. In contrast, SET sees space as a basic variable which must be measured and monitored within the context of any given study of a specific societal system. That is, after their original exposition in the definitional exegis of LST, both space and population size occupy a rather passive role, being mentioned only sporadically, and not given salience in the analysis. In contrast, in SET both space and population are not treated as definitions or "givens," but rather as two of six key variables which must be measured any time a given society is analyzed.

To say it another way, in LST a concrete system is a "nonrandom accumulation of matter-energy" in the context of "a region in physical-space time." Thus, space is relegated chiefly to the level of setting or environment for the system, rather than actually part of the system itself. In SET, space is not the setting or environment for the system, but an integral part of the system itself.

In SET, space includes the physical space (land) within the societal boundaries, as well as all water, and air space over both the land and water area encompassed by the boundaries. As such it encompasses what is often termed the "environment" in social ecology. The term "spatial area" is preferred, not only because it is more descriptive, but also because it allows space to be seen as *within* the system boundaries rather than having the connotation, as the term "environment" implies, of being purely external to the system.

Thus, to clarify, the spatial area (S) within the system is internal to the system rather than external to it. It includes all matter-energy within the system's boundaries. Thus, raw materials such as gold, iron ore, lead, minerals, plus all grain and food grown, are all considered derivative of the spatial area. Thus, when talking of space in SET, *all* matter-energy, of any form, within the boundaries is included in this broad rubric.

This matter must be clarified simply because energy (or matter-energy as Miller terms it), is so central to any sociological systems framework. Energy could be explicitly added to the SET framework if desired, thus yielding seven central variables (*PISTOLE*). However, this is unnecessary as long as readers understand that "space" includes not only distance in SET, but in fact all land, sea, and air reserves within the systems boundary, and all matter-energy of any sort contained therein.

If all land and matter-energy in it constitutes the spatial variable (*S*) of the system, what then is the *environment* of the system? Following the usual systems definition, the environment of SET consists of all components *outside of* (external to) the boundaries of the concrete system which affect the system or are in turn affected by it.

Of the six *PISTOL* variables in SET, population size (*P*) and spatial area (*S*) were first chosen for analysis of points of comparison between SET and LST simply because they are considered so primary or definitional in the concept of a concrete system. As such they are primary or definitional in both SET and LST, as both formulations utilize the concrete system as the genesis of their schema. However, analysis revealed that while *P* and *S* are primary in SET, in LST they have the role of "givens" or even contextual variables. That is, their importance is duly recognized, and they are discussed early in the analysis. After that, however, they are relatively neglected, with focus switching to the twenty subsystems at each of the eight levels. That is, in LST, population size is central simply because a concrete system is comprised of units (e.g., individual organisms) and a scientist studying this system must be able to count its size and analyze the effects of the particular size or change in size whenever deemed relevant, as in the analysis of the society in LST (Miller 1978, pp. 840–42).

Beyond this, population size may be relatively neglected. Similarly, space is an important concept in defining a system, in determining whether the system is concrete or abstracted, and in defining the concept of *structure* of a system. It may be substantively important in certain instances, as in the analysis of spacing behavior (Miller 1978, pp. 564–65). Beyond this, it too, like population size, is relegated to the status of a given while the twenty subsystems are given prominence in the analysis.

Level of Living

Now that the definitionally crucial variables of population size and space have been compared as links between LST and SET, it becomes somewhat arbitrary as to which of the remaining four of the six *PISTOL* variables (*ITOL*) to analyze next. There is a certain logic in turning—once the definitional aspects of population size and space have been treated—to study of the crucial variable of *system state*. It is an elementary, but often neglected, point in systems theory that a scientific approach to systems analysis depends upon the ability to recognize and measure one or more key aspects of the *state of the system*. What is meant by system state? Perhaps the clearest way to make this point is to ask a series of questions. What is the system doing? Is it changing? If so, how? Is it not changing (remaining stable)? If so, how does one know this? These questions are very revealing.

System state refers to any important aspect of the system that can be measured in such a way as to adequately reflect change or lack of change. Ideally, a measure of system state should reflect the whole system, and not just some part of it. For example, though the society is composed of individuals, if the system is the society, the measure of system state should reflect some aspect of the society as a unit rather than some aspect of an individual or subgroup of individuals within the societal system. Thus, if the "whole is greater than the sum of its parts," some mensurable aspect of this greater whole should be chosen as the measure of system state, as this is the truly unique part of the system which provides its systemic nature, thus distinguishing it from components or subsystems.

Perusal of the systems literature in both sociological systems theory and general systems theory will show attention to a myriad of systems issues such as boundaries, information, and energy, but little concentrated discussion of the measure of system state. Yet only a little reflection suffices to show that this is one of the crucial problems of systems theory. If there is no way to measure the "state" of the system (whatever state may be chosen for emphasis), then there is really no way to study the system (as a whole) at all, and little can be known about it. For example, without a measure of system state, it cannot be determined if the system is high or low, good or bad, cold or hot, wet or dry, normal or abnormal, average or nonaverage, and so forth. If no such determination can be made, then in a real sense there can be no system science.

But if the determination of systems state is so fundamental and crucial, why is there not more discussion of it? The primary reason seems to be that there has been an implicit (and often explicit) notion in the systems literature, both in and out of sociology, that the problem of measuring system state has been adequately dealt with. How has this been accomplished? By postulating the concept of a system in equilibrium (or homeostasis or a steady state). Thus, classically, the "state" of the system that has been most crucial has been the state of equilibrium. As long as the system was in equilibrium it was fundamentally sound, and all was right with the world, precluding the need for belaboring the notion of system state. Only if the notion of system state were more problematic would systems scientists need to spend more time analyzing it.

Sociological systems theorists have long been enamored with equilibrium, from Spencer (1864) to Pareto (1935) to Parsons (1951). Both Parsons and James Grier Miller were influenced by Henderson (1935) and Cannon (1932) (see Parsons 1979). Equilibrium was discussed at length in chapter 3, and this discussion need not be repeated. Suffice it to say that emphasis on equilibrium (or homeostasis or steady state—depending on the type of system) greatly simplifies the whole notion of system state, and indeed the whole problem of systems analysis. Potentially, any system might vary from a low or minimum system state to a high or maximum system state. Consider the concept of system integration as a measure of system state. A system could vary from a minimum of completely unintegrated (zero integration) to completely integrated (maximum integration). Any actual system could be measured to see what its actual empirical level of integration was. However, the concept of equilibrium deflects attention from this potential full range of system variation. Rather than expecting a full range of variation to exist among different systems, the system theorist instead emphasizes the fact that system is in "equilibrium." This is the most salient factor regarding the system—that it is in equilibrium—and this is all that needs to be known.

If the system is in equilibrium, and this is *normal*, the classical systems theorist would rarely explore beyond this point to determine the *actual level* of some variable (e.g., integration, or wealth) that determined the "equilibrium." Further, the concept of equilibrium is strictly binary. The system is either "in" or "out" of equilibrium, with no further measure of system state. For additional methodological discussion of the equilibrium concept as a measure of system state, see Bailey 1984a.

LST   How does LST measure system state? Perusal of *Living Systems* reveals the relative neglect of the topic that is characteristic of systems science. Miller (1978, p. 17) discusses system state separately for conceptual and concrete systems. In discussing the state of a conceptual system he says, "This state is the set of values on some scale, numerical or otherwise, which [the system's] variables have at a given instant. This state may or may not change over time." In discussing the state of a concrete system he says, "The state of a concrete system at a given moment is its structure . . . It is represented by the set of values on some scale which its variables have at that instant. This state always changes over time slowly or rapidly."

Since Miller defines system state as structure, his definition of structure should be restated. Miller says (1978, p. 22). "The *structure* of a system is the arrangement of its subsystems and components in three-dimensional space at a given moment in time." Thus, it is clear that Miller does not ignore system state, but does not give it prominence, and does not move past the definitional stage. However, there is emphasis in LST on homeostasis. As Parsons says (1979, p. 47), "Miller, like the present author, was greatly influenced in his younger years by the ideas of Walter B. Cannon and L. J. Henderson. The latter had a direct influence on both of us. Homeostasis is Cannon's conception, but it was broadly shared by Henderson." The emphasis on homeostasis is evident in LST. Miller says:

> At every level of living systems numerous variables are kept in a steady state, within a range of stability, by negative feedback controls. When these fail, the structure and process of the system alter markedly—perhaps to the extent that the system does not

survive. Feedback control always exhibits some oscillation and always has some lag. (1978, p. 37)

He says further:

> All living systems tend to maintain steady states (or homeostasis) of many variables keeping an orderly balance among subsystems which process matter-energy or information. Not only are sub-systems usually kept in equilibrium, but systems also ordinarily maintain steady states with their environments and supra-systems, which have outputs to the systems and inputs from them. This prevents variations in the environment from destroy-ing systems. The variables of living systems are constantly fluctu-ating, however. A moderate change in one variable may produce greater or lesser alterations in other related ones. These alter-ations may or may not be reversible. (Miller 1978, p. 34)

Miller goes beyond the mere statement of homeostasis, however, specifying homeostasis in terms of entropy for living systems. Mil-ler says:

> The living systems are a special subset of the set of all possible concrete systems. . . . They all have the following characteristics:
> (a) They are open systems, with significant inputs, through-puts, and outputs of various sorts of matter-energy and information.
> (b) They maintain a steady state of negentropy even though entropic changes occur in them as they do everywhere else. This they do by taking in inputs of foods or fuels, matter-energy higher in complexity or organization, *i.e.*, lower in entropy, than their outputs. (1978, p. 18)

Miller relates entropy to information and matter-energy. While internal entropy always increases in closed systems in accordance with the second law of thermodynamics, entropy in open systems can remain at a constant level or even decrease. This is because importation of matter-energy and/or information into the system constitutes flows of negative entropy (negentropy) which can coun-teract the natural internal increase of entropy. Information is de-fined in terms of Shannon's H (Miller 1978, p. 13). Further, "Matter-energy and information always flow together. Informa-tion is always borne on a marker. Conversely there is no regular movement in a system unless there is a difference in potential be-tween two points, which is negative entropy of information" (Mil-ler 1978, p. 15). Miller says elsewhere (1978, p. 50) that a nonhomeostatic concrete living system cannot exist, and that each such system must have at least one variable in a steady state. He says (1978, p. 50), "Locating a concrete system which is not homeostatic would create serious doubts about my whole conceptual approach."

To summarize Miller's approach to system state, the concept of system state is clearly recognized but not given a great deal of prominence and, like space and population, is taken somewhat for granted or as a prerequisite rather than as a prime topic of analysis. Regarding the state of the system, LST:

1. Distinguishes between the state of a conceptual system and the state of a concrete system (although in both cases state is defined as a set of values).

2. Equates state of a concrete system with *structure*.

3. Recognizes that without minimum maintenance of system state the survival of the system is in danger.

4. Discusses the need for the system to maintain homeostasis, or values of variables within a certain range, for at least one or perhaps many variables in the system.

5. Recognizes that the maintenance of homeostasis can take the form of the maintenance of low levels of entropy (or high levels of negentropy) within the system.

System State in SET   As in the case of population size and space, there are clear points of congruence between LST and SET with regard to the concept of *state of the system*. Social entropy theory also recognizes the need to maintain system variables above the level of survival. Social entropy theory also stresses the maintenance of entropy internal to the system (below maximum levels of entropy and above minimum levels of negentropy). Social entropy theory also agrees with the general definition of state of the system as used in LST—namely a set of values on some scale at a given point in time (Miller 1978, p. 17). However, there are, as with population and space, a number of significant differences between LST and SET regarding system state, most having to do with degrees of emphasis and with amount of attention given to the concept, rather than with issues of fact or definition. In SET:

1. Measurement of the system state is seen as a major concern—not just a matter of definition or a prerequisite for system analysis.

2. Social entropy theory specifically eschews the degree of reliance on equilibrium (or homeostasis or steady state) seen not only in LST but in other formulations such as Parsons's (1951) theory as well, and discussed at length in chapter 3. Social entropy theory views equilibrium as a legitimate concept which may at times be empirically or theoretically justified or validated, but which at other times may not be found empirically or be justified theoretically. Social entropy theory relies primarily upon entropy. Equilibrium, if it does occur, is a secondary or derivative occurrence, exemplified often by a constant state of entropy (e.g., maximum entropy, as in the second law).

3. Social entropy theory recognizes survival or the "subsistence level" as a system state. But like equilibrium, the survival level is only one of an almost infinite set of system state values that could conceivably occur. Thus, SET does not emphasize the survival level, the equilibrium, homeostatic or steady state level, or any other single level. It prefers instead to recognize that entropy levels in a given society will range between maximum and minimum (zero). As maximum entropy is system death, the system must maintain entropy under the maximum, but generally cannot attain the minimum. The actual entropy level in a given society, at a given point in time, is a matter for empirical determination.

4. In SET, the state of the system is not simply defined or conceptualized as structure at a given point in time (as in LST), but is a primary and important variable in systems theory. As such, its entire *range* must be recognized and analyzed—not just certain values in that range that are either deemed theoretically important (like subsistence level or an equilibrium point) or that are found to exist empirically.

5. In SET, primary operationalization of the full range of system state values is in terms of entropy. The lowest level of system state is maximum entropy. At this level, the system cannot exist. The highest level of system state is minimum (zero) entropy. Here the system has organized to the greatest possible extent. The lowest level (maximum entropy) constitutes *no* available energy or information. The highest level (minimum entropy) constitutes maxi-

mum levels of energy and/or information and/or social organization.

6. In addition to representing system state in terms of entropy, it is also operationalized by the substantive variable of level of living (L). Minimum (zero) level of living is maximum entropy (H_{max}). Maximum level of living (wealth) is minimum entropy (H_0). Thus, entropy H (or S) and L are perfectly (inversely) correlated. In some real sense L can be seen as a measure of energy level in the society. Operationalizations of L either directly represent energy (such as the number of calories available) or represent commodities (such as money) that are easily convertible to energy sources.

For example, in SET (Bailey 1990) one convenient operationalization of (L) for smaller, less developed societies is the total number of calories available to the society. For larger, industrialized societies, perhaps a better measure is the total wealth (in dollars, for example) of the society. Calories are a direct measure of energy, and are directly interpretable as negentropy. Dollars can be used to purchase calories or other energy sources, and as such can be directly converted into energy (negentropy). Thus, in SET, the L variable constitutes an explicit, mensurable recognition of entropy that is left largely implicit in LST.

Thus, in SET the L component, like the other components, is very broad, with a great many potential operationalizations. It is not strictly speaking a single variable, but rather a *component* encompassing many variables which indicate the level of living the society enjoys. Throughout this discussion the distinction between globals and mutables in SET must be kept clearly in mind. A global indicates the *total amount* of that category possessed by the entire society. The mutable (distribution) indicates the allocation of that category among the individuals (P) in the society. Thus, supposing L is operationalized in dollars, the *global L* is the total amount of dollars possessed by the society (everyone in the society). The *mutable L* is the *class system,* or distribution of dollars among the members of the society.

In addition to its significance as a system state measure which is inversely related to the internal entropy level, L also has separate substantive significance. In his original formulation of a set of components similar to *PISTOL*, Ogburn (1951) postulated that a variable similar to *L* was a function of the others. Thus, in SET, one could write $L = f(PISTO)$. However, as is generally true in systems theory, SET does not recognize *L* or any other of the six variables as the single "dependent" variable, but instead recognizes six equations, each showing a different variable in turn as dependent (see chapter 6). Thus, each of the six serves alternatively as a dependent variable (once) and as an independent variable (five times) in the six equations. However, if an exception to this rule were to be made, with the reader choosing only one variable as dependent, it is quite likely that *L* would be the choice, with a society's level of living seen to be a function of its population size, spatial area (and energy sources), information, technology, and organization.

Another point worthy of note is that while *L* may be seen to most directly measure system state, the other five variables (*PISTO*) can also be seen as measures of system state, albeit perhaps indirect or surrogate ones. Since there is such a high degree of intercorrelation between the six components (see Bailey 1990), change in any one directly and rapidly affects all of the other five. Thus, even if *L* is considered the primary measure of system state it must be kept in mind that significant changes in *P, I, S, T,* or *O* will affect *L,* and thus result in changes in the state of the system. It should also be noted that the *L* variable comprises the class of factors often termed social or economic indicators, or measures of quality of life. For further discussion of *L* see chapter 6 and Bailey (1990).

To summarize, to this point in chapter 7 there are clearly points of continuity between LST and SET but clearly differences as well. However, the differences are mainly in terms of emphasis, rather than in terms of fact and definition. That is, LST and SET generally agree on the definitions of space, population size, and system state. However, the differences lie in the fact that LST does not emphasize these three factors or pursue them much beyond their definitions. Any further discussion is somewhat haphazard and unsystematic. It is not found consistently for all eight levels in LST, but is only mentioned (Miller 1978) when one of the variables seems to be of particular salience at a given level. In contrast in SET, *P, S,* and *L* are all central components.

Information

Perhaps information *I* provides a clearer point of contact between LST and SET then do *P, S,* and *L.* Information (*I*) is emphasized in SET as one of the six major components. In contrast to *P, S,* and *L,* it is *also* emphasized in LST. While *P, S,* and *L* (system state) were defined early in *Living Systems* and generally passed by, *I* (information) was much more generally emphasized throughout the volume.

At first glance, LST treats information as it does *P, S,* and system state—by defining it (Miller 1978, pp. 11–51) and then going to other topics. However, further perusal reveals that the continuing coverage of *I* is much more extensive in LST than for the topics previously discussed in this chapter.

Miller (1978, p. 11) makes it clear that he is not conceptualizing information as knowledge, belief, or value, but is using the term in the technical sense. Specifically, he operationalizes information by saying, in effect, that information is what the statistic *H* measures. Miller says:

> Throughout this presentation *information* (*H*) will be used in the technical sense first suggested by Hartley in 1928, and later developed by Shannon in his mathematical theory of communication. It is not the same thing as meaning or quite the same as information as we usually understand it . . . *information* is a simple concept: the degrees of freedom that exist in a given situation to choose among signals, symbols, messages or patterns to be transmitted . . . the amount of information is measured as the logarithm to the base 2 of the number of alternate patterns, forms, organizations, or messages. (1978, p. 11, italics in the original)

Miller explores the relationship between meaning and information further saying:

> In many ways it is less useful to measure the amount of information than the amount of meaning. In later chapters, however, I reluctantly deal more with measurement of the amount of information than of meaning because as yet meaning cannot be precisely measured. (1978, p. 12)

By eschewing the inclusion of concepts which cannot be adequately measured, Miller is pursuing a theme seen constantly in LST. Miller believes that the scientific method must exclude subjective, nonmensurable phenomena from its realm. This includes not only any concept which cannot be properly measured, but also subjective phenomena, including beliefs, ideas, ideology, values, or religious teachings. These are considered appropriate for the scientist to have (personally), but are not considered appropriate topics for scientific study. That is, the scientist can appropriately have religious beliefs, but cannot incorporate them into his or her scientific study—either as influences or methodology or techniques, or as subjective subject matter. For further discussion of this and related matters see Miller (1978, pp. 500–501 and 454–55). Further, although LST specifically eschews nonmeasurable subjective topics for scientific analysis, it does include some discussion of values (see Miller 1978, pp. 802–804). In general, however, one clear difference in emphasis between LST and SET is that the latter much more emphatically and positively embraces the inclusion of topics such as values, ideology, and beliefs (culture in general) as proper topics of study in societal systems theory.

While LST provides preliminary definitions for information just as it does for population, space, and system state, it goes far beyond this for information; information is discussed throughout *Living Systems.* For example, all of chapter 6 (Miller 1978, pp. 121–202) is on information-input overload. Also, in the respective chapters dealing with the different levels there are exhaustive discussions of the subsystems dealing with information. To refresh the reader's memory (from chapter 5 of this volume), there are twelve subsystems dealing with information in LST. Ten of these deal exclusively with information. These are the input transducer, the internal transducer, the channel and net, the timer, the decoder, the associator, the memory, the decider, the encoder, and the output transducer. Two additional subsystems process both matter-energy and information. These are the reproducer and the boundary.

In discussing the society level, Miller (1978, pp. 766–68) gives examples of all of these information-processing subsystems. For example, for the input transducer, he gives numerous examples of institutions which regulate the flow of information into the society (e.g., military border post, weather bureau, foreign news service). Similarly, he provides examples for the other subsystems such as internal transducer (e.g., survey organization), channel and net (e.g., telephone service), decoder (e.g., teaching institution), associator (e.g., teaching institution), memory (e.g., teaching institution), decider (e.g., president), reproducer (e.g., constitutional convention), and boundary (e.g., border guard). Note the degree of imprecision here where a single institution has a function in more than one subsystem. For example the teaching institution serves as decoder, associator, and memory. See Miller 1978 for future discussion.

It is appropriate at this point to summarize the information-processing functions carried out by the twelve subsystems at every level according to LST. These are illustrated at the societal level, as that is the level emphasized in SET, and so makes the analysis most comparable. However, the summary holds for other levels such as the group or organization as well (with different examples, of course). The process begins with the phenomenon (e.g., the society) being reproduced, for example, when created through a charter or constitution. This inevitably involves a great deal of processing and recording of information, including laws and norms of various sorts. This reproducer subsystem has a dual function, also serving to reproduce matter-energy. The other dual-function subsystem is the boundary. Societal boundaries have a great many functions in guarding secrets, and in insuring that needed information flows into the society, while secret information does not flow out.

The other ten subsystems can be seen as input of information, decoding, and recoding, and other types of internal processing, including "throughputs," and output, when the information is sent from the society to another society. Specifically, the input transducer brings in information (past the boundary); the internal transducer generates new information internally; the channel and net carries information within the society; the timer provides time-related information; the decoder transforms information into a form more amenable for internal processing; the associator processes (analyzes) the information; the memory stores information; the decider takes actions based on this information; the encoder creates new codes of information forms (perhaps for transmission to other societies); and the output transducer sends messages out across the societal boundaries on to other societies.

Information in SET   There is a very clear distinction between information in LST and information in SET. Although information is emphasized in each, its emphasis is quite different in the two approaches. As has been seen, information in LST is defined technically as what H measures, and discussion of information in terms of meaning is de-emphasized because of measurement problems. There is relatively little emphasis on what information is (after the early definitional presentation) and on its role in each of the respective levels (e.g., society). A salient exception to this is the long discussion of information-input overload. Most of the analysis is in terms of the concrete subsystems which carry out the various information processing functions (e.g., input transducer, decoder, decider).

In SET the emphasis is quite different. Social entropy theory relaxes Miller's strict adherence to the "scientific" principle that one must be able to adequately measure a phenomena in order to include it in the analysis. Social entropy theory coincides with LST in using H to measure information. The problem is that SET recognizes (while LST fails to) that H is directly and strictly *not* a measure of information, but rather of entropy. Miller (1978, p. 13) recognizes that information is equivalent to negative entropy (negentropy). Thus, information and entropy are inversely related: when entropy increases, information decreases. In spite of this, Miller continues to measure information by H. In reality, H is a measure of entropy (see Theil 1967; Shannon and Weaver 1949). This is clearly seen by looking at the extreme values of maximum H and minimum H. Maximum H is log K, where K is the number of categories (in a table). In statistical terms, this is random assignment of N cases into the K categories. Minimum H is a value of zero (all N cases assigned to only one of the K categories). This means that for maximum H, the scientist has no ability to predict which category a given case is in. For minimum H, he or she has *perfect* predictability. There is no substantive, linguistic, or scientific rationale for equating maximum information with randomization and complete lack of predictability, but that is what Miller does in LST by operationalizing information by H. In reality, H is entropy, and is the inverse of information. For further discussion see Bailey 1990.

As Brillouin (1956) clearly shows, while information cannot be directly measured by H, it can be measured by the inverse of H, such as $H_2 - H_1$, or $1 - H/H_{max}$. The latter is the measure of redundancy and has the advantage of varying between 0 and 1.

Thus, while LST and SET both use H as a measure, LST contends that H is a direct and adequate measure of information, and does not propose a measure of entropy, using S as the measure of entropy, and equating H with $-S$ (Miller 1978, p. 16). Social entropy theory contends that H is a direct measure of entropy and is the inverse of information. Thus, H should be equated with S, not $-S$ as Miller (1978) does. This is demonstrated in Miller's (1978, p. 16) own table 3.1. Among the characteristics of entropy (S), table 3.1 lists randomness and disorder. Thus maximum entropy should be maximum disorder or randomness. Randomness is exemplified by maximum H, and thus H is clearly an entropy measure rather than an information measure.

Another way SET and LST differ with regard to information is that SET uses a much broader definition of information. While LST uses a narrow (and technically incorrect) definition or operationalization of information as H, SET defines information as a broad component encompassing not only the technical pattern of communication symbols (as measured by an analogue of H) but also meaning, values, beliefs, and ideology. In SET the I component embraces all knowledge, values, and beliefs, as well as technical information. In terms of the three-level model in SET (figure 2.2), any concept at the X level is a component of I, and in fact information is included in the indicator (X'') level to the degree that X is mapped into X''.) Thus, any cognitive or conceptual level entity that can be processed by the human brain is included in the component of information in SET. This includes anything commonly known as an element of culture. This is necessary because SET seeks to include all aspects of each of the six components (PISTOL) that affect the other components. Analyzing I, it is clear that technical information affects the other five components. For example, knowledge of how to grow grain or cattle directly affects L as measured in the number of calories. However, beliefs also affect L and must be included in I. For example, in India the taboo against harming sacred cows limits the beef supply and decreases L as operationalized in terms of number of available calories.

Thus, SET uses a very inclusive definition of information, where information includes any cognitive or conceptual elements (X) which affect the *PISTOL* variables. Living systems theory, on the other hand, uses a more restrictive definition of I as that which can be quantitatively measured, eschewing direct inclusion of meaning as part of I until such time as it can be measured quantitatively. It is clear that while LST and SET overlap in their portrayal of *I*, they have some differences. Social entropy theory focuses on the variable "information" (*R*-analysis) while LST focuses on the concrete subsystems which process *I* (*Q*-analysis).

TECHNOLOGY

Technology has major emphasis in SET but is barely mentioned in LST. Technology is not defined as a major system component in LST, unlike all the other components discussed so far (*P, I, S, L*). But while technology per se is not emphasized as a concept in LST, technology does enter rather prominently into the analysis of LST, albeit "through the backdoor" and generally not labeled as technology, but under other names. While Miller (1978) has few explicit discussions of technology, a careful reading of *Living Systems* shows many discussions and examples of tools. For example, Miller 1978 (pp. 856–57) displays sixteen different pictures showing stages in the development ("evolution") of the telephone. According to the terminology of SET, this would be a clear portrayal of the development of a salient form of communications technology. Yet, in LST, the word *technology* is not mentioned in this discussion. Rather, the telephone is referred to as an "artifact" (Miller 1978, p. 855).

Thus, one can see extensive discussion of technology in LST, but under the rubrics of *artifacts* or *prostheses* rather than forms of technology. Miller says:

> An *artifact* is an inclusion in some system, made by animals or man. Spider webs, bird nests, beaver dams, houses, books, machines, music, paintings and language are artifacts—they may or may not be *prostheses*—inventions which carry out some critical process essential to a living system. . . .
>
> Living systems create and live among their artifacts. Beginning presumably with the hut and the arrowhead, the pot and the vase, the plow and the wheel, mankind has constructed tools and devised machines. . . .
>
> A chimpanzee may extend his reach with a stick; a man may extend his cognitive skills with a computer. (1978, p. 33, italics in the original)

Among the examples of prostheses provided by Miller (1978, p. 33) are input transducers which determine type of blood cells and the identity of approaching missiles, photographic and electronic memories which can store masses of information, and computers which carry out mathematical calculations.

In evaluating the role of technology in LST, a number of points can be made.

1. Technology—defined in SET as the tools used by the societal system (computers, telephones, etc.)—is clearly recognized and analyzed in LST.

2. The prominence of this recognition of technology in LST is greatly diminished by generally eschewing the term *technology* (for tools) in place of the terms *artifact* and *prosthesis*.

3. The term *artifact* seems somewhat unfortunate as an indicator of technology for at least two reasons. (*a*) By including such things as music and painting with machines under the rubric *artifact,* Miller is combining art and technology, and thus somewhat diluting the ability of LST to concentrate on the central role of technology in societal systems. (*b*) By treating *artifacts* as *products* of living systems (Miller 1978, p. 34) LST is relegating technology to a needlessly passive role. Thus, while it is true, as Miller (1978, p. 34) says, that artifacts "mirror" the living systems in which they were produced, the other side of the coin is that the living systems "mirror," or are molded by, their artifacts. Thus, while an arrowhead may be a "product" of a certain civilization, it also was instrumental in producing and molding that civilization, so that the civilization, or at least its level of living, was also a "product" of the arrowhead. Thus, artifacts are not in an asymmetrical relationship with the system in which they are always seen (by Miller) as products, but are in a symmetrical relationship so that they produce as well as being produced.

The clarity of the exposition regarding this point in LST is clearly hindered by mixing art with technology under the rubric of "artifact." It may be true, as Miller (1978, p. 34) says, that jewelry of primitive tribes are the products of these systems. In this case the jewelry may be more a product than a producer.

The overall effects of treating technology as artifact is that the central role of technology in the ongoing process of society is obfuscated. Technology as an artifact or product becomes more an epiphenomenon of the functioning society than a prime motor in its development. Technology that is regarded as a *prosthesis* is given a more central role in the ongoing process of system functioning. However, the use of two terms for technology means that the scientist must determine whether a given tool is a prosthesis or merely an artifact. Further, the terminology is unfortunate not only because two terms (prosthesis and artifact) are used when one (technology) would suffice, but also because the terms prosthesis and artifact are much more arcane and have more unfortunate and misleading connotations than the more familiar term *technology.*

Social entropy theory takes a much more direct approach to technology. Technology is defined as tools used by the society to maximize its level of living. Thus, while conceivably such things as art, music, and language (included as artifacts in LST) could be considered aspects of technology, the term is generally limited to tools that are much more explicitly identified with information processing or matter-energy processing functions, such as computers, mining equipment, airplanes, etc. Technology (T) as a global variable consists of the sum total of *all* tools possessed by the society. As a mutable distribution, technology is the allocation of the population of persons into the sum total of available tools, so that as a result some persons have computer skills, others are able to fly airplanes, and others possess still other types of technology. The term *artifact* would not be used in SET because it connotes an epiphenomenon or a phenomenon that has a passive (or even obsolete) role rather than the active role generally assigned to a tool. Further, the term *prosthesis* would not be used in SET, not simply because of its medical connotation, but because it has no clear function in the SET framework. For example, Miller (1978, p. 335) says, "Some prostheses improve the function of normal eyes such as a magnifying glass, a microscope, or a telescope." Social entropy theory would not consider a telescope an "artifact," residue, or epiphenomenon, nor as a prosthesis, but would consider it a tool which could be of help in societal adaptation.

Organization

In comparing LST and SET, this chapter has gone from SET to LST. Population (P), I, S, T, and L (from SET) have been discussed, along with their parallel treatment in LST. Only O remains to be discussed of the six *PISTOL* variables. Organization, as discussed in SET, is a verb, not a noun. Social entropy theory is talking about *organization* as a process and also as an end result or product of that process rather than as an organization (although the organization in this sense is also discussed, see Bailey 1990, chapter 6). In SET, organization means social organization. Social entropy theory defined *organizational structure* (O) as a global property of the society. Both *jobs* ("work roles") and nonwork roles (for example, leisure, retirement, unemployment) are included in the organizational structure (O). Since even the term *role* does not seem sufficiently generic, SET defines the generic *position* as the unit of analysis in O—without defining position narrowly, but with the understanding that most positions are occupations (for example, a position in a bureaucracy). A position is defined as a normatively regulated activity. It will generally have specific goals, and will process and/or extract energy and information to achieve these goals (Bailey 1990). Thus, in SET the global organizational property (O) of the society is the sum of all positions (of all types) in the society (Bailey 1990). As such, O is clearly related to *P, I, S, T,* and *L.* The mutable organizational property O is formed by allocating all *P* persons in the population into the set of O positions possessed by the society. This is often termed the occupational division of labor, or just the *division of labor.* If *P* > *O,* then unemployment exists. If *P* < *O,* there is underemployment in the society. Much of SET is concerned with how the society allocates persons into positions (see especially, chapter 4–6 of Bailey 1990).

The topic of organization does not receive the relative neglect in LST that is accorded some of the other *PISTOL* variables from SET. In contrast it is relatively emphasized in LST, but with some clear differences in emphasis from SET. However, in contrast to SET which analyzes organization of the society as a set of social positions, LST studies organization as a noun, as in "the organization." Among the many topics studied in LST are activities of organizations, life cycles of organizations, and organizational pathology. Perhaps the closest overlap between LST and SET with regard to organization occurs when LST analyzes the industrial classification of organizations (Miller 1978, p. 601–603).

In studying the organization (as a noun) there is also some direct correspondence between SET and LST. Chapter 6 of SET also adopts this approach. In this chapter (Bailey 1990) the organization as a system in itself and also as a subsystem of the large society is discussed at length. Specifically, the six globals and five mutables that were analyzed for the society as the unit of analysis are also studied with the organization as the unit of analysis. While certainly not identical to the discussion in LST, there are some direct parallels between SET and LST. The reader can compare chapter 6 of SET on the organization (Bailey 1990) with chapter 10 of *Living Systems* on the organization (Miller 1978). In addition to devoting all of chapter 10 to the study of the organization in LST, Miller (1978, pp. 747–901) also discusses organizations at length in chapter 11 on the society. This is because when the society is the basic unit of analysis, the next lower level of subsystem is the organization. Thus, Miller discusses the twenty basic subsystems of the society in terms of organization. Examples of some of these twenty subsystems are: reproducer (organization which creates the charter for a new society); boundary (organization which guards the boundary); ingestor (organization which imports matter-energy such as a trucking company); matter-energy storage (organization which determines storage regulations or actually stores, such as the treasury department); internal transducer (organization such as a political party); decoder (organization such as the department of state).

REMAINDER OF SET COMPARED WITH LST

This chapter so far has compared the six *PISTOL* variables of SET with their emphasis in LST. The analysis has clearly shown disparities as well as congruences. To summarize, SET begins with analysis of the society as a concrete system (Q-analysis) but then derives an abstracted system in the form of the six interrelated *PISTOL* variables (R-system). Living systems theory, on the other hand, also begins with the notion of a concrete system, but retains this emphasis more faithfully than SET does throughout the analysis, and does not grow beyond it. Rather than analyzing important system *variables* as SET does, LST emphasizes lower-level system components (subsystems). Generally, these are the next lower level. Thus, for the analysis of the society in LST, the subsystems analyzed are at the level of the community; for the community they are the organization; for the organization they are the group; for the group they are the individual. In contrast, SET emphasizes the micro role of the individual within the macro context of the society (chapter 5). This could be done in LST, but would mean analysis twice removed (from individual through group through organization to society).

Further, while SET is conducive to the analysis of micro-macro links between the individual and the society (through the mutable distributions), LST is ill-suited to micro-macro analysis because of its emphasis on parallel analysis of the levels. Miller (1978) emphasizes cross-level analysis which means, for example, analysis of information processing at the individual and the societal levels. This shows parallel internal processes at the two levels, but keeps them distinct rather than forging linkages between them.

Perusal of SET reveals other topics not covered in depth in LST. These include the global-mutable-immutable distinctions, Q- and R-analyses, the measurement of H, and the analysis of inequality. Likewise, perusal of LST reveals a large amount of material not covered in SET. In addition to the levels of the cell and organ, the supranational is not emphasized in SET. Also, LST emphasizes models and simulations. Further, the matter-energy and information processing subsystems are not covered in detail in SET as they are in LST.

An overall comparison of LST and SET shows similarities between the two approaches, as well as the differences just mentioned. More importantly, the two approaches are clearly complementary. Living systems theory stresses subsystems (Q) while SET stresses the *PISTOL* variables (R). Together the two form a whole. This review sets the stage for chapter 8, which deals with autopoiesis.

COUNTERPOINT

Since counterpoint has already been presented for both LST and SET, this synthesis requires little further comment. This chapter is helpful in emphasizing the contribution that both approaches make to sociology in areas such as the analysis of space, time, organization, entropy, energy, information, and so forth. This analysis has no real parallel in mainstream sociology, at least in terms of depth and richness.

Similarly, the forms of metatheory and the contributions of these approaches were well summarized in the two previous chapters. The reader is directed to them.

CHAPTER 8

Autopoiesis

Autopoiesis is one of the most exciting new notions in systems theory in particular, and in social and behavioral science in general. Despite its appeal, the concept is somewhat controversial in its application to social groups, and heretofore somewhat inaccessible to sociologists. I find the concept of autopoiesis especially appealing for at least two reasons.

1. It is a highly sophisticated model developed in the best sense of classical systems theory. It is first and foremost a systems concept, emphasizing boundary delimitation, the role of the boundary, the internal processes of the system which lead to self-reproduction, and the relationship between the system and its environment (or "medium").

2. Maturana and Varela's (1980) model goes far beyond extant systems models in employing the role of the observer (researcher) in the process of systems analysis. For example, it is emphasized that the distinction between the autopoietic system (e.g., the cell) and its medium is not made by the system itself, and thus not a system property, but is rather made by the observer. One of the most distinctive features of the literature on autopoiesis is indeed its emphasis on epistemology.

One of the freshest things about the autopoietic approach, as outlined by Maturana and Varela, is its modern and perhaps revolutionary epistemology. This is in the best interests of my stated goal of developing a "positive positivism." The positive positivism as displayed by Maturana and Varela reveals a distinctive and perhaps radical epistemology which, though not labeled as such as Maturana and Varela, is clearly reflective of postmodern thought and displays a certain amount of deconstructionism (see the discussion in chapter 9).

As discussed by Varela (1984) the changes come in emphasizing closure and internal coherence of the system as opposed to the traditional emphasis on input. While the input perspective values the environment as the major guideline to understanding transformations in the system, the closure-type perspective sees the internal coherence of the system as the key to its transformation. Varela (1984, p. 29) says that the input perspective has been dominant for fifty years (at least in Anglo-Saxon science). Varela says:

> I firmly believe that there is a major change, or trend of a change in our contemporary sensibilities and scientific epistemology in the sense that we are becoming more and more interested in an epistemology which is not concerned with the world-as-picture, but with the *laying down* of a world, where a unit and its world co-arise by mutual specification. It could be said that the notion of self-organization serves as a clear symptom that differentiates between input-type machines (whether we call them Turing, state-transition, or simply functionalist theories), and biological autonomy and human understanding. This is so because self-organizing behavior depends on a history of coupling and it is based on a mechanism which is explicitly distributive and hermeneutic: it is interpretive, precisely in the sense of laying down of being. (1984, p. 31, italics in the original)

Classical positivism emphasized objectivity and eschewed the intrusion of values and subjectivity into the scientific enterprise. The emphasis was on the phenomenon being studied, and the relationship between this phenomenon and the observer was deemphasized (except to insure lack of bias). Postmodern thought emphasizes subjectivity, and the "constructed" nature of so-called "reality" (see Featherstone 1988). The parallels between the postmodern perspective and autopoiesis are striking, although Maturana and Varela do not use either the terms "positivism" or "postmodernism." Maturana and Varela's approach is postmodern in its emphasis on the hermeneutic, the interpretive, the role of the observer in the system, and the fact that autopoiesis is not functionalist. Yet, the approach retains recognition of homeostasis, and in many ways is just traditional science.

In terms of the preceding chapter (chapter 7), autopoiesis can be seen as an approach intermediate to living systems theory and social entropy theory. It shares a biological approach to living systems with the former, but departs from LST with its interpretive approach and its emphasis on the role of the observer. In this respect it has more in common with SET, whose three-level model forges a link between the two approaches. All three approaches were also derived during the 1970s and 1980s, with LST being launched first, autopoiesis second, and SET third.

Despite the excitement generated by this approach, there are some hindrances which have kept it from spreading more rapidly. Perhaps the chief among these is the complex writing style of Maturana and Varela, and their lexicon of jargon which is largely new (to all fields) and some of which contradicts standard usage (such as their definition of structure). Among the terms which need to be understood by the student of this approach are: autopoiesis, unity, organization, medium, allopoiesis, closure, coupling, structural coupling, structural plasticity, structure, structure-determined system, observer, property, space, and interaction. As noted by Maturana (1978, p. 360) autopoiesis is a Greek word composed of the Greek words for "self" and "to produce."

I will first present the various terms as defined by Maturana and Varela, and then pursue some secondary analysis for added clarity. Maturana and Varela, like many systems theorists, have broad interests, being interested not only in systems theory and biology, but also in philosophy and linguistics. Unfortunately, it almost takes a linguist to decipher some of their complex definitions. These definitions are as follows.

Unity is a "network of interaction of components which constitute a living system as a whole" (Varela, Maturana, and Uribe 1974, p. 187). In other words, "a unity is an entity, concrete or conceptual, dynamic or static, specified by operations of distinction that delimit it from a background and characterized by the properties that the operations of distinction assign to it" (Maturana 1978, pp. 32–33).

The second important term is *organization*.

> . . . A complex system is defined as a unity by the relations between its components which realize the system as a whole, and its properties as a unity are determined by the way the unity is defined, and not by particular properties of its components. It is these relations which define a complex system as a unity and constitute its organization. Accordingly, the same organization may be realized in different systems with different kinds of com-

ponents as long as these components have the properties which realize the required relations. (Varela, Maturana, and Uribe 1974, p. 188)

Now I turn to the definition of *autopoiesis*.

The autopoietic organization is defined as a unity by a network of productions of components which (*i*) participate recursively in the same network of productions of components which produced these components, and (*ii*) realize the network of productions as a unity in the space in which the components exist. (Varela, Maturana, and Uribe 1974, p. 188)

They say further, "in contradistinction, mechanistic systems whose organization is such that they do not produce the components and processes which realize them as unities and, hence, mechanistic systems in which the product of their operation is different from themselves, we call allopoietic" (Varela, Maturana, and Uribe 1974, pp. 188–89).

Structure "refers to the actual components and to the actual relations that these must satisfy in their participation in the constitution of a given composite unity" (Maturana 1978, p. 32). Another salient term is *property*. "A property is a characteristic of a unity specified and defined by an operation of distinction. Pointing to a property, therefore, always implies an observer" (Maturana 1978, p. 33). What is an observer?

An *observer* is a human being, a person, a living system who can make distinctions and specify that which he or she distinguishes as a unity, as an entity different from himself or herself that can be used for manipulations or descriptions in interactions with other observers. An observer . . . is able to operate as if he or she were external to (distinct from) the circumstances in which he or she finds himself or herself. (Maturana 1978, p. 31)

As for space: "*Space* is the domain of all the possible interactions of a collection of unities (single, or composite that interact as unities) that the properties of these unities establish by specifying its dimensions. . . . Once a unity is defined, its space is defined" (Maturana 1978, p. 33).

Another term is *interaction*. "Whenever two or more unities, through the interplay of their properties, modify their relative position in the space that they specify, there is an interaction" (Maturana 1978, p. 33). The *medium* is the host, context, or environment in which the system exists (or "domain of interactions," Maturana 1978, p. 41; Maturana and Varela 1980, p. xxi). *Closure* refers to organizational closure or autonomy. This is distinct from the analysis of input and output or transfer properties. As a closed network, the system operates by generating relations of activity determined by its structure, not by environmental circumstances (Maturana 1978, p. 41). Yet another important term is coupling. *Coupling* refers to a relationship or interaction between a system and its medium (environment).

Structure-determined systems experience only changes determined by their organization and structure that are either changes of state or disintegration.

> The organization and structure of a structure-determined system, therefore, continuously determine: (a) the domain of states of the system, by specifying the states that it may adopt in the course of its internal dynamics or as a result of its interactions; (b) its domain of perturbations, by specifying the matching configurations of properties of the medium that may trigger its disintegration. (Maturana 1978, p. 34)

Another important term is *structural coupling*.

> The outcome of the continued interactions of a structurally plastic system in a medium with redundant or recurrent structure, therefore, may be the continued selection in the system of a structure that determines in it a domain of states and a domain of perturbations that allow it to operate recurrently in its medium without disintegration. I call this process 'structural coupling'. (Maturana 1978, pp. 35–36; see also Maturana and Varela 1980, pp. xx–xxi)

But what is structural plasticity?

> Now, if a structure-determined system, as a result of its interactions, undergoes changes of state that involve structural changes in its components (and not only in their relations), then I say that system has a second-order plastic structure, and that it undergoes plastic interactions. . . . If the medium is also a structurally plastic system, then the two plastic systems may become reciprocally structurally coupled through their reciprocal selection of plastic structural changes during their history of interactions. (Maturana 1978, pp. 35–36)

Now that I have presented the basic nomenclature, I may proceed to further discussion of these complex concepts. As the reader has by now guessed the rather intractable nature of some of these concepts, it behooves me to discuss, and even redefine, them in some detail so that their basic meanings become sufficiently clear. I will also follow the practice of quoting extensively from both primary and secondary sources to insure that meanings and nuances are accurately transmitted. After that is done, I can move the discussion to such matters as the relationship of autopoiesis to other parallel approaches such as living systems theory, and the efficacy of applying autopoiesis to social systems. Due to the complexity and importance of the term autopoiesis, it may be wise to present another definition by Maturana, which overlaps the one previously given, but expands upon it a bit.

This second definition of an autopoietic system is:

> A dynamic system that is defined as a composite unity as a network of productions of components that (a) through their interactions recursively regenerate the network of productions that produced them, and (b) realize this network as a unity in the space in which they exist by constituting and specifying its boundaries as surfaces of cleavages from the background, through their preferential networks within the network, is an autopoietic system. (Maturana 1980b, p. 29).

Notice that this statement has two main parts. The first says that the system reproduces itself, the second says that it constructs its own boundaries.

As presented by Mingers (1989), Maturana and Varela's basic task is to show what distinguishes living systems from equally complex nonliving systems (e.g., how does a Martian distinguish a horse from a car?). In discussing the living cell, Mingers says:

> What characterizes this as an autonomous, dynamic living whole? What distinguishes it from a machine such as a chemical factory which is also dynamic and also consists of complex components and interacting processes of production forming an organized whole? It cannot have to do with any function or purposes that the cell might fulfill in a larger whole since it can live by itself, nor can it be explained in terms of particular components of the cell. The difference must stem from the way the parts are organized together as a whole. (Mingers 1989, pp. 161–62)

Continuing, Mingers says that:

> . . . a cell produces its own components, which are therefore what produces it. A factory, in contrast, produces chemicals which are used elsewhere, and is itself produced or maintained by other systems. It produces and is produced by something other than itself. This simple idea is all that is meant by auto-poiesis. The word means "self-producing" and that is what the cell does: it continually produces itself. Living systems are autopoietic—they are organized in such a way that their processes produce the very components that are necessary for the continuance of these processes. (1989, p. 162)*

At this point, the basic elements of the concept of autopoiesis as outlined by Maturana and Varela seem fairly clear. An autopoietic system is a living system which can be distinguished from its nonliving counterpart (such as a machine) of equal complexity by at least two features of the former that are not shared by the latter: the autopoietic living system reproduces itself (produces the processes which produce it) and produces and maintains its boundaries.

Perhaps there is relative consensus at this point, in the sense that most readers will agree that a cell can (in a sense) reproduce itself and can also produce its boundaries (in certain respects) while a nonliving system cannot do either of these things. However, at this point the consensus seems to end, as the seminal work of Maturana and Varela touches upon a number of intellectual issues in thought-provoking ways, but also in sometimes controversial or tentative ways. Perhaps the biggest issue for sociology, and one that has been tangentially addressed by Maturana and Varela, is the question of whether (any) social systems such as groups, organizations, or societies qualify as autopoietic systems—and if so, how. There seems to be little agreement at this point, with authors such as Beer (1975), Luhmann (1982, 1984, 1986, 1989), and Robb (1989a, 1989b) being apparently convinced that social systems are autopoietic, while other authors such as Maturana (1981), Varela (1980), and Mingers (1989) stop short of saying

*From John Mingers, "An Introduction to Autopoiesis." *Behavioral Science* 2:159–180. Copyright © 1989 Plenum Publishing Corporation. Reprinted with permission of the author and the publisher.

that social systems are autopoietic. Mingers (1989, p. 177) concludes that "overall it seems difficult to sustain the idea that social systems *are* autopoietic, at least in strict accordance with the formal definition" (italics in the original).

It is easy to see why there is no consensus on this issue because the notion of autopoiesis, which seems to be simple in its nucleitic formulation, actually spawns a host of ancillary intellectual controversies. Thus, anyone who jumps right into the fray and declares that a social system either is or is not autopoietic is probably acting prematurely. Only by embarking on an odyssey through the prerequisite ancillary debates can one arrive at an informed opinion regarding the ultimate question of whether some social systems are indeed autopoietic.

Fortunately, we have a number of analytic tools in the form of extant systems approaches which allow us to clarify the complex issue of autopoiesis and thus to arrive at a final answer as to whether a social system is autopoietic. By relating the concept of autopoiesis to other systems approaches, we not only help answer the primary question, but also further the general systems goals of integration and cumulation by providing bridges among approaches that were independently derived.

Autopoiesis and Living Systems Theory

There are several clear links between living systems theory and autopoiesis. Since the former has been considered here in detail, development of these links should aid in explication of the latter. One link between autopoiesis and living systems theory is that both focus on the distinction between living and nonliving systems from a biological perspective (see Miller 1978, pp. 18, 85, 302, 355, 482, 583 for discussions of nonliving systems). Another link is that both approaches focus on a cell as a primary example, offering a ready point of comparison. In addition, Miller (1978) presents a systematic discussion of social systems such as the group, organization, and society using the same framework that he used to analyze the cell. This provides a ready bridge from Maturana and Varela's study of the cell to the study of social systems which will prove helpful in deciding if such social systems are autopoietic. Maturana and Varela also analyze social systems, but in a much less structured and less comprehensive fashion than does Miller.

A third link is that both autopoietic theory and living systems theory focus on the self-reproductive and boundary functions of the living system. In living systems theory, these are but two of the twenty basic subsystems common to all living systems at all levels. However, the reproducer and the boundary, alone of the twenty subsystems, receive special attention in living systems theory as the only two subsystems which process *both* matter/energy and information. The fact that these two particular subsystems are stressed in both living systems theory and autopoietic theory not only attests to their importance in systems theory, but also provides a clear link between the two approaches.

A fourth link between autopoietic theory and living systems theory is that both approaches generally take the individual (cell) as the unit of analysis, and thus analyze concrete systems (in Miller's terms). Further, while Miller also recognizes conceptual and abstracted systems (1978), Maturana (1975) recognizes both concrete and conceptual systems, thus generating the notion of an autopoietic conceptual system (see Mingers 1989, p. 17).

It is premature at this time to address the ultimate question of whether social groups are autopoietic. Also, living systems theory offers no explicit answers to the question. Miller (1978) does not use the term autopoiesis. However, one logical supposition is that according to the principles of living systems theory, systems at all eight levels (cell, organ, organism, group, organization, community, society, and supranational) would be autopoietic in some sense. The chief basis for this supposition is that cells are autopoietic, and they are, in Miller's (1978) terms, the basic building blocks for the next seven levels. That is, the components of social groups are individual persons; the components of persons are organs; the components of organs are cells.

Clearly, then, all social groups, according to living systems theory, would contain cells (in organs which are in individuals) and thus all social groups contain autopoietic subsystems (twice removed). This fact in itself does not dictate that groups are themselves autopoietic, but at least says that they provide a context or "medium" (Maturana's term) for autopoietic systems (cells), a point that Maturana apparently agrees with. While not conceding that a social system is autopoietic, Maturana does term the individual human as autopoietic, and defines a social system as a medium in which the individual realizes this autopoiesis (Maturana 1980a, p. 13).

In other words, while Maturana stops short of saying the social system itself is autopoietic, he says (1980a, p. 13) that "It is constitutive of a social system that its components should be autopoietic . . ." Living systems theory implies at least this much, and probably more. Not only do cells reproduce (both asexually and sexually—see Miller 1978, p. 224), but "in organisms the fertilized egg divides repeatedly by mitosis as the tissues grow and differentiate. Cells of some tissues retain the capacity to divide as long as the organism of which they are part survives" (Miller 1978, p. 224). This shows that not only does the cell reproduce, but that cell mitosis is crucial in the reproduction of the organism, and in that sense the organism can be said to be autopoietic. Even further, the reproducer is clearly a subsystem in all eight levels of living systems (not merely in the cellular level) thus implying that all eight levels are autopoietic in some sense.

The whole issue can be approached from the other direction—from the "top down." From this perspective, the social group (male-female dyad) is required for organism reproduction of the individual human; the human individual is in turn required for organ reproduction; and the organ for cell reproduction (in the case of cells in human cell tissue). Thus, in the case of human cells, the social group is *required* before cell production (autopoiesis) can occur. Thus, social groups are at least hierarchically autopoietic.

More directly, most of the controversy over whether social groups are autopoietic centers around the question of what is the appropriate component or basic unit of the social system. Although there is no consensus on this matter, the position of Miller (1978) in *Living Systems* is clear—the system at all eight levels is concrete. The basic unit of the group is the individual; the unit of the organization is the group; the unit of the society is the community; and the unit of the supranational system is the society. Inasmuch as each higher level is required for the reproduction of its component, the system is autopoietic, or at the very least is a medium or context for an autopoietic system at some link in the chain.

Thus, analysis of living systems informs the study of autopoiesis, and leads, at least tentatively, to the conclusion that social groups, organizations, and societies are autopoietic in some sense. However, in addition to the obvious congruences between auto-poietic theory and living systems theory, there is one fundamental difficulty that may only indicate slight displacement of scope, or may indicate more serious ontological problems. To illustrate, number Miller's hierarchical levels: $L1$, the cell; $L2$, the organ; $L3$, the organism; $L4$, the group; $L5$, the organization; $L6$, the community; $L7$, the society; and $L8$, the supranational system. When Miller (1978) discusses reproduction at any of the eight L levels, he is referring to L level reproduction by subsystems at the L or ($L - 1$) levels or below. In contrast, students of autopoiesis often concentrate first on production by the L level of the $L - 1$ level components, and only subsequently, reciprocal production of the L level by the $L - 1$ level components.

For example, consider level $L5$ (the organization). In discussing autopoiesis, Mingers (1989, p. 175) says: "Autopoiesis is concerned with processes of production—the *production of those components* which themselves constitute the system" (italics in the original). The language is a little misleading, as components cannot constitute a system (they can only constitute components), but can *produce* a system through their relations. At another point, Mingers (1989, p. 162) says, "so a cell produces its components, which are therefore what produces it" (this is much clearer). To generally code Mingers's last statement, level L (the system) produces level ($L - 1$) (components or subsystems) which in turn produces level L (the system). Returning to the organizational level of the living system, $L5$ (organization) should produce $L4$ (components, which are groups), which should in turn produce $L5$ (the system, which is the organization).

Miller (1978, p. 606) describes the reproducer of $L5$ (the organization) as follows: "*Reproducer.* Any organization or group that produces an explicit or implicit charter for a new organization may be downwardly dispersed to a single person." Thus, $L5$ (the organization) can be produced either by another organization ($L5$), a group ($L4$), or even a single person ($L3$).

Autopoiesis as described by Mingers (1989, p. 162) would entail:

$$L5 \rightarrow L4$$

then

$$L4 \rightarrow L5.$$

Miller discusses

$$L5 \rightarrow L5$$

$$L4 \rightarrow L5$$

$$L3 \rightarrow L5.$$

Thus Miller does two things which hamper comparisons with the autopoietic theorists. He focuses only on the last half of the autopoietic cycle (production of the L level by the $[L - 1]$ level or component), and does not describe production of the $(L - 1)$ level by the L level as emphasized by the autopoietic theorists (e.g., production of groups $[L4]$ by organizations $[L5]$). He also discusses "downward dispersion" (Miller 1978, p. 606) in which component levels below $(L - 1)$ (or $L4$ in this case) such as $(L - 2)$ (which is $L3$, the individual), are seen to produce $L5$. These points of difference in the explications of the two approaches do not demonstrate contradictions between autopoiesis and living systems theory, but only show a lack of parallelism, which makes comparison difficult.

Epistemology: Autopoiesis and the Three-Level Model

The disparities just discussed probably represent only slight inconsistencies in maintenance of scope (level of analysis) rather than fundamental ontological difficulties. Nevertheless, there are clear ontological and epistemological issues in autopoietic theory that must be addressed before an understanding of autopoiesis can be achieved. This is facilitated by analyzing autopoiesis in the context of the three-level model (Bailey 1984c, 1990). One epistemological concern in autopoiesis derives from the relation of the observer to the system being observed. I will follow Mingers's (1989) excellent discussion of this topic. As quoted above, Mingers (1989, p. 161) says that all descriptions or explanations are made by the observer. One must not confuse properties of the observer with properties of the system. An observer can perceive both the system and its environment and relate them. A system or its reacting components cannot do this.

Further, Mingers says,

> . . . all descriptions and explanations are made by *observers* who distinguish an entity or phenomenon from the general background. Such descriptions will always depend in part on the choices and purposes of the observer, and may or may not correspond to the true domain of the observed entity. That which is distinguished by an observer, Maturana calls the *unity,* that is, a whole distinguished from a background. In making the distinction, the properties which specify the unity as a whole are established, e.g., in calling something a "car" certain basic attributes are specified. (1989, p. 163, italics in the original)

Another distinction is between the organization of a unity and its structure (Maturana 1980a; Mingers 1989). The organization is the generalized model showing relations between components and the properties of components which define the unit as a member of a general class. Structure, in contrast, refers to the specific components and relations of an actual empirical example or case. For example, a car may be defined as an organization by having wheels, steering, transmission, etc. A *specific* car has a structure of an engine of a certain size, an actual kind of wheel, etc. Thus a unity has both an organization and a structure. My car has an *organization* in that it has the general properties that define it as being in the class labeled car. It also has a structure of specific empirical properties (blue color, large size, etc.). These specific properties of the unity are not specified by the organization, but are specific to the structure. Thus, structure as used by Maturana and Varela is essentially opposite of the way it is generally used (it is commonly used to refer to general phenomena, and they use it to refer to specific phenomena).

Another epistemological aspect of autopoietic systems is that the observer's domain of description is inherently relative and subjective. This means that the observer is always affected by his or her nervous system, and can never have access to an absolute, objective reality (Mingers 1989, p. 171). To clarify further, the observer is describing an autopoietic system. But the observer *is also* an autopoietic system himself or herself, and so his or her description is always a product of the organization and structure of his or her autopoietic system, and so is subordinate to the action of the nervous system and the autopoiesis of the organism.

Thus, if you as a researcher observe an independent autopoietic system, the independent events can trigger a response or description in your nervous system. However, the resulting description that you provide can never be merely a product of the independent triggering events, but must always be structurally determined by the specific autopoietic processes of your particular nervous system. This means that your description will always be subjective in the sense that it is structurally determined, and it may even be unique to you. This does not mean that descriptions are always unique to the individual, as observers develop similar cognitive structures through cultural experience and language interaction (Maturana 1974; Mingers 1989).

To summarize, there are three main points here:

1. The distinction between the domain of the autopoietic system and the domain of description of it by the observer (researcher) must be maintained.

2. A *unity* is a whole distinguished from a background (environment). The act of making this distinction can only be accomplished by the observer (not by the system itself). The description made by the observer may or may not coincide with the reality of the autopoietic system. This is because the observer's description is subjective, and because the description will depend on the choices and purposes of the observer.

3. The organization is the general abstract class of autopoietic systems, while the structure is the set of specific characteristics or the actual properties of an actual, empirically existing autopoietic system.

These important distinctions can be analyzed through the three-level model (see chapters 2, 6, and 7, and also Bailey 1984c; 1988; 1990). For an illustration of the three-level model refer to figure 2.2. It is immediately clear from perusal of the three-level model that description of an autopoietic system by an observer involves more than just two levels. It is clear that the structure of the autopoietic system exists on level X' (the empirical level), while the description of the system by the observer entails the observer's mental image or perception, and is on the conceptual level (X).

The beauty of the three-level model is that it easily facilitates maintenance of the clear distinction between the autopoietic system as empirical entity (X') and the perception of the system in the mind of the observer (X). This distinction can be quite difficult to maintain in a strictly verbal analysis, but is easy in the three-level model. The autopoietic system, its structure, and its environment all exist on the empirical level (X').

The three-level model also clarifies Maturana's point that the observer's perception is always relative and subjective. What this means is that the perception (X) of the empirical system (X') may not be totally accurate, but may involve some measurement error (path a of figure 2.2). It means further that three perceptions $(X_1, X_2,$ and $X_3)$ by three different observers of the same empirical system (X') may all be different, and may differ in their amount of measurement error and thus in the degree to which they adequately represent X. That is, paths $a_1, a_2,$ and a_3, which represent observations by three different observers of the same path a of figure 2.2, may all be different.

It is also clear that while both the system and its environments can exist empirically in level X', they cannot distinguish themselves, but can only be distinguished from each other and labeled as environment and system by the observer at the conceptual level (X). Thus, only the observer can distinguish a unity, and only at the X or conceptual level. Further, structure in Maturana's terms is on the empirical level (X'), while organization, or the general class of systems, is on the conceptual level (X). I can conceive of a car as a general combination of components and their relationships on an abstract level (X). This includes all cars, and is what is meant by organization. I can observe a single car which has one specific state (out of many) for each component. This individual car exists on the empirical level (X') and is a structure.

To summarize, the observer's perception (X) of the system (X') is subjective, and measurement or perception may be faulty (path a). Further, the concepts of unity and organization, description, and explanation all exist at the conceptual level (X) and reside in the mind of the observer. The entities of structure, components, system, and environment, all exist at the empirical level (X'). The link between the former (unity, organization, description, explanation) and the latter (structure components, system, environment) is path a.

Although conducted within the context of the three-level model, the discussion to this point has involved only two levels (X and X'). Maturana's distinction between the observer and the system, and his distinction between organization and structure, only require two levels. However, it has been pointed out elsewhere (Bailey 1990) that two-level discussions are incomplete. There is nothing wrong with a two-level analysis, it is only that the actual research process entails all three levels—the conceptual (X), the empirical (X'), and the operational or indicator level (X'').

While the concepts of unity and organization are possible at the conceptual level (X) in the two-level model, as are the concepts of explanation and prediction, in reality, unity, organization, explanation, and prediction are only possible at the conceptual level in relatively simple cases, and generally require utilization of the indicator level (X'') as well.

To illustrate, if I observe a small and simple system within a small and simple environment, I can possibly perform four operations in my head. I can distinguish the system as a whole within its environment (unity), I can generalize the concept of organization as a class (not necessarily from observation of this case), and I can describe and explain, all at the conceptual level (X). All of these operations can be performed mentally and stored as perceptions in my mind.

Presume, however, that I wish to construct a complex functional explanation of how the system operates. This will entail analysis of the parts (components) and of relations among them, part/whole analysis, and analysis of system/environment interchanges. It is doubtful that I can adequately formulate such a complex explanation solely within my mind (X) without the aid of a marker such as paper, computer diskette, etc. Any use of such materials entails the indicator level (X''). Even if I could formulate an adequate explanation solely in X, I probably could not remember it all very long without storage in X''. If the system and environment (X') are complex, the situation is even more difficult. I probably could not even formulate a unity, organization, or description at the conceptual level (X), but require utilization of the indicator level (X'').

In actuality, the observer utilizes not only the perceptual or conceptual level (X) and the empirical level (X'), but also the indicator level (X''). First consider the concept of unity. All of the components in the system are interrelated, but it is also true that the system has relations with its environment. How can the identity of the system be distinguished from the larger environment of which it is a part? This is done by the observer, and is not determined by the system itself. However, in all but the simplest cases of the most primitive systems and ideal observational conditions, the process of determining unity probably is a multistage process, utilizing all three levels of the three-level model, but especially relying on the indicator level for memory, or information storage.

As Luhmann points out, since observing systems are autopoietic, observation only comes about as an operation of autopoietic systems. Thus:

> If an autopoietic system observes autopoietic systems, it finds itself constrained by the conditions of autopoietic self-reproduction, and it includes itself in the field of observations, because as an autopoietic system observing autopoietic systems, it cannot avoid gaining information about itself. (Luhmann 1986, p. 186)

Initially, the observer perceives a system and begins describing its components and their interrelations, including the details of the autopoietic process. However, if a system is complex, its environment is complex, and its boundaries are problematic, all identification and analysis of unity can only be established by perceiving aspects of the system and environment (X') as mental images at the conceptual level (X), and then recording them at the indicator level (X''). This is a path a, path b endeavor. Further, details can be coded into the model along path c when additional data are gathered, and the model is refined.

The process of research (using X' and X) results in a model (X'') of autopoiesis at the indicator level (e.g., as stored in a book such as Maturana and Varela 1980). In the next stage, this model (X'') is revised through further conceptualization (X), e.g., through further reflection on the unity and through data gathering (X'). Thus, the research process is an endless combination of the process-structure-process-structure cycles, and the three-level model. That is, the process-structure cycle is accomplished through use of all three levels (X, X', X'') of the three-level model (see chapters 6 and 7 and Bailey 1990).

The concept of organization is similar to unity in its derivation.

It may be inductively derived through generalizations from observation (X) of a large number of actual empirical examples (structures) (X'), as in the case of living systems such as organisms. Conversely, the organization (X) may proceed the structure (X'), as in the case of a machine where the inventor first sketches out the organization (X) at the conceptual level, then builds a structure or prototype, which is an actual empirical example (X'). In the latter case, the indicator level is particularly salient, as the inventor generally will not go from organizational conceptualization (X) to prototype (X') without the intermediate stage of developing plans or blueprints (X'') or a working scale model (X''), both at the indicator level. This process is X (organization) $\rightarrow$ X'' (scale model or blueprint) $\rightarrow$ X' (empirical example such as prototype or structure). In terms of paths, this is path b to path c. If the organization is derived by induction through observation of the structure, as in Glaser and Strauss's (1967) grounded theory, then the sequence is coded as X' (structure) is observed, leading to general derivation of the abstract organization at the conceptual level (X), leading to preliminary sketching of a model on paper or stored on a computer disk at the indicator level (X''). The model can then be revised both through further reflection (path b) or further data analysis (path c). Thus, the two common sequences are either path b to path c ($X \rightarrow X'' \rightarrow X'$), or else path a to path b ($X' \rightarrow X \rightarrow X''$). The former represents going from organization to structure, while the latter represents going from structure to organization.

Open and Closed Systems

Are autopoietic systems open or closed? There is no debate over the fact that, in terms of the classical distinction, all autopoietic systems are open. Only living systems can be autopoietic, as defined by Maturana and Varela, and it has been established earlier in this volume, as elsewhere in system theory, that all living systems are open systems (see Miller 1978; Klapp 1978; Bailey 1990). By saying that living systems are open, one is simply saying that they exchange matter/energy and information with their environment. This is in opposition to isolated systems, which exchange neither matter/energy nor information with their environment, and closed systems, which exchange energy, but not matter (see Hall and Fagen 1956).

Thus, there is consensus that, as with all living systems, auto-poietic systems are technically open systems in the classical sense. For example, Mingers (1989, p. 168) notes that autopoietic systems have interactions with the environments, including the intake of resources and the output of wastes. Luhmann also considers autopoietic systems as open in the classical sense, saying that:

> The theory of autopoietic systems, however, has been invented for a situation in which the theory of open systems has become generally accepted. It does not return to the old notion of "closed [versus open] systems." (Luhmann 1986, p. 183)

But given that autopoietic systems are open in the classical sense of exchanging matter/energy and information across their boundaries with their environments, they are organizationally closed. Luhmann says that:

> Given the historical context, the concept of autopoietic closure has to be understood as the recursively closed organization of an open system. . . . The problem, then, is how autopoietic closure is possible in open systems. The new insight postulates closure as a condition of openness, and in this sense the theory formulates limiting conditions for the possibility of components of the system. Components in general and basic elements in particular can be reproduced only if they have the capacity to link closure and openness. (1986, p. 183)

But in what sense, then, can autopoietic systems be considered as closed if in fact they are open? What is meant by the "organizational closure" of autopoietic systems? Luhmann (1986, p. 174) expresses it very well, saying:

> Thus, everything which is used by the system is produced by the system itself. This applies to elements, processes, boundaries and other structures, and last but not least to the unity of the system itself. Autopoietic systems, of course, exist within an environment. *But there is no input and no output of unity.* (1986, p. 174, italics in the original)

Mingers (1989, p. 167) says that autopoietic systems are organizationally closed in the sense that the product of the organization is itself. All possible states that the organization can enter will maintain autopoiesis. The state of the structure at a given time will determine the actual changes that the structure undergoes, and this is known as being *structurally determined* (Mingers 1989, p. 168;

Maturana 1974, p. 34). Both internal and external actions can trigger such changes. The internal structure determines what changes can occur (only those which maintain autopoiesis), and thus which environmental interactions will trigger changes in the system.

Thus, environments do not determine internal changes of the system. The environment merely selects states from among those determined by the internal structure. This is known as *structural coupling*. This is a form of adaptation except that the environment does not specify internal changes—these are specified by the internal structure, and the environment only triggers such changes as the internal structure specifies (Mingers 1989, pp. 168–69); Maturana 1981, p. 29). Organisms can become structurally coupled not only to their environment but also to other organisms, such that behavior in one organism can trigger behavior in the second.

Autopoiesis and the Law of Requisite Variety

As has just been shown, an autopoietic system can be viewed in terms of a finite set of states that the system can enter. Further, all of these states must be autopoietic, or maintain autopoiesis. Any change in states must be a change which maintains autopoiesis. The specific states that the structure undergoes depend upon the state of the structure itself at a given point in time (Mingers 1989, p. 168; Maturana 1978, p. 34).

Ashby's law of requisite variety (Ashby 1956) can be viewed in a parallel fashion. Any system faces a certain number of spontaneous fluctuations triggered by the environment (just as the environment can trigger changes in the structure of autopoietic systems). The number of possible fluctuations represent a certain amount of disturbance, disorder, or entropy. In a cybernetic or controlled system, there are a certain number of alternative control actions (leading to a certain subsequent number of system states). The law of requisite variety states that the number of control actions must be at least equal to the number of spontaneous fluctuations to be corrected, if the control system is to function effectively. "*Only variety in* R *[the regulator] can force down the variety due to* D *[the disturbance]; only variety can destroy variety*" (Ashby 1956, p. 207, italics in the original).

The similarity of the parallel development between autopoietic theory and requisite variety theory is clear. In both cases the system faces a set of N disturbances (triggers) from the environment, as both are open systems. The difference is that in autopoietic systems, only a specific set of states (that are structurally determined by the internal structure) can be triggered by environmental conditions. In the law of requisite variety, the set of internal control states must match the number of externally triggered disturbances.

The thrusts of the two theories are different. Autopoietic theory says that only a set of states can be altered—disturbances will not result in others outside of this set. The law of requisite variety says that disturbances can cause problems with the system unless matched one-for-one by controls. Thus, the autopoietic system can be seen as one in which controls are adequate, and the variety of controls does match the variety of disturbances. Then, the autopoietic system is effective. However, the law of requisite variety allows for the possibility that the variety of control actions does *not* match the variety of disturbances. In such a case control is not effective, with the consequence that internal entropy increases. If continued long enough, this would result in serious impairment or extinction of the system.

Autopoiesis and Entropy

The law of requisite variety says that if the internal controls do not match the variety of external disturbances, then the system will not function effectively (internal entropy will increase). The theory of autopoiesis says that controls will be adequate in autopoietic systems. For one thing, internal changes are not selected or caused by external changes from the environment, but are specified by the internal structure. Only a specified set of changes can be triggered by environmental influences, and all of these facilitate or maintain autopoiesis. Thus, large entropy increases are precluded.

What is the relation of autopoiesis to entropy? While autopoietic theory does not deal with this point in detail, a number of inferences can be made. To begin with, the autopoietic system has adapted to its environment, and is maintaining a level of system efficiency. Thus, homeostasis is maintained. Robb (1989a, p. 53) says, "In classical positivist cybernetic terms, an autopoietic system is a homeostatic system in which the variables to be maintained constant are those which ensure the maintenance of its own unity, its organization." This homeostasis of organization is characterized by negentropy (Miller 1978, p. 18). Thus, an autopoietic system is one in which entropy does not increase to a maximum, but is maintained within certain organizational levels.

The concept of entropy also enters into the epistemological discussion of autopoiesis concerning the subject of how the boundary of the system is identified by the observer. One basic tenet of autopoietic theory is that the system constructs and maintains its own boundaries (Robb 1989a, p. 53). Another tenet is that only an observer can perceive the autopoietic system and its environment and relate them—the interacting components cannot do this (Mingers 1989, p. 161).

Thus, while boundaries are constructed by the autopoietic system and are real, the system itself cannot distinguish elements such as components, boundaries, and the larger environment. A crucial epistemological question then is, how can the observer distinguish the autopoietic system from its environment? This question has probably not yet been answered by autopoietic theory. I can give some partial answers. For one thing, it is clear that distinction of system from environment is not a random choice, although it may appear to be at first glance. This is because interactions exist not only internally between components within the system, but also externally, between the system components and their environment. Suppose that components A and B are in the system, and component C is not. Rather, C is external to the system (is a part of the environment). Suppose the observer perceives interactions between A and B (the system), but also between A and C, and B and C (nonsystem). How can boundary determination be made, so that the observer correctly specifies that A and B are in the system and are separated from C by a boundary?

Part of the answer lies in the perception of entropy changes. The entropy level of an autopoietic system will generally be different from the entropy level of its environment. This is because of the internal production of negentropy (through the importation of matter/energy and information across the boundary) which maintains entropy levels of the living system at lower levels than the entropy of its nonliving environment.

To see that this is so, imagine a case where it were not so. In this case, entropy levels within the living system would equal those outside of the system. What would be the function of the boundary here, in Miller's (1978) sense of a subsystem processing matter/energy and information? Obviously, it would have no function, and would not meet the definition of a boundary. It could be removed without any change in the internal entropy levels of the system, because they already equal those of the environment, and negentropy is not being constructed and maintained internally. There is no self-regulation. The boundary in this case is useless, and can be removed. The system becomes swallowed up by the larger system of which it is a part, and its boundaries become the boundaries of the larger system.

In the usual case of living systems, the boundary does function so that the system maintains a state of organization (negentropy) that does not follow the second law of thermodynamics. This is accomplished through the Prigogine entropy law (see chapter 4, and also Prigogine 1955; Bailey 1990; Laszlo 1986). In this case there is a clear "entropy break" (Bailey 1990) at the boundary, with the entropy level on one side of the boundary (within the system) being markedly different from the entropy level on the other side of the boundary (in the environment).

It would seem, then, that the observer could easily perceive boundaries simply by observing the point at which there was a clear break in entropy levels. By calculating entropy within the system and entropy outside the system, the observer could empirically establish the existence of the boundary. However, this leads to a fundamental paradox: the observer *can* establish system boundaries by calculating internal entropy for the system and comparing it with external entropy to document an entropy break, but those calculations can only be made if the location of the boundary is known. So, to document the boundary we must know it. The boundary can be perceived if entropy levels are perceived, and entropy levels can be perceived if the boundary is perceived.

How does one escape this paradox (which is not limited to systems, but applies to all empirical groups, including classes or types)? This cannot be done with the two-level model. Only the three-level model is sufficient. If one works with a simple two-level model of observer (X) and empirical system (X'), then an impasse is reached. The observer can perceive a conceptual set of boundaries and calculate entropy for them, but these may be entirely hypothetical. For example, I can imagine two hypothetical cells A and B, and two hypothetical cell entries, 100 and 0. This is a case of minimal (zero) entropy, but is

100	0
A	B

hypothetical, existing only in my mind and just being a product of the observer's perceptions. At the empirical level (X') entropy levels exist, but cannot be computed by the components, as the process of entropy calculation is a mental one which must be conducted by a conscious external observer. To this point the two-level model consists of entropy calculation (without data) (X) and data (X') without entropy calculation. Only by

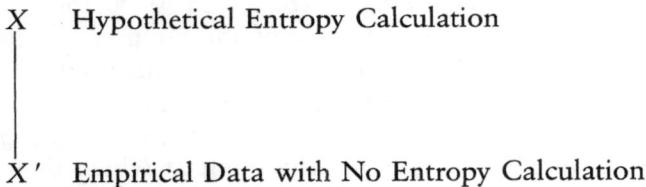

X Hypothetical Entropy Calculation

X' Empirical Data with No Entropy Calculation

adding the third level (indicator level, X'') can entropy be calculated, especially for larger complex systems. This entails mapping the conceptual categories from the conceptual level (X) onto the indicator level (X'') and storing them in some physical marker such as a computer or paper. This is the path b mapping (see figure 2.2). The next step is to code empirical data from the empirical level (X') into the corresponding categories representing them (path c mapping). Only when both X and X' are mapped into X'' can entropy be calculated in such a way as to determine the boundary of the system.

A corollary to the relation of autopoiesis to entropy is the relation of autopoiesis to equilibrium. For living (open) systems, the correct concept is homeostasis, not equilibrium (see Buckley 1967; Bailey 1990). As mentioned above, autopoiesis is compatible with the notion of homeostasis, meaning that structurally determined changes in internal states that further autopoiesis also maintain homeostatic (negentropy) levels.

ARE SOCIAL SYSTEMS AUTOPOIETIC?

It is now time for the big question: Are social systems autopoietic? Before jumping to conclusions one needs to ask whether the question is even appropriate. As discussed in chapters 2 and 3, many systems concepts have come to sociology from physical and biological science. Some, such as equilibrium, have been used inappropriately. Is it desirable for sociological systems theorists to import yet another concept (autopoiesis) from science, or will it simply engender the problems caused by equilibrium? Further, what is the rationale or logical mechanism for applying the concept of autopoiesis to social systems? Is there a clear line of deductive reasoning which leads to the conclusion that autopoiesis applies also to social systems? Conversely, if autopoietic social systems are not derived deductively, could one inductively study the social system and derive the concept of autopoiesis (à la the grounded theory of Berger and Luckman 1967)?

There are some good arguments for favoring the latter approach. Sociologists should study the social system directly (without the concept of autopoiesis) with a view toward identifying its characteristics. If it is seen to be autopoietic, then this concept can be used. If, on the other hand, its characteristics do not fit the model of autopoiesis, the model should not be "bent" to fit the social system, but a new model which is isomorphic with the reality of the social system should be constructed. This was the goal of *Social Entropy Theory* (Bailey 1990), and also of *Living Systems* (Miller 1978).

Despite these reservations, the model is appealing, and it is tempting to see whether it can be applied to social systems. As indicated above, there is no consensus on the matter. This is a disturbing sign, because one would think that if the model clearly fit the social system, researchers could agree on this. However, some theorists such as Luhmann (1982, 1986) and Robb (1989a, 1989b) seem convinced that it applies, while the theorists who are not convinced generally do not say that it *cannot* be applied, but only that it is not clear at this point that this is the case (Maturana 1981; Varela 1979; Mingers 1989).

Much of the disagreement centers around the familiar issue of the proper unit of analysis for the social system: the individual or the social role (or some alternative). This has been discussed at length by Parsons (1979), Parsons and Shils (1951), Miller (1978) and Bailey (1981, 1990). It is still not resolved, with Miller (1978) continuing to espouse the concrete system with the individual as the unit, while Parsons defended the abstracted system with the role as the unit almost to his dying day (Parsons 1979). This debate resurfaces as a crucial issue in the discussion over whether social systems are autopoietic.

Luhmann (1982, 1984, 1986, 1990) is convinced that social systems are autopoietic, and that communication is the basic unit of the social system. Luhmann says that:

> Social systems use communication as their particular mode of autopoietic reproduction. Their elements are communications which are recursively produced and reproduced by a network of communications and which cannot exist outside of such a network. Communications are not "living" units, they are not "conscious" units, they are not "actions." Their unity requires a synthesis of three selections: namely, information, utterance, and understanding (including misunderstanding). (1986, p. 174)

Elsewhere, Luhmann says:

> Die Gesellschaft ist ein autopoietisches System auf der Basis von sinnhafter Kommunikation. Sie besteht aus Kommunikationen, sie besteht nur aus Kommunikationen, sie besteht aus allen Kommunikationen. Sie reproduziert Kommunikation durch Kommunikation. (1984, p. 311)

Loosely translated, this says that a social system is autopoietic on the basis of communication. It consists of communication, only of communication, and of all communication. It reproduces communication through communication.

Luhmann (1986, p. 175) also says, "The elementary, undecomposable units of the system are communications of minimal size." He says further:

> For a theory of autopoietic systems, only communication is a serious candidate for the position of the elementary unit of the basic self-referential process of social systems. Only communication is necessarily and inherently social. Action is not. Moreover, social action implies communication. . . . Therefore the theory of autopoietic social systems requires a conceptual revolution within sociology; the replacement of action theory by communication theory as the characterization of the elementary operative level of the system. (1986, pp. 177–78)

Elsewhere (1986, p. 177) Luhmann says that action is generally considered the basic unit of analysis; the role or the individual can be so considered, but communication is a kind of action.

Mingers (1989, p. 175) has difficulty with this view. He finds social autopoiesis problematic because of the difficulty of determining whether a term (such as organization) refers to an empirically existing entity or a construct (objectivism versus subjectivism). Further, he is unsure what the basic units of the system are. Humans may be the components of social systems, but they are *not* reproduced by the system but by physical biological processes. If humans are not the components, what are? (Mingers 1989, p. 175). He also questions the emphasis in autopoietic theory on physical space and the self-determined boundary, as humans choose to belong or not to a particular organization, and can belong to many at the same time. What then can be the boundary of an autopoietic system that would distinguish components from noncomponents? How can the social group act as a unit, as only the individual people act? (Mingers 1989, pp. 175–76).

Beer (1975, p. 70) agrees with Luhmann that human societies are autopoietic, saying, "any cohesive social institution is an autopoietic system. . . . As examples I list: firms and industries, schools and universities, clinics and hospitals, professional bodies, department of state and whole countries." Further, Facheux and Markidakis (1979, p. 216) say that social systems are autopoietic; "Varela's views are valid for all living systems, but take particular significance for social systems . . . ." Zeleny (1989) says that all "spontaneous" social systems are autopoietic.

Robb (1989a) supports the view that "suprahuman" autopoietic social systems may be developing. He distinguishes between colonial forms of organization and autopoietic forms. In the colonial form, the subsystems maintain their autonomy and their individual autopoietic processes. There is no subordination of the unity of the subsystem to its larger host system. Such colonial organizations are at the same "logical level" as the humans that order them. In contrast, suprahuman autopoietic systems are at a *higher* logical level than the humans and human organizations which form their components. They are logically closed, and not open to control by lower-level components (humans). Robb (1989b, p. 344) says that not all social systems are autopoietic, but that some are or may become so. He says that not the whole human, but only certain human *properties,* are the components of the social system. These properties include "mind-sets," such as goals, ideas about organizational structure, culture, and actions set in the organizational context (Robb 1989b, p. 345).

So where do we stand at this point? Robb (1989a, b, c), Beer (1975), Luhmann (1982, 1984, 1986), Zeleny (1989), and Facheux and Markidakis (1979) all support the notion that at least some social systems (at various of Miller's eight levels) are autopoietic. Mingers remains unconvinced. He does concede (Mingers 1989, p. 177) that the concept of autopoiesis might be applied to the social system as a metaphor, or perhaps in a more general version.

Mingers is not alone in his reservation. The originators of autopoietic theory, Maturana and Varela, are also unsure that it applies to social systems. Varela (1980, p. 38) says that he does not see how autopoiesis can *directly* be transported to social analysis, but does offer a less specific version called organizational closure (Varela 1979, p. 55) which lacks the physical process of component production and thus may apply to social systems. Maturana (1981) also stops short of saying that social systems are autopoietic, but does say that they serve as mediums for autopoietic systems, and also applies the idea to concepts (Maturana 1975).

Can (or should) a concept from the world of cells and animals, which refers to specific physical processes of component production in cells, be transported to social systems? What can one conclude from all of this? Strictly speaking, the same types of biological *production* that have been so carefully described by Miller, Maturana, and Varela *do not exist* in social systems. I thus agree with Varela (1980) that biological autopoiesis as developed for the cell, with emphasis on production of components, is not present in social systems, and probably cannot be documented in the strict, technical sense. Neither should one necessarily try to transport all of the technical jargon of autopoietic theory (e.g., structural coupling, plastic structure) to social systems theory. Thus, to this extent, I side with Varela, Maturana, and Mingers in having reservations about strictly and literally applying the concept and theory of biological autopoiesis, as derived for cells, directly to social systems.

However, I believe that there may be more consensus in the positions than originally recognized. Thus, I again agree with Varela (1979, p. 55) that a much more generalized version, such as his concept of organizational closure, can be applied to social systems, *"but without the particular specification of physical processes of component production"* (Mingers 1989, p. 176, italics added). In reality (although I cannot speak for all authors), there is probably more consensus on this position than appears at first glance. While Robb says that autopoiesis can be applied to social systems and Mingers says it cannot, Robb does not mean that *all* social systems are autopoietic, and Mingers does not mean that some more generalized version of autopoiesis cannot be adequately applied to social systems. Further, probably all authors agree that autopoiesis has never been documented for an *actual* social system in the painstaking detail that it has for the actual cell.

Thus, the issue is not over whether autopoiesis in terms of physical component production can be transported to social systems (the answer is no). The issue is that social systems undergo sweeping structural changes and complete turnover of members, yet still maintain their autonomy and identity. Thus, an organization can grow, divide, move geographically, and have all of its old members replaced by new ones, while still maintaining its identity and autonomy. These are the same kind of phenomena that autopoietic theory explains in cells and animals (Mingers 1989, p. 172). Thus, it would seem a crime not to attempt to generalize and expand the basic concepts of autopoietic theory in such a way as to explicate these factors, while avoiding a narrow, technical transport of emphasis on physical production.

Thus, it is agreed to this point that *physical*, biological autopoiesis will *not* be applied to social systems, but that we will continue the effort to develop a theory of social autopoiesis which parallels the theory as developed for cells and physical living systems. Luckily, there is a major precedent for this strategy in the form of Miller's (1978) living systems theory. Living systems theory clearly discusses reproduction not only for the cell, organ, and organism, but also for the group, organization, community, society, and supranational system. However, Miller does not naively apply descriptions about physical production processes in the cell to the social system levels. He merely begins with the premise that the twenty subsystems exist, and can be described, for all eight levels. This is the strategy taken here. It is assumed that autopoietic-type reproduction occurs at all levels—cell, individual, group. However, this process at the higher levels will appear much different than it will at the lower levels.

The genius of Miller's approach is that he simultaneously presents the twenty subsystems at all eight levels. If he had merely worked them out in great detail for the cell (as was the case with autopoiesis), then doubtlessly the same debate would have occurred in living systems theory over whether these twenty subsystems (which had been found at the cellular level) applied to social systems or not. By working at all eight levels simultaneously, Miller (1978) avoided such controversy. For example, Miller (1978), p. 224) in discussing the reproducer subsystem at the level of the cell, writes of asexual reproduction in which the parent cell retains its identity, while its DNA is duplicated and a small cell breaks away. How can this analysis ever be transported to social systems? The secret is that it is *not* transported, but the parallel processes at the social system level are explored independently, and *not* as extensions of cell reproduction.

However, having arrived at this point (which is perhaps a breakthrough given the apparent controversy over social autopoiesis), there is still a major obstacle in the form of the debate over the proper unit of analysis: individual, role, action, event, or communication. Mingers (1989) would choose the individual as the basic unit of the social system, as would Miller (1978). Luhmann (1986, p. 177) says that most sociologists would choose action, but that sometimes roles or human individuals are preferred. He would choose communications. He said earlier (Luhmann, 1982, p. 131) that social systems use communication to constitute and interconnect *events* (actions) which build up the system, with events being reproduced and serving as components of the system.

Using communication as the basic element of the social system, Luhmann (1986, p. 172) says that if social systems are living systems, it would seem to follow that they are autopoietic, but problems immediately arise in precisely defining the components. He prefers to use general autopoietic theory as merely a foundation for deriving an autopoietic social theory, with the concept of social autopoiesis *abstracted* from biological connotations (Luhmann, 1986, p. 173).

In Luhmann's theory, the network of events (communications) reproduces itself, and structures are required for the reproduction of events by events. This is accomplished through the synthesis of information, utterance, and understanding (Luhmann 1986, p. 174) and requires self-reference. Communication cannot use the environment, but requires references to previous and future communications. Information, utterance, and understanding are co-created within the process of communication and cannot exist independently of the system. Information is not something external (in the environment) to be picked up by the system, but is produced in the system. The synthesis of information, utterance, and understanding is an undecomposable unit, engaged in autopoiesis only as an element of the system.

Reflexive communication is not an occasional event, but an ongoing process that is co-reproduced by autopoiesis, with each communication undergoing recursive elaboration (questioning, denial, or correction) as well as adaptation to future events (Luhmann 1986, p. 178). Societies also form and use their own boundaries. The social system is a self-referentiated system operating within its own world, and constituting a world of its own. Societies, as autopoietic social systems, communicate about themselves, and this entails distinguishing themselves from the environment. Communication can never exist outside the social system's own boundaries, which are components of the system and formed by the system (Luhmann 1986, p. 179).

Luhmann's (1986, pp. 180–81) discussion clearly shows one fundamental difference between biological autopoiesis and social autopoiesis. While biological autopoiesis reproduces elements (e.g., molecules in cells) in order to stave off decay (an increase in internal entropy according to the second law), social systems produce their own decay. Their thought and communications are events which vanish as soon as they occur. If a social system were to store all its communications it would soon be drowned in complexity so that communication patterns could not be established and chaos would result (Luhmann 1986—this is similar to Miller's 1978 concept of information-input overload).

Luhmann says:

> The solution is to renounce all stability at the operative level of elements and to use events only. Thereby the continuing dissolution of the system becomes a necessary cause of its autopoietic reproduction. The system becomes dynamic in a very basic sense. It becomes inherently restless. The instability of its elements is a condition of its duration.
>
> All structures of social systems have to be based on this fundamental fact of vanishing events, disappearing gestures, or words that are dying away. Memory, and then writing, have their

function in preserving not the events, but their structure-generating power. The events themselves cannot be saved, but their loss is the condition of their regeneration. Thus, time and irreversibility are built into the system not only at the structural level, but also at the level of its elements. (1986, p. 180)

Luhmann says that autopoiesis involves an important shift from self-referential structural integration to self-referential constitution of elements. Maintenance is not simply replication, cultural transmission, or reproduction of patterns. It is processing the production of next elements which are different from previous ones (or else they cannot be recognized as new events, Luhmann 1986, p. 181). Thus, the system maintains itself not through storing patterns but by producing elements. Information is an internal aspect and not something to be derived from the environment. In describing social autopoiesis two dichotomies are necessary—system/environment, and event/situation.

It is clear from this brief summary that Luhmann's social autopoietic theory is far from a literal and inappropriate transportation of biological autopoiesis to social systems. To the contrary, it differs markedly from biological autopoiesis (as when Luhmann notes that social systems produce decay while biological systems do not), and goes well beyond biological autopoiesis in many ways (see his discussion of reentry in Luhmann 1986, p. 183). Luhmann presents a viable theory of social autopoiesis which uses the general theory of biological autopoiesis only as a foundation on which to build, in the best Comtean tradition.

Object Versus Role (Action, Communication, Event) as the Basic Unit

Luhmann presents such a vital, viable, and sophisticated theory of social autopoiesis that it would be presumptuous to attempt to extend it here. However, it is fruitful to return to the question of whether the individual object or the role (or action, event, or communication) is the proper unit of the social system. Although this question has been dealt with at length elsewhere (Miller 1978; Parsons 1979; Bailey 1981, 1990), as well as in chapter 5, 6, and 7 above, it is still a "loose end" in the context of autopoiesis. Luhmann (1986, p. 180) stresses that memory and writing have their functions not in preserving events, but in their structure-generating power. This is a clear tie-in to social entropy theory, which emphasizes the nexus between action (process) and structure (structure, as used by Bailey 1990, not by Maturana and Varela), and the role of social markers in the three-level model).

Thus, SET complements Luhmann's autopoietic theory by combining object-variable (Q-R) analysis in the score matrix (table 2.1), and by combining process-structure. I have emphasized in SET that the human individual serves as a marker in carrying information. Further, the score matrix of table 2.1 shows objects (individuals) in the rows and variables (including "roles" such as husband, electrician, etc.) in the columns. Thus, this matrix allows one to easily move from the domain of the individual (Q-analysis) to the domain of the variable (R-analysis). This gets past the "individual-role as the basic unit of analysis" dilemma.

Next, SET shows how individuals serve as social markers (physical information carriers) so that when two or more people interact, they react in terms of properties of each other, such as the immutable properties (race, sex, age) and mutable properties (region, occupation, etc.). These interactions involve communication and events in a situated context. This is a clear tie-in to Luhmann's autopoietic theory. This interaction, over time, as process, uses structure. Structure includes norms and rules, and particularly the written-down rules and laws, and culture of all sorts in law books, dictionaries, etiquette manuals, etc. By interacting, two individuals use written rules, and also reproduce and revise these written rules. They use structure as a basis for (through actions) changing structure. This is done through the three-level model, where each individual has a mental expectation and goal of the outcome of the structure (X), an empirical reality counterpart (X'), and a mediated indicator counterpart (X'') consisting of pictures, resumes, applications, etc. For further details see chapters 6 and 7 of this volume, and also Bailey 1990.

This chapter has shown that biological autopoietic theory *cannot* be directly transported to social systems, but that a separate, independent, unique theory of social autopoiesis can be derived using general autopoietic theory as a foundation—and Luhmann has done this in a very valuable and sophisticated fashion. I now turn to the task of summing up the volume by charting the path for the future of social systems theory.

COUNTERPOINT

What are the points of comparison between mainstream sociological theory and autopoietic theory? There is little overlap with Alexander's neofunctional concepts of action, order, equilibrium, and idealism/ideation. Autopoiesis has a very unique jargon which has relatively little overlap with sociological terms. Thus action, order, and idealism are not autopoietic concepts. However, "action" could occur within systems boundaries and in mediation with the environment. Further, the result of autopoiesis can be seen as "order" in the technical sense of entropy below the maximum, and a resulting state of organizational complexity. That is, order in the sense of negentropy would exist in the autopoietic system. Further, although the concept of equilibrium is not generally used, some autopoietic theorists do use the related notion of homeostasis.

Turning next to concepts derived from Collins's conflict theory, I can analyze autopoiesis in terms of conservatism, conflict, age, sex, and hypostatization. Age and sex have generally not emerged as salient variables in the sociological applications of autopoiesis. Also, autopoiesis is not clearly a conflict model. It is perhaps more of a consensus model, but only to the extent that "self-organization" or "self-reproduction" or "self-regulation" infer a sort of consensus. This aspect has not really been analyzed to my knowledge.

As to "conservatism," inasmuch as the model comes from biology and admittedly embraces (at least to some extent) the notion of homeostasis, it may have some vulnerability to the charge of conservatism. However, a little reflection shows that Maturana, Varela, Luhmann, and other adherents of autopoiesis are not the usual "conservatives," but are definitely innovative and highly sophisticated and epistemologically aware theorists. As highly innovative and nontraditional thinkers, if we were to label them as "conservative" we would certainly be widening the scope of this term. I personally certainly do not regard them as conservative.

The issue of hypostatization is a little difficult to judge. It seems to me that the element of autopoiesis that is most vulnerable to the charge of reification is the notion of a system that is empirically "open" in terms of boundary, but is organizationally "closed." It seems to me that if any part of autopoietic theory is to be hypostatized, it is the notion of organizational closure. Does closure consist of empirically observable recursive action, or is it rather more in the nature of a construct?

I have not seen the question posed in this way before, and thus not directly answered. I would assume that adherents of auto-poiesis would want to make the case that organizational closure is a concrete phenomenon, and is not subject to reification. I think that this case can probably be made for cells. Whether it can be made for human systems is part and parcel of the whole question of whether autopoiesis can be applied to such systems. In any event, I suspect that in the absence of further elaboration of this point, autopoietic theory is more vulnerable to charges of hyposta-tization than living systems theory or social entropy theory, but can probably be successfully defended against such a charge.

Turning next to concepts derived from Giddens's structuration theory, we see much more overlap with autopoiesis than for the other two theories. These concepts are: agency, structure, system, structuration, system integration, time, and space. Time and space are used in a straightforward fashion in autopoietic theory, in the sense that autopoiesis occurs over time in the context of a spatially bounded system. Agency has no direct parallel in autopoiesis, ex-cept as process within the system, or conceivably as interaction with the medium (environment). The terms structure, structura-tion, system, and system integration, as defined by Giddens, can all be related to autopoietic theory. As a prelude, note that Giddens (1979, pp. 66–77) defines *system* as "reproduced relations be-tween actors and collectivities," while *system integration* is "reci-procity between groups or collectivities." Thus, integration refers to the degree of interdependence of "systemness."

Throughout this book, with the salient exception of the discus-sion in chapter 1 and of my extended critique of equilibrium and other aspects of Parsonian functionalism, we have not encountered another analysis of mainstream sociological theory by a "new" systems theorist. Another example is at hand now in Mingers's (1989) analysis of social autopoiesis, in which he comments on parallels between autopoiesis and structuration, saying that Gid-dens:

> . . . has developed a social theory based on a distinction between "system" and "structure" and what he calls the process of "structuration" by which social structures maintain and produce

themselves over time. Although there are significant differences, his work has definite resonances with Maturana's. For example, Giddens' distinction of structure/system is similar to Maturana's organization/structure, although the usage of the term "structure" is reversed. For Giddens, system refers to the actual, observable social systems of interaction between actors and collectivities as they are realized in various social practices. Structure, on the other hand, refers to an unobservable "virtual" set of rules and resources which actors draw on and which enable the actual activities which occur within social systems. Thus Giddens' system is the same as Maturana's structure, while Giddens' structure plays a similar role to Maturana's organization. Structuration is then the dual process whereby actors draw on the existing structure in their interactions and the social structure itself is continually and recursively reproduced through its involvement in these interactions. Giddens, too, stresses that purposes and reasons cannot be ascribed to systems and that explanation must be nonfunctionalist and nonteleological. (Mingers 1989, p. 174)

I would hesitate to agree with Mingers that "Giddens' system is the same as Maturana's structure." While Giddens (1979, p. 66) speaks of a system as "reproduced relations between actors or collectivities," Maturana (1978, p. 32) defines structure as "actual components and actual relations . . . in the constitution of a given composite unity." I would call these definitions similar, but would stop short of calling them "the same," especially since Giddens does not define the term "composite unity." Nevertheless, Mingers's analysis shows once again that there are clear parallels between mainstream sociological theory and the new systems theory.

Giddens (1979, p. 75) also comments on autopoiesis, concluding that ". . . it is probably too early to say just how close the parallels with social theory might turn out to be. The chief point of connection is undoubtedly recursiveness, taken to characterise autopoietic organisation."

I agree with Mingers and Giddens that there are clear parallels between mainstream sociological theory (especially structuration theory) and autopoietic theory, and am glad that this has been recognized "from both sides of the gulf." I also feel that these parallel developments are compatible and complementary, and proceeding in the same direction. I would encourage further explication of these parallels—a task that I must leave unfinished for lack of time and space, as this volume must close.

As a caveat, anyone attempting this analysis must continually remember that autopoietic theorists define structure roughly *opposite* the way most sociologists define it. This is a good case in point. I feel that often critics view systems theorists' attempts to standardize terminology as rather unimportant quibbling or as "chiefly semantic." The fact is that such standardization is imperative if the goals of synthesis and integration—shared by both mainstream sociological theorists and by systems theorists—are to be achieved. It is difficult enough to understand and synthesize two complex theories such as structuration theory and autopoietic theory without first having to translate between two very divergent notions of structure. But anyone who blithely tries to integrate structuration and autopoiesis without being aware of the different conceptions of structure is surely doomed to failure.

Turning to Ritzer's (1990a) discussion of metatheory, what forms of metatheory are used in autopoiesis? The most evident form is M_P (metatheorizing as a prelude to theory development). Also prominent is a variant of the first form of metatheorizing designed to achieve a deeper understanding of theory (M_U). This is the internal-intellectual approach, and is evident not only in the observer-observed distinction, but also in terms of the discussion of the difference between input models and closure models such as autopoiesis. Also, of limited visibility, is the third variant of M_U (external-intellectual), as when Mingers (1989) turns to Giddens's structuration theory for analytic tools with which to study autopoiesis.

What contributions has autopoietic theory made that mainstream sociological theory has not made, and that thus adds breadth and richness to the mainstream? This question is more difficult to answer for autopoietic theory than for other approaches in the new systems theory such as living systems theory and social entropy theory. In order to answer this question fully we really need to agree on whether autopoiesis applies to social systems, but as has been seen, there is a decided lack of consensus on this issue. Given this fact, certain limited statements can be made.

First, while somewhat arcane, the jargon of autopoiesis is complex, rich, and challenging. Second, the notion of an open system with organizational closure is fascinating, and potentially very valuable. Third, the recursive theory of self-reproduction is rich and challenging. Fourth, the notion of structural coupling is a notable contribution. Although still somewhat of a theoretical enigma, autopoietic theory is challenging and compelling. There is something about it that is very appealing. Its ties with social theory are clear, and it may yet be heralded as a major theoretical break-through in sociology when its applications to social phenomena are better understood. It deserves careful attention.

CHAPTER 9

The Dual Synthesis

It is now time to conclude this volume by formulating a dual synthesis of the new systems theory. As it stands, the new systems theory consists chiefly of living systems theory, social entropy theory, autopoietic theory, and the "new sociocybernetics" of Geyer and van der Zouwen (1986) and their associates. These approaches form a coherent package, but display some heterogeneity. The links between LST and SET have already been forged in chapter 7, but autopoiesis remains somewhat "unlinked" to these two. All of these rest on a foundation formed by the "old" systems approaches of general systems theory, information theory, and cybernetics (and some aspects of reconstructed functionalism—but without equilibrium).

The first task is thus to synthesize all of the systems approaches—the old and the new—but with an emphasis on the new. The second goal (the duality) is to forge the synthesis in such a way that links to the mainstream of sociological theory are emphasized and strengthened. Thus, the final synthesis will be a synthesis of the new systems theory, and will be constructed in such a way that the complementarity to and compatibility of systems theory with the mainstream will be maximized. This second goal—heightening congruence between systems theory and the mainstream, will be accomplished chiefly through the use of the comparison terms culled from the analysis of mainstream theory in chapter 1, and utilized in the counterpoints of the respective chapters.

Theoretical Presuppositions

Prior to the construction of the integrated model there are a number of preconditions for it that can be laid down.

1. The synthesis will be first and foremost a contribution to systems theory, but within that constraint will mesh with mainstream sociological theory as much as is possible.

2. The resulting model will *not* preclude or exclude any intellectual approaches (in the manner that equilibrium excluded emphasis on social change).

3. Equilibrium will *not* be defined as part of the systems definition. The model will not assume that a given system is in equilibrium. In line with condition number 2, equilibrium itself will not be precluded by the model, but will not be assumed nor emphasized. However, the final model should be able to successfully model an empirical system which, if found, was in a state of "equilibrium" of some sort. Similarly, it will be able to accommodate other notions such as conflict or social change.

4. Time and space *will be* an integral part of the model.

5. Synchronicity and diachronicity will be seen in intersection in the model.

6. The model will be as general as possible, but will apply to living human social systems only.

7. The role of the boundary in systems analysis will be emphasized.

8. Social and cultural factors will be emphasized, but due attention will also be paid to the analysis of flows of matter-energy and information.

9. The notion of system will be broadly defined.

10. Some measure of system state (e.g., entropy) will be defined and operationalized.

11. The model will be capable of generating testable hypotheses.

12. The model will not be strictly positivistic in Alexander's (1982, p. 5) terms.

Given numbers 10 and 11 (which certainly may appear positivistic to some), it is important to clarify the notion of positivism for the synthesis. The model will not be strictly positivistic according to Alexander (1982, p. 5). Alexander presents three postulates central to positivism: a "radical break between empirical observations and nonempirical statements," more general philosophical or metaphysical issues have no fundamental significance, and this elimination of nonempirical references results in a "scientific self-consciousness."

Since the discussion to this point in the volume has been replete with metatheoretical, philosophical, and metaphysical statements, it seems clear that the final model will have them too. Thus the model will not be "strictly positivistic" in Alexander's terms. It will require operationalization, formalization, and empirical observation. But *in addition* it will require philosophical analysis, and a *blending* of empirical and nonempirical statements. It will also emphasize values, culture, and ideational factors. Thus, the resulting model will also not be as "conservative" as functionalist models. I would hope that the final synthesis could be seen as an example of the modified, philosophical (and perhaps even liberal or humanitarian), positivism described earlier in this volume as "positive positivism."

In summary, the model will be a synthesis, and thus a hybrid—not only of systems theory, but of mainstream sociological theory. From systems theory, it will have to accommodate a number of features such as isomorphism, hierarchy, information theory, cybernetics, energy and information processing, entropy, and boundary analysis. From mainstream sociological theory, it will have to accommodate notions such as action, order, agency, structure, conflict, culture, norms, values, ideology, and micro-macro analysis. Such a two-pronged model will retain the metatheoretical focus of systems theory and be true to systems notions and ideals (such as multidisciplinary integration and synthesis) while also striving to include the topics that are important to mainstream sociology. Such a model will be broad and must *include*—not exclude (and certainly not preclude or prohibit). Obviously, the model cannot be broad enough to include *everything of interest to everyone,* but it must have the *potential* to include any topic, and must not be constructed in a biased fashion where potential inclusion is prohibited. Thus the model will be a skeleton. It will not be fleshed in to include everything, but must have a potential place for every intellectual notion, and further it should be fairly clear from looking at the systems skeleton just where in the model the given topic would fit.

This is clearly an ambitious project. What sort of strategy might prove efficacious in constructing such a broad synthesis? The strategy that I propose is to select a broad and relatively neutral model with wide generality as the initial "bare-bones" skeleton or nucleus for the synthesis. This model will then be progressively fleshed in by borrowing needed intellectual elements from the various systems approaches (the first prong) and then from mainstream sociological theory (the second prong). It should come as no surprise to readers of this volume that I will select the general social entropy model as the bare-bones carrier or nucleus. There are a number of reasons for selecting this model (besides my obvious bias in favor of it).

1. Social entropy theory is the only systems model since functionalism (except for Buckley 1967) that was designed for sociology by a sociologist. Thus, it should (and does) share more goals with mainstream sociological theory than the other new systems theories such as living systems theory and autopoiesis.
2. Social entropy theory has roughly the same goals as this new synthesis—to be general, and to not preclude any intellectual notions. Thus, it should have the qualifications to be a good "carrier" of other ideas.
3. Social entropy theory is one of the newest and most up-to-date systems theories.

The basic strategy in building the dual synthesis is first to sketch the bare-bones skeleton as derived from the SET model. After this is done, I will proceed through systems theory (both old and new) adding new elements for each of the respective perspectives as they seem appropriate. After the systems prong of the model has been sketched, I will proceed through the mainstream sociological theories, both old and new—beginning with Parsonian functionalism, and proceeding through neofunctionalism, structuration theory, and conflict theory. The volume will then conclude by assessing the model (both as a synthesis of systems theory and in terms of its congruence with mainstream sociological theory). This discussion will include a brief assessment of how the model meshes with other mainstream perspectives not emphasized in this volume (such as ethnomethodology, symbolic interactionism, network analysis, etc.). The last discussion will be a statement of needed future research in this area.

SYSTEMS THEORY

To begin, I will stress, as I did in SET, that the model of complex society must be as complex as the phenomena it is modeling, and must be isomorphic with it. That is, there must be a point-by-point congruence between the society and the systems model. Since modern society is very large and complex, the generic model must be very general. Also, since the model is designed as a contribution to systems theory (above and beyond its application to sociology), the systems definition should be extremely general. To this end, I repeat the generic model presented in chapter 2:

> A system is a bounded set of interrelated components that has an entropy value below the maximum.

This general definition can then be immediately narrowed down to apply to social systems. To this end, we specify that components are individual humans (or collectivities). This is recognized as a concrete rather than an abstracted system in Miller's (1978) terms. The case for choosing concrete systems has been made repeatedly both in this volume and in Bailey (1990) and also in Miller (1978). The boundaries referred to in the system will be political borders such as national, state, county, or city borders. Thus, the concrete system is located in space and time as Giddens (1979) suggests. Note further that Giddens's (1979) definition of a system can be recognized as a concrete system also. Actually, my definition, when applied to society, is very much like his, except that mine specifies boundaries and the entropy level, while his does not.

Now that the definition has been specified, the remaining components can be loaded in. As entropy has already been specified, I next need an operationalization for it. As noted above and in Bailey (1990), Shannon's H will suffice. I will first apply the general model at the level of society. Within the society, there is a set of interrelated (interacting) humans of population size P. The boundary is the national border. We thus have two macro variables—S (space) and P (population). The human actors, through their mutual interrelations, *organize* themselves (O) in such a way as to maximize (or optimize) their level of living (L). They do this by using information (I) and technology (T). Thus, the social system possesses six macro level variables (*PILOTS*). These are the *globals.* The human actors living in the society all possess certain *immutable* micro variables or characteristics such as skin color, sex, and date of birth. These are essentially permanent. They also tend to be very visible, and are easily determined by strangers, with little prior information and without advanced technology. (I can generally tell by looking at you that you are black or female without having to first "run you through the computer.") Being fixed, permanent, and highly visible, they are perfect bases for manipulation and discrimination by those in power.

It is given that each person in the society represents one population unit (*P*). Beyond this, each person should have some position (or zero) on the other five macro variables—*ILOTS* or *LOTIS*. It is not surprising to assume that some persons will occupy more advantageous positions on *each* of these five variables than others (a better site or residence [*S*], more education [*I*], more money [*L*], a better job or organizational position [*O*], and more sophisticated technology [*T*]). The immutables are important bases for determining where each person in the society is allocated in terms of these five variables. In lieu of additional information, the strategy is simple—just determine the immutable characteristics of the persons in power (the decider subsystem in Miller's terms), then predict that persons with these immutable characteristics will be allocated into the choicest positions (first choice), with persons with the other (nonpower or nondecider) immutable characteristics being allocated into the less desirable position.

In American society, it is easy to see that favored immutables are *white* skin color, *male* gender, and age range *twenty-five to seventy-five* (approximately). Thus, while males aged twenty-five to seventy-five will be predicted to remain in more favorable overall positions on all allocations than any other group. This does not mean that women, blacks, and persons under twenty-five and over seventy-five will not ever get favorable statuses. It only means that, all other things (e.g., mutables) being equal, a higher percentage of *white males aged twenty-five to seventy-five* will get favorable positions on space (*S*), information (*I*), technology (*T*), and organization (*O*) levels and level of living (*L*), or (*LOTIS*), than the percentage of any other group. Thus, white males aged zero to twenty-four and seventy-six-plus will place a lower percentage of their group in favorable positions, as will white females of any age, and all blacks of any age.

Bear in mind that this is the prediction that one would make about an individual's probable allocation by possessing knowledge *only* about that individual's immutable characteristics (and bear in mind that salient immutables in addition to these three examples could be specified). This knowledge represents, in statistical terms,

FIGURE 9.1.

The Opportunity Matrix Comprising the Immutable (*GRA*)
and the Mutable Distributions (*LOTIS*)

Mutables

		I	S	T	O	L
Immutables						
Skin color	White					
	Nonwhite					
Gender	Male					
	Female					
Age	25–75					
	<25, >75					

the *column marginals* in the opportunity matrix (figure 9.1). But where an individual is allocated in the opportunity structure depends *not only* on immutable characteristics, but also on the individual's prior levels on each mutable variable prior to allocation.

The six macro variables (*PILOTS*) are *globals*. They are true macro properties and cannot be defined for individuals, but are characteristics of the society *as a whole*. They represent, respectively, the *totality* of each variable in the system. That is, P is the total population size, S is the total space in the system, L is the sum total of all life sustenance however operationalized (e.g., as money, in terms of total number of dollars possessed by the system), I is the total information possessed by the system, T is the total technology possessed by the system, and O is the sum total of organizational positions possessed by the system.

In contrast, the immutables are micro properties belonging only to the individuals possessing them. Each person in the system has a set of three basic, salient immutables. These are his or her gender, race, and age (*GRA*). While the *GRA* is probably salient in most every system, there are other immutables (some correlated with the *GRA*) that may also be deemed salient in a given society in a given point in time, and used by those in power to allocate persons to positions in the society. These include a host of physical features such as height, weight, hair color, physical disabilities, etc.

But while immutable properties, particularly the *GRA*, form a convenient basis for allocation to societal positions, the person's original position in the allocation opportunity structure at a given point in time is also highly relevant. That is, each person in the population P also has a *position,* at any given point in time, on the five other variables, *ISTOL* or *LOTIS.* Thus, each person has an occupation or organizational position (O), spatial location or *residence* (S), *level of living* such as wealth (L), level of *information* such as education (I), and certain *technological* skills such as ability to pilot an airplane (T).

How can the *LOTIS* variables, which were said to be properties of society (globals) be properties of individuals also? Because they are *mutables* or achieved properties. But then are mutables and globals the same? No, because the *PISTOL* or *PILOTS* variables that are globals are the *sum total* of each variable for the society as a whole. Mutables, on the other hand, are *not sum totals,* but rather are *distributions.* Thus, when I say mutable variables, I am using a shorthand, because it is technically correct to always say *mutable distributions.* The difference then is methodologically and theoretically clear. The six globals (*PILOTS*) are total sums. In contrast, there are *not* six mutable distributions, but only five (*LOTIS*) which are formed by distributing the total population (P) across the remaining five variables (level of living, organization, technology, information, and space).

The five mutable distributions have a special quality. While the globals (*PILOTS*) *can only be defined for societies,* and the immutables (such as the *GRA*) *can only be defined for individuals* (societies do not possess race or gender, but they do have age of course), the mutable distributions can be defined both for societies (macro) and individuals (micro). How can this be? It is simple—the total distribution (e.g., occupational distribution or division of labor) is a *macro property of the society,* while the individual's position in this total distributional structure (e.g., the individual's job) is a *micro property.*

How can one tell a global such as O from the occupational mutable distribution O_m? It is very simple. The global O is the sum total of all occupations *without incumbents!* That is, the population size is not needed to define the global O, but is needed to define the mutable distribution O_m. The mutable O_m is defined by *distributing* all P individuals in the population of the system into the O global positions, to form an O_m mutable distribution. Thus, while population size (P) is not needed to define *LOTIS* variables as global variables, it is needed to define *LOTIS* as mutable distributions. Thus, the global O (or I or S or T or L) can be defined without the population P, but the mutable distribution O_m (or I_m or S_m or T_m or L_m) cannot be, and that is the difference, and that is how one can tell the globals and the mutables apart. Thus, for occupation, we can define three *separate and distinct variables:* (1) the total number of *occupational positions O* (e.g., 1,000,000) (this is a *global* and a *macro* variable); (2) the occupational division of labor (O_m, this is a *mutable distribution* and a *macro* variable); (3) an individual's occupation O_{im} (e.g., teacher) (this is a *mutable* and a *micro* variable). Thus, there are three different types of occupational variables (O, O_m and O_{im}), as well as three types of spatial variables, three types of technological variables, three types of level of living variables, and three types of information variables, in each system. These are in addition to the population global and the immutables (GRA).

Thus, the opportunity matrix of figure 9.1 essentially shows marginals, or the prediction we would make about where a person will be allocated into the mutable distribution *without* knowing his or her current position in the five mutables. In reality, I can make a much better guess about allocation if I know where a person is currently located in the mutable structure. I can conceive of the mutable structure as a five-dimensional space ($LOTIS$) and each person in the system is located in this space, meaning that each person has a position on all five dimensions. Obviously, where each person ends up in the five-dimensional matrix depends to some extent on where he or she started. The transition states are not independent.

I can then envision a transition matrix, where an individual's position in the five-dimensional *LOTIS* matrix at any given point in time is a function of his or her prior position (that is, her or his prior *LOTIS* mutables), and his or her immutables (such as the *GRA*). If the individual is dissatisfied with his or her position, he or she has to change the mutables (education, residence, occupation, wealth, or technology), as the immutables cannot be changed. However, while the immutables *cannot* be changed, how the system treats the immutables *can* be changed, so that while a person's *GRA* is not changed, its impact on his or her position in the five-dimensional *LOTIS* structure is changed.

At a given point in time, the person does not have power over the given set of five mutables that he or she possesses, and the given set of salient immutables. In other words, these are "givens." However, he or she *can* change the mutables (possibly) for the future, and *may* be able to influence the norms governing the immutables. While *theoretically any* person can achieve *any* position in the five-dimensional structure, the reality is that persons who initially have low positions on the five mutables, and who *also* have GRA immutable characteristics not held by those currently in power, will have a much more difficult time reaching their desired positions in the five-dimensional *LOTIS* mutable structure than will those persons with high initial *LOTIS* values (from birth) and with the same GRA immutable characteristics as those in power (the decider subsystem in Miller's terms). Those persons with low *LOTIS* values and favorable GRA characteristics have a good probability of upward mobility, as do those with high *LOTIS* values and unfavorable GRA characteristics.

Notice that the five-dimensional *LOTIS* model is basically a stratification model except that it is a much more general one than is normally found in sociology. Most stratification models deal with income or wealth (L) and perhaps with education (I) and occupation (O). Most *do not* deal with technology (T) and residence (S), and none that I have seen deal with all five simultaneously. Yet, *all* are needed for an adequate analysis, because each person in the social system has a value or score on each dimension. Notice also that all five are related. If a person has a low value on any one of the five *LOTIS* variables (not just on L) it is difficult to get high values on the others. Also, if a person has high values on one, he or she can expect high values on the others, but there may be normative forces against this.

For example, a person with a lot of money (L) *may* elect to eschew education (I) and to live in a slum (S), but the normative pressure against such behavior is likely to be enormous, and the behavior will most likely be considered deviant. Thus there will be a high degree of correlation among the five dimensions. Persons with a low value on one dimension will tend to have low values on the other four; persons with a high value on any given dimension will tend to have high values on the other four. Instances will occur where persons will have a mix of low and high values for various reasons. These can be labeled as a general form of "status inconsistency" (Lenski 1954).

Note that to a certain degree, a person's position in the five-dimensional structure is within his or her control, and to a certain degree it is not. Not only are the immutables (*GRA*) beyond individual control, but the mutables (*LOTIS*) might be, as when a manufacturing plant closes (*O*), or a person's job is transferred to another state (*S*). Also, there is considerable correlation between the immutables and the mutables, so an "achieved" or mutable variable such as education (*I*) might be much more difficult to achieve if one has unfavorable *GRA* values than if he or she has favorable ones.

Now that the structure of SET has been laid out, how about the "motor" or action? How does this model work over time, on a day-to-day basis? In order to understand this, it is necessary once again to consider the interaction of the Q-R model and the three-level model (see chapter 2).

The Q facet (objects) is represented in SET by P in the globals, and P is also represented in the distribution, as cell frequencies or percentages in each category (e.g., the number of teachers in the U.S.). The Q facet is also represented by individual actors having micro immutable characteristics. Thus Q is prevalent at all levels. The R (variable) facet is also prevalent at all levels. The *LOTIS* characteristics are variables (R) at both the global and mutable levels, as are the *GRA* immutable characteristics. Perceptions about *both* objects and variables are stored in the memory of those in power (X) (the conceptual level of figure 2.2) and also are stored on information-carrying markers such as employee personnel files, computer diskettes, police records, etc. (X''). They are also observed empirically (X').

Giddens (1979) advocates an interplay of synchrony and diachrony in sociological theory. This is seen clearly in SET in the Q-R "spin" or "paddle-wheel" effect—an ongoing dialectic where $Q_1 \rightarrow R_1 \rightarrow Q_2 \rightarrow R_2 \rightarrow Q_3 \rightarrow R_3 \rightarrow$ etc. This can be seen as a dialectic between action and order, or between agency and structure. The agency or action takes the form of Q, or persons or objects acting over time diachronically (process) to accomplish their various goals within a societal context. The structure or order is in the form of the resulting R symbol structure. This dialectic takes place at all three levels of the three-level model (X', X, X'').

I have described the Q-R (process-structure) sequence several times in this volume, as well as in Bailey 1990, but will review it briefly again in slightly different terms. The objects (actors or the population [*P*]) in the system engage in action (singly) or interaction (with other actors). They do this in an organized way (*O*) in a spatial context (*S*). They utilize information (*I*) and technology (*T*) to achieve certain goals, often (but not always) to maximize their level of living (*L*). This action is sometimes "rational" and strictly goal oriented, and sometimes not. It is generally regulated by norms. The result of this action is often *replicated action*. This *replicated action* is defined in SET as *order*. Order (and predictability) results *anytime* actions are replicated, and for whatever reason. The replication may result from mass acceptance of certain values, from internalization of norms (à la Parsons), from ritualization, custom, or coercion. Whatever the mechanism, as long as action is replicated, social order results. The particular mechanism is empirically determined in a given society at a given time and place.

This human action, over time, is the *Q* portion of the *Q-R* sequence. The *R* portion represents variables of whatever form. They may be globals (*PISTOL*), mutables (*LOTIS*), or immutables (*GRA*). As variables, these *R*-elements are simply characteristics or properties, of persons or collectivities. Thus, they cannot act themselves, but serve to guide action, and are in turn shaped by it. The *Q-R* sequence represents a relationship between the diachronic process of human action (*Q*-relationships) and the synchronic symbol structure (*R*-relationships). The diachronic process action (*Q*) shapes the nonliving symbol structure (*R*). The *R*-structure then guides the diachronic action for the next time period, etc. This cycle of diachronic action, synchronic symbol structure, diachronic action, etc., is continuous and never ending. The synchronic symbol structure can be altered by altering human action. Conversely, human action can be altered by altering the synchronic symbol structure that guides it. The synchronic symbol system (*R*) contains the rules, norms, laws, definitions, etc., as written in law books, guidebooks, rule books, etiquette books, dictionaries, encyclopedias, etc. For further discussion see chapters 6 and 7 of this volume, and Bailey (1990, pp. 171–209).

The cyclical and permanent relationship between the process of interaction among human actors, the resulting symbolic structure, and the action once again in a new sequence at a new time period, is clearly facilitated by the use of the three-level model.

Virtually all human action, particularly orderly action, utilizes both the empirical (X') and conceptual levels (X). In complex society, much interaction is guided by information stored in markers (X'') such as rule books, personnel manuals, etc. There are many possible action sequences in the three-level model, such as X'' to X to X'; or X' to X'' to X. One of the most popular and common is the basic X to X'' to X' sequence. As an example, the sequence begins with one actor conceptualizing some goal (X). These expectations will entail certain mutable and immutable characteristics. The next stage is to map this goal (X) (perhaps a new position in a company) into the company's markers, in the form of a job description in the personnel manual (X''). The last stage is to search for an applicant (X') who fits this position (whose mutables such as education and immutables such as age are isomorphic with the job description in the X'' marker).

The action-symbol (diachronic-synchronic) interplay takes place over all three levels of the three-level model, causing a "spin" or constant interplay over time. Time and space preclude further elaboration of this complex topic. For a detailed discussion see Bailey 1990. A basic conclusion to be reiterated here is that *the relationships between variables (R) that are the focus of the abstracted system are generated by interaction among concrete human actors (Q) which are the focus of the concrete system. Thus, the holistic SET model deals with both concrete and abstracted systems, and shows their nexus* (Bailey 1990, p. 186). Further, replicated action by actors (Q) results in order (correlations) among symbolic variables (R). This order can be measured and operationalized by the entropy measure H, as well as by various correlation and contingency coefficients.

Thus, SET is a broad synthesis, and is highly integrated. As examples, SET:

1. Integrates theory and method.
2. Discusses order in both verbal and statistical (entropy) terms, and operationalizes order in terms of H.
3. Integrates the Q- and R-facets.
4. Integrates abstracted and concrete systems.
5. Presents the three-level model.
6. Illustrates several levels (individual, group, society).

7. Includes values, norms, and culture as well as materialistic variables such as energy. Both empirical and nonempirical factors are included in the model.

8. Presents the globals (*PILOTS* or *PISTOL*), mutables (*LOTIS*), and immutables (*GRA*).

9. Links diachronic process and synchronic symbol structure.

10. Shows how replicated human action leads to correlations between variable symbols.

11. Presents a broader than average model of opportunity structure (the five-dimensional *LOTIS* structure).

12. Analyzes the allocation process based on mutables and immutables.

It is now time to begin the synthesis in earnest. I will first deal only with the systems prong, then go back through the sociology literature to complete the second prong of the synthesis. The strategy is to use SET (both as hastily sketched in this chapter and as presented in greater detail in chapters 6 and 7, and in Bailey 1990) as a skeleton, and to flesh it in with concepts from other systems theories. Since SET has its roots in functionalism, general systems theory, cybernetics, information theory, and nonequilibrium thermodynamics, it is not necessary to review these systems approaches, as they are already somewhat incorporated in SET. Thus, I will begin the synthesis with the "new" sociocybernetic theory of Geyer and van der Zouwen (1986), Aulin (1986), and others, and will continue through living systems theory and autopoiesis.

Of special interest in the new sociocybernetics is the concept of self-steering (Aulin 1982, 1986; Baumgartner 1986; Klabbers 1986). The notion of self-steering versus outside steering is often applied to planning and political analysis. In self-steering, actors have control of their destiny. In outside steering, the power is in the hands of politicians or outside decision makers. This is readily applied to the allocation process in SET. I said that persons are allocated to positions in the five-dimensional *LOTIS* structure (at all levels—city, state, nation, etc.) on the basis of both their current mutable characteristics (*LOTIS*) as well as their immutables (*GRA*). Inasmuch as *GRA* is fixed, self-steering is impossible. Inasmuch as the mutables are fixed, self-steering is impossible. While the mutables are essentially what are often called "achieved" variables, the connotation is that they can all be changed. In practice, however, this is often not feasible, and may be impossible.

Self-steering is possible, however, in many cases in dealing with the mutables. It may be possible for a person to change one or all of his or her mutables: income level (L), organizational position (O), technological skills (T), education (I), and residence (S). In addition, self-steering is also possible (in some cases) in terms of changing the laws or norms (symbolic synchronic structure) that govern how immutables such as GRA are treated. Examples include civil rights laws and changes in retirement laws. Beyond such changes in mutables and changes in the synchronic symbol structure governing immutables, the steering is not self-steering, but is outside steering by those in power, and is beyond the individual's immediate control.

Moving to living systems theory, we see that several key elements of LST need to be added to the SET skeleton for our ongoing synthesis of systems theory. One obvious addition is the analysis of six levels. Of the eight levels of LST (cell, organ, organism, group, organization, community, society, and supranational), the first two are excluded by our decision (stated earlier) to study only human systems (and not subsystems such as the cell and organ). This leaves six key levels for analysis—individual, group, organization, community, society, and supranational system. Social entropy theory (Bailey 1990) focused upon the society, individual, and organization, and generally neglected (except in passing) analysis of the group, the community, and the supranational levels.

Since the globals ($PILOTS$) and the mutable distributions (LO-TIS) are group (macro) properties that cannot be defined for individuals (see the discussion of the one-level lag or "drop-back" in chapter 5), there are five basic levels of systems analysis—the group, organization, society, community, and supranational, with the individual being the subsystem components for the group, and each succeeding level forming the subsystem for the next higher level (groups are components of organizations; organizations are components of communities; communities are components of societies; and societies are components of supranational systems).

The other chief element to be added to our synthesis from LST is the twenty basic information and energy processing subsystems, with emphasis on the dual subsystems of the reproducer and the boundary. The respective terms (distributor, motor, input transducer, decider, etc.) are easily incorporated into our basic model, and add a breadth and richness of detail not present in the original formulation of SET. Another needed contribution to the synthesis from LST is the notion of information-input overload, as well as the richness of the systems definitions. For further explanation of the relationship between LST and SET, and further additions to the synthesis, see chapter 7 of this volume.

The last of the new systems theories to be added to the synthesis is autopoietic theory. This is a little more complicated than the additions from LST. The additions of LST were relatively straightforward, because both SET and LST begin with a concrete model using the individual as the unit of analysis. As applied to social groups by Luhmann (1986, pp. 177–78), the unit of analysis for autopoiesis is communication. The obvious link to SET is through the emphasis in SET on information, and the information component in *PISTOL* (*PILOTS*). There is also a clear relationship between Luhmann's (1986, p. 178) analysis of decision points, and Miller's (1978) decider subsystem. The "selection of communication" (Luhmann 1986, p. 178) is a vital decision in autopoietic theory, and is made by the "decider subsystem" (Miller 1978).

While synthesizing SET and LST was relatively straightforward, autopoiesis raises a number of problems. For one thing, there is a lack of consensus (as we have seen) over whether autopoiesis can be applied to human social groups. If it cannot be, how can we synthesize it into SET? Secondly, if it *can* be applied to humans, what is the basic unit of analysis? If autopoiesis cannot be applied to humans, then we cannot synthesize it. If it can be applied to humans, but only with communication as the basic unit of systems analysis, then we can synthesize it, but it may be difficult. The optimum state of affairs is if it can be applied to human concrete systems, with the individual as the basic unit of analysis. Then synthesis is relatively straightforward once again.

Utilization of a concrete system seems to be part of the problem in gaining consensus about the efficacy of autopoiesis for human groups. The logic seems to be that "self-reproduction" means reproducing units, and if the units are humans, then clearly the organization does not reproduce them, as reproduction is a biological and not an organizational process.

The key to this dilemma lies in the relation of Q and R elements discussed in chapter 2 and chapter 6, and throughout this volume. One of the chief reasons for advocating the use of concrete systems is that it is relatively easy to go from concrete to abstracted systems, but not vice versa. Since Q and R are two sides of the same coin (see chapter 2), if we begin with concrete (Q) systems based on objects, we can generate abstracted (R) systems based on variables (symbolic structure).

The end result is that the SET framework can easily accept Luhmann's communication units. In fact, communications are generally stored in level X'' on markers such as computer diskettes, books, videotapes, etc. The communications form part of the symbolic synchronic structure that has been discussed in this volume and in Bailey 1990. They play a significant role in the diachronic process—synchronic symbolic structure cycle, as they are part of the latter. Thus, the interplay between human action (Q) and the symbolic structure (R) that guides it and is shaped by it is a crucial concept of SET. If one takes the act or the communication (process) as the systems unit, or the symbol product carried on a marker, the analysis fits well into SET.

Thus, my conclusion is that autopoiesis *can* be synthesized into our integrative model. I will accept Luhmann's choice of communication as the unit. Then, the concrete human actors generate the communication within the system. It is the communications (not the living organisms) that are self-produced by the society. It is the communication system that possesses organizational closure. The concrete (Q) system remains an open system, but the communication system (R) that it generates is organizationally closed. It is the *latter* that is self-reproduced and thus autopoietic.

In addition to self-reproduction of reflexive communication, there are other aspects of autopoietic theory that need to be synthesized into the SET model. One of these is the autopoietic emphasis that the system's boundary is not clearly visible to the system, but is in fact a determination of the observer. As Mingers puts it:

> All explanations or descriptions are made by observers and one must not confuse that which pertains to the observer with that which pertains to the observed. Observers can perceive both an entity and its environment and relate the two together. Interacting components cannot, however, do this. (1989, p. 161)

As Morgan puts it:

> Thus, a system's interaction with its 'environment' is really a reflection and part of its own organization. It interacts with its environment in a way that facilitates its own self-reproduction,

and in this sense we can see that its environment is really part of itself . . . living systems close in on themselves to maintain stable patterns of relations, and . . . it is this process of closure or self-reference that ultimately distinguishes a system as a system. (1986, pp. 236–37)

While all of this may seem confusing, it is relatively easily summarized, and easily synthesized with SET. The autopoietic system is working first and foremost to maintain its autonomy and identity. It does so by engaging in circular patterns of interaction, or interaction that is reflexive, recursive, and organizationally closed. To this extent, the concept of boundary is meaningless to the organization, as it interacts with the environment (as well as within its boundaries). All interaction is geared to achieving and maintaining organizational autonomy and identity.

From the standpoint of systems theory, autopoietic theory is not unusual. Autopoiesis specifies that the system produces its own boundaries (Mingers 1989, p. 164). This is part and parcel of SET (see Bailey 1990). Autopoietic systems also interact across boundaries (Morgan 1986, p. 236), but remain organizationally closed. General systems theory is based on the notion that there are relations between the system and the environment. In fact, while entropy may decrease in the open system through importation of energy into the system, *the open system plus environment is a closed system subject to the second law of thermodynamics.* In a sense this is what Maturana and Varela are saying in different words, and with a little different emphasis. The (concrete) system is open, but the (system plus environment) is closed, and the system is organizationally closed in reflexive, recursive relations to maintain its autonomy and identity.

All of these statements are easily integrated into our synthesis. I will now summarize the three basic aspects of autopoietic theory to be integrated into SET. Bear in mind that not only are these statements compatible with GST (and with homeostatic theory) but to some extent may be just a reiteration of features *already found in our model,* but in different words and with different emphases. The contributions from autopoietic theory are:

1. Social systems recursively and reflexively reproduce communications in an organizationally closed manner. These self-reproduced communications *represent, maintain, and preserve* basic organizational autonomy and identity.

2. Social systems produce boundaries but operate both within and across them to maintain organizational closure. All explanations are made by observers who can perceive the system and its environment.

3. Through *structural coupling* systems interact with the environment in such a way that only those stimuli from the environment which maintain autopoiesis are utilized.

The point about the role of boundaries and the observer can be illustrated from SET. Boundaries, according to SET (Bailey 1990) are the "chicken and the egg." Boundaries come first but also last. By this I mean that social systems construct boundaries, and then boundaries construct (constrain) social systems. A boundary can be set arbitrarily, or by random start, in addition to being carefully and purposefully constructed. After this is done, systems work within the boundary. The existence of a boundary can be determined through entropy calculation, because a boundary is an entropy break, meaning that entropy values are different on different sides of the boundary. However, as Maturana and Varela stress, all "explanation" (which would include the calculation of entropy), is not the role of the system, but of the observer. The systems theorist computes entropy—the system does not. In fact, entropy *cannot* be calculated solely on the empirical level (X'), but requires the X and X'' (conceptual and indicator) levels. The calculation of entropy is a symbolic exercise on the X'' level which requires specifying boundary categories (done by the observer) and then mapping empirical frequencies into them, then making calculations.

Suffice it to say that these three elements from autopoiesis (self-reproduction, organizational closure, and structural coupling) fit well into our synthesized SET model, although they may be more difficult to understand than other approaches. To this end we will eschew most of the complicated jargon of autopoietic theory, and will not incorporate it into our synthesis.

MAINSTREAM SOCIOLOGICAL THEORY

Now that all of the new systems theory has been synthesized, it is time to synthesize mainstream sociological theory in the form of the three comparative theories of Alexander, Giddens, and Collins. As in the counterpoints, the comparison terms will be used to structure the analysis.

Before turning to these contemporary theorists I shall deal briefly with Parsons. Parsons is the foundational nexus between the new systems theory and mainstream sociological theory, as he contributed to *both* systems theory *and* mainstream sociology. It is an oversimplification, but I can almost say that *without equilibrium* and without the *cybernetic hierarchy*, almost all of Parsons's systems theory is acceptable to the new systems synthesis. As I have shown throughout this volume and in Bailey 1990, equilibrium is unacceptable inasmuch as all societies are postulated to be in equilibrium, and it is made part of the systems definition. Equilibrium caused a myriad of problems. It blocked the study of change and conflict, was conservative, and was in large measure responsible for problems of teleology, tautology, and determinism in functionalism. I also have some difficulty with the notion of a "hierarchy" of energy, information, culture, etc., although this is not so damaging as the notion of equilibrium. The problem is solved by eschewing the notion of an abstracted system (at least initially) and beginning with a concrete system. Space, time, energy, and information flows are all basic, and culture is then derived through human interaction. The notion of an elaborate "hierarchy" is unnecessary, although without equilibrium it is probably not very harmful either.

By removing equilibrium, using a concrete rather than an abstracted system (the individual rather than the role as the basic unit) and de-emphasizing the cybernetic hierarchy, Parsons's contribution to the new synthesized systems theory is great. I cannot elaborate all of it here. Suffice it to say that his work on socialization and the internalization of norms, and his work on culture and personality, as well as on differentiation theory, all fits right into the SET synthesis, and is a very welcome addition. For further details of his contribution see, for example, Parsons (1951) and Parsons and Shils (1951).

Alexander

If Parsonian thought—sans equilibrium and abstracted systems—can be integrated into the new systems synthesis, it is clear that neofunctionalism can also be. Since neofunctionalism has basically rejected (or at least reconstructed) equilibrium (see Alexander and Colomy 1990, p. 45), it fits right into the synthesis. Alexander and

Colomy also critique the cybernetic hierarchy (p. 45). The neo-functionalist program involves a reconstruction of Parsons's material and idealist presuppositions, as well as a marked revision of orthodox functionalism's approach to change (see Alexander and Colomy 1990, pp. 52–53). Many aspects of the neofunctionalist program of elaboration, revision, and reconstruction are easily synthesized into my model. For example, the emphasis on order is welcomed, as is the call by Alexander for a "return to the more concrete, group-oriented . . . phase . . . in which the institutional context of a particular social system was clearly differentiated from its abstract mechanisms" (Alexander and Colomy 1990, p. 47). In addition to the neofunctionalist emphasis on material factors, the new systems synthesis also welcomes the return to dynamism in systems analysis (Alexander and Colomy 1990, pp. 46–48). In short, while I obviously cannot offer a blanket endorsement of all neofunctionalist research, it appears that most ongoing research programs, dealing with issues such as idealism-materialism, differentiation, and institutionalization (Alexander and Colomy 1990, pp. 52–56) not only mesh well with my synthesis of the new systems theory, but are welcome additions, as they fill large gaps not traditionally addressed by systems theory.

Giddens

The comparison list of terms from neofunctionalism includes action, order, equilibrium, and idealism/ideational. All except equilibrium have a role in our synthesis. The corresponding comparison terms from structuration theory are agency, structure, system, systems integration, time, and space. Can these be synthesized into our model? The answer is obviously yes. Most of these points of correspondence have already been discussed in the respective counterpoints, and apply also to the dual synthesis. I have already discussed how time and space are integrated into the synthesis, and this need not be repeated. The way this is done (with space as one of the *PISTOL* variables and one of the mutable distributions [*LOTIS*] and time used in an interplay between diachronic action and synchronic symbol structure) seems consistent with the usages advocated by Giddens (1979) for social theory.

I have also already commented on Giddens's notions of systems and systems integration, and will only note here that they are consistent with the dual synthesis. The central feature of Giddens's (1979) definition of system that makes it readily compatible with my synthesis is his use of the actor as the unit of analysis. Thus, his system is a concrete rather than an abstracted system (in Miller's 1978 terms). He is thus departing from Parsons's strategy, but fortuitously his concrete system meshes well with my synthesis. In fact, his definition is readily seen as a specific application of my more general definition (my definition subsumes his, but both are compatible). Giddens's notion of system integration as "degree of interdependence" (Giddens 1979, pp. 76–77) is directly compatible with my use of entropy, which in a sense measures "degree of interdependence" (the lower the entropy or the higher the negentropy, the higher the degree of interdependence).

Giddens's concept of structure also is very compatible with the dual synthesis, and can be incorporated as long as it is remembered that Maturana and Varela's use of structure is very different. Notice that I eschewed inclusion of their use of structure into the dual synthesis. My own notion of synchronic symbol structure is very close in some ways to Giddens's notion of structuring properties as rules and resources (Giddens 1979, p. 68) in the sense that I emphasized that the symbolic structure (such as a rule book) guides action.

This leaves the terms action-order, agency-structure, and structuration. While the dual synthesis has no concept of structuration, the concept of agency will mesh well into our integration. Social entropy theory, as has been seen, does deal with the cycle between diachronic action (process) and synchronic symbol structure (product). There are very close parallels between the action-order notions of neofunctionalism and the agency-structure relations of structuration theory. I see them all as compatible, and thus all as partners in the dual synthesis.

Collins

This brings me to conflict theory, specifically to conservatism, age, sex, conflict, and hypostatization. I am very happy to say that all of these *have already been dealt with to some degree* in this discussion of the new dual synthesis. The notions of age and sex are directly included in the *GRA* (gender-race-age) analysis of the immutables. As for hypostatization, the concrete system is as real as

Collins's "real people" and is not subject to hypostatization. Thus, SET and the first prong of the dual synthesis (the systems synthesis) is not particularly vulnerable to hypostatization. It is only in the second (mainstream) prong that some problems of hypostatization may arise through the inclusion of Parsonian theory (sans equilibrium). However, Alexander and other neofunctionalists have taken steps to increase the concreteness of this approach, and this should decrease vulnerability to hypostatization. Also, to turn the tables a bit, I should note that the rigor provided in SET by the *PILOTS* globals, the *LOTIS* mutables, and the *GRA* immutables provides a concrete framework (in terms of space, population, age, sex, race, technology) that far exceeds the rigor found in the average conflict theory. Thus, compared to SET, conflict theory seems in danger of floating away on a cloud of ideological hypostatization. It too needs a framework to lend concreteness to its analysis.

Since the dual integration deals with such factors as race, age, and sex, and social change, it is not conservative as was functionalism. This is further evidenced by the analysis of conflict in SET. Conflict fits right into SET in the allocation model. I have already stressed how those in power use the immutable *GRA* to discriminate against individuals that they have power over, and this is in a real sense a conflict analysis. Thus, the new systems integration is not "conservative" as it deals with conflict. In fact, as I have pointed out, entropy is generally defined in terms of change over time rather than as a synchronic quantity. Thus, the final analysis is that even conflict theory is compatible with the dual synthesis, and deserves a place in it. As a further example, conflict can easily be expressed in terms of the new sociocybernetic notions of self-steering and outside steering. Conflict occurs when self-steering does not produce results to the actor's satisfaction, and outside steering by a greater power overrides his or her desires.

CONCLUDING REMARKS

What has the dual synthesis accomplished?

1. It has synthesized systems theory. Specifically, the "new systems theory" was synthesized, using the "old" systems theory of GST, information theory, and cybernetics as a foundation.

2. It demonstrated the compatibility of mainstream sociological theory to the new systems theory, and integrated certain elements of mainstream theory into the systems model. Specifically, neofunctionalism, structuration theory, and conflict theory—are all seen to be compatible with the synthesis, and all are integrated into it to some degree.

In concluding, I do not claim to have merged mainstream sociological theory and systems theory, nor to have even brought the new systems theory into the mainstream of sociology for that matter. However, I hope that I have shown the considerable overlap and parallels between systems and the mainstream, and have shown that both are headed in the same direction.

A still unanswered question is how the new systems theory relates to the many variants of mainstream sociological theory (such as interpretive sociology, exchange theory, rational choice theory, Marxist theory, etc.) that have not been explicitly examined. One reason for choosing the theories of Alexander, Giddens, and Collins was that they are all broad theories which are themselves syntheses, and which are concerned with micro-macro linkages. Thus, their inclusion enabled me to parsimoniously cover a lot of theoretical ground. While there may not be as many points of congruence between every mainstream theory and the new systems theory as was evidenced in the theories chosen, I am confident that the breadth of systems theory and my determination that SET is a theory which does not logically preclude other approaches has resulted in a state of affairs where most mainstream theories, be they micro or macro, network, ethnomethodology, etc. will find a basic compatibility and complementarity with the new systems synthesis.

Some of these complementarities are evident in only a cursory review of SET. For example, perusal of SET shows basic complementarity not only with human ecology and demography (e.g., in the *PISTOL* model), but also with network analysis, rational choice theory, symbolic interactionism, and ethnomethodology, among others. There are even some clear points of intersection with Marxism (for example, the parallel between self-reproduction in autopoiesis and the Marxian notion of reproduction). While the new systems theory cannot be (and does not want to be) all things to all people, it is important that the compatibilities between the new systems theory and mainstream sociological theory be stressed. There are many approaches that stress differences (indeed the whole notion of academic "specialization" is based on this). Systems theory stresses synthesis and integration, and thus takes its place in the third stage of postwar theory (see Alexander and Colomy 1990).

By being broader than sociology, systems theory is in a rather unique position to add to the sociological imagination. Concepts such as equilibrium which may be reified in sociology can be seen from the broader perspective of systems theory to be used differently (or to be nonexistent) in other systems. Thus, systems theory has known for some time that equilibrium was not an adequate concept, and has turned to nonequilibrium analysis. I would guess that the contemporary Parsons (as both a systems theorist and sociologist) would also realize the inadequacies of equilibrium if he were alive today, although when he was first developing his theory the concept remained largely unchallenged in (pre-Prigogintan) systems theory, as well as in sociology and the other social sciences. Thus, systems theory was instrumental in the reconstruction of equilibrium which is now prevalent in sociology and the social sciences.

The lesson from systems theory is: do not build equilibrium as a given into your presuppositions, but leave it to be determined as an empirical matter on a case-by-case basis. The new systems theory does not want to preclude any intellectual notion, including equilibrium, but it certainly does not want to assume its existence or treat it as a given, or incorporate it into its definitions and presuppositions as Parsons did. The lesson from systems theory to sociology is: use equilibrium if you want to and if you can make the case for it, but do not build it into your theory presuppositionally. Thus, one benefit of a very broad approach such as systems theory is that it allows sociologists to expand their understanding and their sociological imagination by taking what had been givens or constants (such as equilibrium) and turning them into variables.

In a sense this volume has covered a lot of ground. However, much remains to be done. It is frustrating from a systems standpoint that there is still so much difference in the meaning of terms, particularly in the autopoietic jargon. It would be nice if at least everyone shared the same definition of structure. Thus, there is still no common language, and translating between "structuration" and "structural coupling" remains a formidable task. Nevertheless, a start has been made. This attempt at dual synthesis is offered in the spirit of the systems principle of multidisciplinary integration, and the principle of the accumulation of knowledge. I hope that readers find this volume true to this spirit, and will enlarge upon my efforts.

REFERENCES

Aberle, David. F., A. K. Cohen, A. D. Davis, M. J. Levy, Sr., and F. X. Sutton. 1950. "The Functional Prerequisites of a Society." *Ethics* 60: 100–111.

Ackoff, Russell L. 1974. *Redesigning the Future: A Systems Approach to Social Problems*. New York: Wiley Interscience.

Ackoff, Russell L. and Fred E. Emery. 1972. *On Purposeful Systems*. New York: Aldine.

Alexander, Jeffrey. 1982. *Theoretical Logic in Sociology*. Vol. 1, *Positivism, Presuppositions, and Current Controversies*. Berkeley: University of California Press.

———. 1983. *Theoretical Logic in Sociology*. Vol. 4, *The Modern Reconstruction of Classical Thought: Talcott Parsons*. Berkeley: University of California Press.

———. 1984. "The Parsons Revival in German Sociology." In *Sociological Theory*, ed. Randall Collins, pp. 394–412. San Francisco: Jossey-Bass.

———. 1985. "Introduction." In *Neofunctionalism*, ed. Jeffrey Alexander, pp. 7–18. Beverly Hills: Sage.

———. 1987. "Action and Its Environments." In *The Micro-Macro Link*, ed. Jeffrey Alexander, Bernhard Giesen, Richard Munch, and Neil J. Smelser, pp. 289–318. Berkeley: University of California Press.

———. 1988. *Action and Its Environments*. New York: Columbia University Press.

Alexander, Jeffrey and Paul Colomy. 1990. "Neofunctionalism Today: Reconstructing a Theoretical Tradition." In *Frontiers of Social Theory: The New Synthesis*, ed. George Ritzer, pp. 33–67. New York: Columbia University Press.

Alexander, Jeffrey, Bernhard Giesen, Richard Munch, and Neil J. Smelser, eds. 1987. *The Macro-Micro Link*. Berkeley: University of California Press.

Althusser, Louis. 1970. *For Marx*. New York: Pantheon.

Archer, Margaret. 1985. "Structuration versus Morphogenesis." In *Macro-Sociological Theory*, ed. S. N. Eisenstadt and H. J. Helle, pp. 58–88. London: Sage.

Ashby, W. Ross. 1954. "The Application of Cybernetics to Psychiatry." *Journal of Mental Science* 100: 114–24.

———. 1956. *An Introduction to Cybernetics.* New York: Wiley.

Aulin, Arvid. 1982. *The Cybernetics Law of Social Progress.* Oxford: Oxford University Press.

———. 1986. "Notes on the Concept of Self-Steering." In *Sociocybernetic Paradoxes: Observation, Control and Evolution of Self-Steering Systems,* ed. R. F. Geyer and J. van der Zouwen, pp. 100–18. London: Sage.

Bailey, Kenneth. 1972. "Polythetic Reduction and Monothetic Property-Space." In *Sociological Methodology 1972,* ed. Herbert L. Costner, pp. 83–111. San Francisco: Jossey-Bass.

———. 1975. "Cluster Analysis." In *Sociological Methodology 1975,* ed. David L. Heise, pp. 59–128. San Francisco: Jossey-Bass.

———. 1981. "Abstracted versus Concrete Sociological Theory." *Behavioral Science* 26: 313–23.

———. 1982. "Post-Functional Social Systems Analysis." *Sociological Quarterly* 23: 18–35.

———. 1983. "Sociological Entropy Theory: Toward a Statistical and Verbal Congruence." *Quality and Quantity* 17: 251–68.

———. 1984a. "Equilibrium, Entropy, and Homeostasis: A Multidisciplinary Legacy." *Systems Research* 1: 1–18.

———. 1984b. "Beyond Functionalism: Toward a Nonequilibrium Analysis of Complex Social Systems." *British Journal of Sociology* 35: 1–18.

———. 1984c. "A Three-Level Measurement Model." *Quality and Quantity* 18: 22–45.

———. 1985. "Entropy Measures of Inequality." *Social Inquiry* 55: 200–11.

———. 1987. "Restoring Order: Relating Order to Energy and Information." *Systems Research* 4: 327–37.

———. 1990. *Social Entropy Theory.* Albany: State University of New York Press.

———. 1993. "Living Systems Theory and Functionalism." Paper presented at the 37th Annual Meeting of the International Society for the Systems Sciences, Sydney, Australia.

Ball, Richard A. 1978. "Sociology and General Systems Theory." *The American Sociologist* 13: 65–72.

Banathy, Bela H. 1988. "Systems Inquiry in Education." *Systems Practice* 1: 193–212.

Banathy, Bela H., et al., eds. 1985. *Proceedings of the Society for General Systems Research International Conference.* Vol. 1 and 2. Seaside, Calif.: Intersystems Publications.

Bates, Frederick L. and Clyde C. Harvey. 1975. *The Structure of Social Systems*. New York: Gardner Press.

Baumgartner, Thomas. 1986. "Actors, Models and Limits to Societal Self-Steering." In *Sociocybernetic Paradoxes*, ed. R. F. Geyer and J. van der Zouwen, pp. 9–25. London: Sage.

Beauchamp, Murray. 1989. "Chaos in Sociology." *Perspectives* (American Sociological Association Newsletter) 12: 1–2.

Beer, Stafford. 1975. "Preface." In *Autopoietic Systems*, ed. H. R. Maturana and F. G. Varela. Urbana: University of Illinois.

Behavioral Science. 1980. Vol 25: 65–87.

Berger, Peter L. and Thomas Luckman. 1967. *The Social Construction of Reality*. New York: Doubleday-Anchor Books.

Berrien, F. Kenneth. 1968. *General and Social Systems*. New Brunswick: Rutgers University Press.

Berry, Brian J. L. 1964. "Cities as Systems Within Systems of Cities." *Papers and Proceedings of the Regional Science Association*: 34.

Berry, Brian J. L. and W. L. Garrison. 1962. "Alternate Explanations of Urban Rank-Size Relationships." *Annals of the Association of American Geographers* 48: 83–91.

Bertalanffy, Ludwig von. 1956. "General System Theory." *General Systems* 1: 1–10.

———. 1962. "General System Theory: A Critical Review." *General Systems* 7: 1–20.

———. 1967. *Robots, Men and Minds: Psychology in the Modern World*. New York: George Braziller.

———. 1968. *General System Theory*. New York: George Braziller.

Blalock, Hubert M., Jr. 1968. "The Measurement Problem: A Gap Between the Languages of Theory and Research. In *Methodology in Social Research*, ed. Hubert M. Blalock and Ann B. Blalock, pp. 5–27. New York: McGraw-Hill Book Co.

Blau, Peter M. 1977. *Inequality and Heterogeneity: A Primitive Theory of Social Structure*. New York: The Free Press.

Blumer, Herbert. 1969. *Symbolic Interactionism: Perspective and Method*. Englewood Cliffs, N.J.: Prentice-Hall.

Boulding, Kenneth. 1956. "General Systems Theory: The Skeleton of Science." *Management Science* 2: 197–208.

———. 1978. *Ecodynamics: A New Theory of Social Evolution*. Beverly Hills: Sage.

Brillouin, Leon. 1949. "Life, Thermodynamics, and Cybernetics." *American Scientist* 38: 594–668.

———. 1956. *Science and Information Theory*. New York: Academic Press.

Buckley, Walter. 1967. *Sociology and Modern Systems Theory*. Englewood Cliffs, N.J.: Prentice-Hall.

———. ed. 1968. *Modern Systems Research for the Behavioral Science*. Chicago: Aldine.

Burns, Tom R. and Walter Buckley, eds. 1976. *Power and Control: Social Structures and Their Transformation*. Beverly Hills: Sage.

Busch, John and Gladys Masih Busch, eds. 1984. *Issues in Sociocybernetics: Current Perspectives*. Seaside, Calif.: Intersystems Publications.

———. 1988. *Sociocybernetics: Rethinking Social Organization*. Salinas, Calif.: Intersystems Publications.

Butler, E. W. and S. N. Adams. 1966. "Typologies of Delinquent Girls: Some Alternative Approaches." *Social Forces* 44: 401–07.

Cannon, Walter B. 1929. "Organization for Physiological Homeostasis." *Physiological Reviews* 9: 399–431.

———. 1932. *The Wisdom of the Body*. New York: Norton.

Cavallo, Roger, ed. 1979. "Systems Research Movement: Characteristics, Accomplishments and Current Development." *General Systems Bulletin* Special Issue 9: 1–131.

Cavallo, Roger and George J. Klir. 1978. "A Problem-Solving Basis for General Systems Research." In *Applied General Systems Research: Recent Developments and Trends*, ed. George J. Klir, pp. 53–60. New York: Plenum.

Checkland, Peter. 1981. *Systems Thinking, Systems Practice*. Chichester: Wiley.

———. 1985. "From Optimizing to Learning: A Development of Systems Thinking for the 1990s." *Journal of the Operations Research Society* 36: 757–67.

Churchman, C. West. 1968. *The Systems Approach*. New York: Dell.

Clausius, R. 1850. "On the Mechanical Theory of Heat." Berlin: Poggendorff's Annalen.

———. 1879. *The Mechanical Theory of Heat*, trans. Walter R. Browne. London: Macmillan.

Collins, Randall. 1975. *Conflict Sociology: Toward an Explanatory Approach*. New York: Academic Press.

———. 1988. *Theoretical Sociology*. San Diego: Harcourt Brace Jovanovitch.

———. 1990. "Conflict Theory and the Advance of Macro-Historical Sociology." In *Frontiers of Social Theory*, ed. George Ritzer, pp. 68–87. New York: Columbia University Press.

Comeau, Larry R. and Leo Dreidger. 1978. "Ethnic Opening and Closing in an Open System: A Canadian Example." *Social Forces* 57: 600–02.

Comte, Auguste. 1830–1842. *Cours de Philosophie Positive*, 6 vols., trans. and condensed by Harriet Martineau and published as *The Positive Philosophy of Auguste Comte*, 2 vols. 1953. London: J. Chapman.

Contemporary Sociology: A Journal of Reviews. 1979. Vol. 8, pp. 687–715.

Costner, Herbert L. 1969. "Theory, Deduction, and Rules of Correspondence." *American Journal of Sociology* 75: 245–63.

Davies, Lynda J. 1988. "Understanding Organizational Culture: A Soft Systems Perspective." *Systems Practice* 1: 11–30.

Davis, Kingsley. 1949. *Human Society*. New York: MacMillan.

———. 1959. "The Myth of Functional Analysis as a Special Method in Sociology and Anthropology." *American Sociological Review* 24: 757–72.

Davis, Kingsley and Wilbert E. Moore. 1945. "Some Principles of Stratification." *American Sociological Review* 10: 242–49.

Deutsch, K. 1951. "Mechanism, Teleology and Mind." *Philosophy and Phenomenological Research* 12: 185–223.

Dore, Ronald D. 1967. "Function and Cause." *American Sociological Review* 26: 843–53.

Duncan, David. 1908. *Life and Letters of Herbert Spencer*. Two volumes. New York: Appleton.

Duncan, Otis Dudley. 1966. "Path Analysis: Sociological Examples." *American Journal of Sociology* 72: 1–16.

Duncan, Otis Dudley and Leo F. Schnore. 1959. "Cultural, Behavioral, and Ecological Perspectives in the Study of Social Organization." *American Journal of Sociology* 65: 132–46.

Durkheim, Emile. 1982. *The Rules of Sociological Method*. London: Macmillan.

Easton, David. 1965. "Limits of the Equilibrium Model in Social Research." *Behavioral Science* 1: 96–104.

Facheux, Claude and Spyros Markidakis. 1979. "Automation or Autonomy in Organizational Design." *International Journal of General Systems* 5: 213–20.

Fararo, Thomas J. 1989. *The Meaning of General Theoretical Sociology: Tradition and Formalization*. New York: Cambridge University Press.

Featherstone, Mike. 1988. "In Pursuit of the Postmodern." *Theory, Culture and Society* 5: 195–216.

Forrester, J. W. 1973. *World Dynamics*, 2nd ed. Cambridge: Wright-Allen.

Foster, C., A. Rapoport, and E. Trucco. 1957. "Some Unsolved Problems in the Theory of Non-Isolated Systems." *General Systems* 2: 9–29.

Galtung, Johan. 1975. "Entropy and the General Theory of Peace." In *Essays in Peace Research*, vol. 1, ed. Johan Galtung, pp. 47–75. Atlantic Highlands NJ: Humanities Press.

———. 1980. *The True Worlds: A Transnational Perspective.* New York: Free Press.

Garfinkel, Harold. 1967. *Studies in Ethnomethodology.* Englewood Cliffs, N.J.: Prentice-Hall.

Geyer, R. F. and J. van der Zouwen, eds. 1978. *Sociocybernetics: An Actor-Oriented Social Systems Approach*, vol. 2. Leiden Holland; Martinus Nijohff.

———. 1982. *Dependence and Inequality: A Systems Approach to the Problems of Mexico and Other Developing Countries.* Oxford: Pergamon Press.

———. 1986. *Sociocybernetic Paradoxes: Observation, Control and Evolution of Self-Steering Systems.* London: Sage.

Gibbs, Jack P. 1989. *Control: Sociology's Central Notion.* Urbana: University of Illinois Press.

Gibbs, J. Willard. 1874–1877. "On the Equilibrium of Heterogenous Substances." *Transactions of the Connecticut Academy of Arts and Sciences*, III.

Giddens, Anthony. 1979. *Central Problems in Social Theory: Action, Structure and Contradiction in Social Analysis.* Berkeley: University of California Press.

———. 1982. *Profiles and Critiques in Social Theory.* Berkeley: University of California Press.

———. 1984. *The Constitution of Society: Outline of the Theory of Structuration.* Berkeley: University of California Press.

———. 1987. *Social Theory and Modern Sociology.* Stanford: Stanford University Press

Glaser, Barney and Anselm Strauss. 1967. *The Discovery of Grounded Theory.* Chicago: Aldine.

Gleick, J. 1987. *Chaos: Making a New Science.* New York: Viking.

Gouldner, Alvin W. 1959. "Reciprocity and Autonomy in Functional Theory." In *Symposium on Sociological Theory*, ed. L. Gross, pp. 241–70. New York: Harper and Row

———. 1970. *The Coming Crisis of Western Sociology.* New York: Basic Books.

Haken, Herman. 1983 *Synergetics*, third revised and enlarged edition. Berlin: Springer-Verlag.

Hall, A. D. and R. E. Fagen. 1956. "Definition of System." *General Systems* 1: 18–28.

Hempel, Carl G. 1959. "The Logic of Functional Analysis." In *Symposium on Sociological Theory*, ed. Llewellyn Gross, pp. 271–307. New York: Harper and Row.

Henderson, L. J. 1928. *Blood*. New Haven: Yale University Press.

———. 1935. *Pareto's General Sociology: A Physiologist's Interpretation*. Cambridge: Harvard University Press.

Homans, George C. 1964. "Bringing Men Back In." *American Sociological Review* 29: 809–18.

Homans, George C. and C. P. Curtis. 1934. *An Introduction to Pareto*. New York: Knopf.

Horowitz, Irving Louis. 1962. "Consensus, Conflict and Cooperation: A Sociological Inventory." *Social Forces* 41: 177–88.

Jantsch, Erich. 1975. *Design for Evolution: Self-Organization and Planning in the Life of Human Systems*. New York: George Braziller.

Johnson, Benton. 1975. *Functionalism in Modern Sociology: Understanding Talcott Parsons*. Morristown, N.J.: General Learning Press.

Klabbers, Jan H. G. 1986. "Improvement of (Self) Steering through Support Systems." In *Sociocybernetic Paradoxes: Observation, Control and Evolution of Self-Steering Systems*, ed. R. F. Geyer and J. van der Zouwen, pp. 64–88. London: Sage.

Klapp, Orrin E. 1975. "Opening and Closing in Open Systems." *Behavioral Science* 20: 251–57.

———. 1978. *Opening and Closing: Strategies of Information Adaptation in Society*. New York: Cambridge University Press.

Klir, George J. 1969. *An Approach to General Systems Theory*. New York: Van Nostrand.

———. 1978. *Applied General Systems Research: Recent Developments and Trends*. New York: Plenum.

Krippendorff, Klaus. 1986. *Information Theory: Structural Models for Qualitative Data*. Beverly Hills: Sage.

Kuhn, Alfred. 1963. *The Study of Society: A Unified Approach*. Homewood, Ill.: Irwin.

———. 1974. *The Logic of Social Systems*. San Francisco: Jossey-Bass.

———. 1979. "Differences versus Similarities in Living Systems." *Contemporary Sociology* 8: 691–696.

Kuhn, Thomas. 1962. *The Structure of Scientific Revolutions*. Chicago: University of Chicago Press.

Laszlo, Ervin. 1972. *Introduction to Systems Philosophy*. Chicago: University of Chicago Press.

———. 1986. "Systems and Societies: The Basic Cybernetics of Social Evolution." In *Sociocybernetic Paradoxes: Observation, Control and Evolution of Self-Steering Systems*, ed. R. F. Geyer and J. van der Zouwen, pp. 145–71. London: Sage.

Lazarsfeld, Paul F. 1937 "Some Remarks on the Typological Procedures in Social Research." *Zeitschrift für Socialforschung* 6: 119–39.

———. 1958. "Evidence in Social Research." *Daedalus* 8: 99–130.

Le Chatelier, H. 1888. "Recherches Experimentales et Theoriques sur les Equlibres Chimiques." *Annales des Mines, Huitieme serie, Memiories,* XIII. Paris: Dunod.

Lenski, Gerhard E. 1954. "Status Crystalization: A Nonvertical Dimension of Social Status." *American Sociological Review* 19: 405–13.

Lewin, Kurt. 1936. *Principles of Topological Psychology,* trans. F. Heider and G. M. Heider. New York: McGraw-Hill.

Lewis, Gilbert Newton, and Merle Randall. 1923. *Thermodynamics and the Free Energy of Chemical Substances.* New York: McGraw-Hill.

Light, Ivan. 1983. *Cities in World Perspective.* New York: MacMillan.

Lilienfeld, Robert. 1978. *The Rise of Systems Theory: An Ideological Analysis.* New York: Wiley Interscience.

Lingoes, J. C. 1968. "An IBM 360/67 Program for Guttman-Lingoes Multidimensional Scalogram Analysis—III." *Behavioral Science* 13: 52–53.

Lockwood, David. 1956. "Some Remarks on the 'Social System'." *British Journal of Sociology* 7: 134–46.

Lopreato, Joseph. 1971. "The Concept of Equilibrium: Sociological Tantalizer." In *Institutions and Social Exchange: The Sociologies of Talcott Parsons and George C. Homans,* ed. Herman Turk and Richard L. Simpson, pp. 309–43. New York: Bobbs-Merrill.

Lotka, Alfred J. 1925. *Elements of Mathematical Biology.* New York: Dover.

Luhmann, Niklas. 1982. "The World Society as a Social System." *International Journal of General Systems* 8: 131–38.

———. 1984. "Die Wirtschaft der Gesellschaft als Autopoieteisches System." *Zeitschrift für Soziologie* 13: 308–27.

———. 1986. "The Autopoiesis of Social Systems." In *Sociocybernetic Paradoxes: Observation, Control and Evolution of Self-Steering Systems,* ed. R. F. Geyer and J. van der Zouwen, pp. 172–92. London: Sage.

———. 1989. *Ecological Communication.* Cambridge: Polity Press.

———. 1990. *Essays on Self-Reference.* New York: Columbia University Press.

McFarland, David D. 1969. "Measuring the Permeability of Occupational Structures: An Information-Theoretic Approach." *American Journal of Sociology* 75: 41–61.

Malinowski, Bronislaw. 1948. *Magic, Science and Religion and Other Essays,* Glencoe, Ill.: Free Press.

Margaleff, D. Ramon. 1958. "Information Theory in Ecology." *General Systems* 3: 36–71.

Markovsky, Barry. 1987. "Toward Multilevel Sociological Theories: Simulations of Actor and Network Effects." *Sociological Theory* 5: 101–17.

Maruyama, Magoroh. 1963. "The Second Cybernetics: Deviation Amplifying Mutual Causal Processes." *American Scientist* 51: 164–79.

Mason, William, George Y. Wong, and Barbara Entwisle. 1983. "Contextual Analysis through the Multilevel Linear Model." In *Sociological Methodology 1983–1984*, ed. Samuel Einhardt, pp. 72–103. San Francisco: Jossey-Bass.

Maturana, Humberto. 1974. "Cognitive Strategies." In *Cybernetics of Cybernetics*, ed. H. von Foerster. Urbana: University of Illinois.

———. 1975. "Communication and Representation Functions." In *Encyclopedie de la Pleiade*, ed. J. Piaget. Paris: Gallimard.

———. 1978. "Biology of Language: The Epistemology of Reality." In *Psychology and Biology of Language and Thought: Essays in Honour of Eric Lenneberg*, ed. G. Miller and E. Lenneberg, pp. 27–63. New York: Academic Press.

———. 1980a. "Autopoiesis: Reproduction, Heredity and Evolution." In *Autopoiesis, Dissipative Structures and Spontaneous Social Orders*, ed. M. Zeleny. AAAS Selected Symposium 55, Boulder Colo.: Westview Press.

———. 1980b. "Man and Society." In *Autopoietic Systems in the Social Sciences*, ed. F. Bensler, R. P. Hejl, and W. Kock, pp. 11–31. Frankfurt: Campus Verlag.

———. 1981. "Autopoiesis." In *Autopoiesis: A Theory of Living Organization*, ed. M. Zeleny, pp. 21–23. New York: Elsevier, North-Holland.

Maturana, Humberto R. and F. G. Varela. 1980. *Autopoiesis and Cognition: The Realization of the Living*. Dordrecht: Reidel.

Meadows, D. H., D. L. Meadows, J. Randers, and W. W. Behrens. 1972. *The Limits to Growth*. New York: Universe Books.

Menzies, Ken. 1977. *Talcott Parsons and the Social Image of Man*. London: Routledge and Kegan Paul.

Merton, Robert K. 1949. *Social Theory and Social Structure*. Glencoe, Ill.: The Free Press.

Micklin, Michael and Harvey M. Choldin, eds. 1984. *Sociological Human Ecology: Contemporary Issues and Applications*. Boulder: Westview Press.

Miller, James Grier. 1937. "Whitehead: History in the Grand Manner, Part I." *Harvard Guardian* 1: 23–29.

————. 1955. "Toward a General Theory for the Behavioral Sciences." *American Psychologist* 10: 513–31.

————. 1978. *Living systems.* New York: McGraw-Hill Book Co.

————. 1984. Personal communication.

————. 1985. "The Application of Living Systems Theory to 41 U.S. Army Battalions." *Behavioral Science* 30: 1–50.

————. 1986. "Can Systems Theory Generate Testable Hypotheses? From Talcott Parsons to Living Systems Theory." *Systems Research* 3: 73–84.

————. 1987. Personal communication.

Miller, James Grier and Jessie L. Miller. 1980. "The Family as a System." In *The Family: Evaluation and Treatment,* ed. Charles Hofling and Jerry Lewis, pp. 141–83. New York: Brunner/Mazel.

————. 1983. "General Living Systems Theory and Small Groups." In *Comprehensive Group Psychotherapy,* second edition, ed. H. I. Kaplan and B. J. Saddock, pp. 33–47. Baltimore: Williams and Wilkins.

Miller, Jessie L. and James Grier Miller. 1992. "Greater Than the Sum of its Parts I. Subsystems which Process Both Matter-Energy and Information." *Behavioral Science* 37: 1–38.

Mingers, John. 1989. "An Introduction to Autopoiesis—Implications and Applications." *Systems Practice* 2: 159–80.

Moore, Wilbert E. 1981. "Can the Discipline Survive Its Practitioners?" *The American Sociologist* 16: 56–58.

Morgan, G. 1986. *Images of Organization.* Beverly Hills: Sage.

Odum, Howard. 1983. *Systems Ecology.* New York: John Wiley and Sons.

Ogburn, William F. 1951. "Population, Private Ownership, Technology and the Standard of Living." *American Journal of Sociology* 56: 314–19.

Olsson, Gunnar. 1967. "Central Place Systems, Spatial Interaction, and Stochastic Processes." *Papers and Proceedings of the Regional Science Association* 18: 44.

Pareto, Vilfredo. 1935. *The Mind and Society,* vol. 4. New York: Harcourt, Brace.

Parra Luna, Francisco. 1990. "Sociological Systems Theory." In *Sociology in Spain,* ed. Salvador Ginerand Luis Moreno, pp. 353–56. Madrid: Instituto De Estudios Sociales Avanzados.

Parsons, Talcott. 1937. *The Structure of Social Action.* Glencoe, Ill.: The Free Press.

————. 1951. *The Social System.* Glencoe, Ill.: The Free Press.

————. 1961a. "An Outline of the Social System." In *Theories of Society,* vol. 1, ed. Talcott Parsons, Edward Shils, Kaspar D. Naegle, and Jesse R. Pitts, pp. 30–79. Glencoe, Ill.: The Free Press.

———. 1961b. "The Point of View of the Author." In *The Social Theories of Talcott Parsons*, ed. Max Black, pp. 311–63. Englewood Cliffs, N.J.: Prentice-Hall.

———. 1966. *Societies*. Englewood Cliffs, N.J.: Prentice-Hall.

———. 1967. "Boundary Relations Between Sociocultural and Personality Systems." In *Toward a Unified Theory of Human Behavior*, second ed., ed. R. R. Grinker. New York: Basic Books.

———. 1979. "Concrete Systems and 'Abstracted Systems.'" *Contemporary Sociology* 8: 696–705.

Parsons, Talcott and E. A. Shils, ed. 1951. *Toward a General Theory of Action*. New York: Harper and Row.

Pask, Gordon. 1975. *Conversation, Cognition and Learning: A Cybernetic Theory and Methodology*. New York: Elsevier.

Pattee, Howard H., ed. 1973. *Hierarchy Theory: The Challenge of Complex Systems*. New York: George Braziller.

Pickler, A. G. 1954. "Utility Theories in Field Physics and Mathematical Economics (I)." *British Journal of the Philosophy of Science* 5: 47–58.

———. 1955. "Utility Theories in Field Physics and Mathematical Economics (II)." *British Journal of the Philosophy of Science* 5: 313–16.

Pollner, Melvin. 1987. *Mundane Reason: Reality in Everyday and Sociological Discourse*. New York: Cambridge University Press.

Prigogine, Ilya. 1955. *Introduction to Thermodynamics of Irreversible Processes*. Springfield, Ill.: Charles C. Thomas.

———. 1962. *Non-equilibrium Statistical Mechanics*. New York: Interscience Publishers.

Prigogine, Ilya and Isabelle Stengers. 1984. *Order Out of Chaos*. New York: Bantam Books.

Rapoport, Anatol. 1956. "The Promises and Pitfalls of Information Theory." *Behavioral Science* 1: 303–09.

Rhee, Yong-Pil. 1982. *The Breakdown of Authority Structure in Korea in 1960: A Systems Approach*. Seoul: Seoul National University Press.

Ritzer, George. 1975. *Sociology: A Multiple Paradigm Science*. Boston: Allyn and Bacon.

———. 1983. *Sociological Theory*, first ed. New York: Knopf.

———. 1988. *Sociological Theory*, second ed. New York: Knopf.

———. 1990a. "The Current State of Social Theory: The New Synthesis." In *Frontiers of Social Theory: The New Synthesis*, ed. George Ritzer, pp. 1–32. New York: Columbia University Press.

———. 1990b. "Micro-Macro Linkage in Sociological Theory: Applying a Metatheoretical Tool." In *Frontiers of Social Theory: The New Syn-*

thesis, ed. George Ritzer, pp. 347–70. New York: Columbia University Press.

Ritzer, George and Richard Bell. 1981. "Emile Durkheim: Exemplar for an Integrated Sociological Paradigm?" *Social Forces* 59: 966–95.

Robb, Fenton. 1989a "Cybernetics and Supra Human Autopoietic Systems." *Systems Practice* 2: 47–74.

———. 1989b. "The Application of Autopoietic Systems to Social Organizations: A Comment on John Minger's 'An Introduction to Autopoiesis: Implications and Applications.'" *Systems Practice* 2: 349–51.

———. 1989c. "The Application of Autopoietic Systems to Social Organizations: A Comment on John Minger's Reply." *Systems Practice* 2: 353–60.

Robertson, Ian. 1987. *Sociology*. New York: Worth Publishers.

Rossi, Peter. 1980. "Report of the President: Rossi Expresses Concern About Diversity in Sociology." *American Sociological Association Footnotes* 8: 1–7.

Rothstein, Jerome. 1958. *Communication, Organization and Science*. Indian Hills, Colo.: Falcon's Wing Press.

Russett, Cynthia. 1966. *The Concept of Equilibrium in American Sociological Thought*. New Haven: Yale University Press.

Samuelson, Paul A. 1983 [1947]. *Economics*, eleventh ed. New York: McGraw-Hill.

Schrodinger, Erwin. 1945. *What is Life?* Cambridge: Cambridge University Press.

Schuster, Heinz Georg. 1984. *Deterministic Chaos*. Weinheim, Germany: Physik-Verlag.

Seyle, H. 1956. *The Stress of Life*. New York: McGraw-Hill.

Shannon, C. E. and W. Weaver. 1949. *The Mathematical Theory of Communication*. Urbana, Ill.: University of Illinois Press.

Simon, Herbert A. 1955. "On a Class of Skew Distribution Functions." *Biometrika* 42: 425–40.

———. 1964. *Models of Man*. New York: Wiley.

Small, Albion and George E. Vincent. 1894. *An Introduction to the Study of Society*. New York: American Book Co.

Sneath, Peter H. A. and Robert A. Sokal. 1973. *Numerical Taxonomy: The Principles and Practice of Numerical Classification*. San Francisco: W. H. Freeman.

Sokal, Robert A. and Peter H. A. Sneath. 1963. *Principles of Numerical Taxonomy*. San Francisco: W. H. Freeman.

Spencer, Herbert. 1892 [1864]. *First Principles*. New York: Appleton.

Spilerman, S. 1972. "Extensions of the Mover-Stayer Model." *American Journal of Sociology* 78: 599–626.

Stinchcombe, Arthur L. 1965. "Social Structure and Organizations." In *Handbook of Organizations*, ed. James G. March, pp. 142–93. Chicago: Rand McNally.

———. 1968. *Constructing Social Theories*. New York: Harcourt, Brace, Jovanovich.

Stryker, Sheldon. 1979. "The Profession: Comments from an Interactionist's Perspective." *Sociological Focus* 12: 175–86.

Szilard, Leo. 1929. "Uber die Entropieverminderung in Einem Thermodynamischen System bei Eingriffen Intelligehter Wesen." *Zeitschrift fur Physik* 53: 840–56. Trans. A. Rapoport and M. Knoller as "On the Decrease of Entropy in a Thermodynamic System by the Intervention of Intelligent Beings." *Behavioral Science* 9 (1964): 301–10.

Theil, Henri. 1967. *Economics and Information Theory*. Chicago: Rand McNally.

Troncale, L. R. 1978. "Linkage Propositions Between Fifty Principal Systems Concepts." In *Applied General Systems Research: Recent Developments and Trends*, ed. George J. Klir, pp. 29–52. New York: Plenum Press.

———. 1985. "On the Possibility of Empirical Refinement of General Systems Isomorphies." In *Systems Inquiring: Theory, Philosophy, Methodology*. Proceedings of the Society for General Systems Research. International Conference, vol. 1, ed. Bela Banathy, et al., pp. 7–13. Los Angeles, May 27–31, 1985.

Turner, Jonathan H. 1990. "The Past, Present and Future of Theory in American Sociology." In *Frontiers of Social Theory: The New Synthesis*, ed. George Ritzer, pp. 371–91. New York, Columbia University Press.

———. 1991. *The Structure of Sociological Theory*, fifth ed. Belmont, Calif.: Wadsworth.

Turner, Jonathan H., and Alexandra Maryanski. 1979. *Functionalism*. Menlo Park, CA: Benjamin/Cummings.

van den Berghe, Pierre. 1963. "Dialectic and Functionalism: Toward Reconciliation." *American Sociological Review* 28: 695–705.

Varela, F. G. 1979. *Principles of Biological Autonomy*. New York: Elsevier North-Holland.

———. 1980. "Describing the Logic of the Living. The Adequacy and Limitations of the Ideas of Autopoiesis." In *Autopoiesis, Dissipative Structures and Spontaneous Social Orders*, ed. M. Zeleny. AAAS Selected Symposium 55. Boulder, Colo.: Westview Press.

———. 1984. "Two Principles of Self-Organization." In *Self-Organization and the Management of Social Systems*, ed. J. Ulrich and G. Probst, pp. 25–32. Frankfurt: Springer.

Varela, F. G., Humberto R. Maturana, and R. Uribe. 1974. "Autopoiesis:

The Organization of Living Systems, Its Characterization and a Model." *BioSystems* 5: 187–96.

Vickers, Geoffrey. 1959. "The Concept of Stress in Relation to the Disorganization of Human Behavior." In *Stress and Psychiatric Disorder*, ed. J. M. Tanner, pp. 3–10. Oxford: Blackwell.

von Foerster, H., Margaret Mead, and H. L. Teuber. 1949–1957. *Transactions of Conferences on Cybernetics*, 5 vols. New York: Josiah Macy, Jr. Foundation.

von Neumann, John. 1958. *The Computer and the Brain*. New Haven: Yale University Press.

Wagner, Helmut. 1964. "Displacement of Scope: A Problem of the Relationship Between Small Scale and Large Scale Sociological Theories." *American Journal of Sociology* 69: 571–85.

Wallerstein, Immanuel. 1974. *The Modern World System: Capitalist Origins of the European World Economy in the Sixteenth Century*. New York: Academic Press.

Weber, Max. 1947. *The Theory of Social and Economic Organization*, trans. A. M. Henderson and Talcott Parsons, ed. Talcott Parsons. New York: Oxford University Press.

———. 1949. *The Methodology of the Social Sciences*, ed. Edward Shils and Henry Finch. New York: Free Press.

Wheeler, Lynde Phelps. 1951. *Josiah Woodard Gibbs: The History of a Great Mind*. New Haven: Yale University Press.

White, H. C., S. A. Boorman, and R. L. Brieger. 1976. "Social Structure From Multiple Networks. I. Blockmodels of Roles and Positions." *American Journal of Sociology* 81: 730–80.

Wiener, Norbert. 1948. *Cybernetics*. New York: Wiley.

———. 1950. *The Human Use of Human Beings: Cybernetics and Society*. Boston: Houghton Mifflin Co.

———. 1961. *Cybernetics*, second ed. Cambridge: MIT Press; and New York: Wiley.

Zeleny, Milan. 1989. "Are Social Systems Autopoietic?" In *Proceedings of the Annual Meeting of the International Society for Systems Sciences*, vol. 3, ed. P. W. J. Leddington, pp. 148–52. Edinburgh, Scotland.

Zetterberg, H. L. 1965. *On Theory and Verification in Sociology*. Totowa, N.J.: Bedminster Press.

AUTHOR INDEX

SUBJECT INDEX

Abstract systems, 47–48; Living
Systems Theory, 168–169, 208,
209, 259–260; Social Entropy
Theory, 223–227, 244, 327;
three-level model, 55–56, 60–62,
73
Action, 30, 33, 243–244, 250, 233–
235, 243–244, 334
Agency, 11, 19, 32, 36, 76
American Cybernetics Society, 130
Autopoiesis, 1, 3, 15, 50, 58–59, 62,
131, 285–322; basic terms, 287–
290; boundaries, 285, 290–291,
306–307, 311, 318, 339–340,
341; communication, 310–311,
314–316, 339–340; conflict
theory, 318–319; contributions of,
321–322; definition, 288, 290;
entropy, 305–308; epistemology,
285–287, 288, 296–302, 306;
homeostasis, 305–306, 308;
interaction, 288, 339–340; law of
requisite variety, 304–305; Living
Systems Theory, 292–296, 313–
314; medium, 285, 287–288;
metatheory, 321; organization,
287–288, 297, 301–302, 320;
open versus closed systems, 302–
304, 340; organizational closure,
289, 303–304, 318–319; self-
reproduction, 285, 290–291, 318,
338, 335, 340; social systems,
291–292, 293–296, 309–317;
structural coupling, 289, 304,
structuration theory, 319–320;
structure, 288, 297, 303–304, 320;
three-level model, 296–302, 307–
308, 317; unit of analysis, 293,

294–296, 309–311, 314–317;
unity, 287, 301–302

Bell Telephone Laboratories, 135–
136
Boundaries: autopoiesis, 285, 290–
291, 306–307, 311, 318; entropy,
122–124; functionalism, 109–110;
Living Systems Theory, 171, 177–
178; Social Entropy Theory, 23,
227, 241–242, 264–265; systems,
84–85; theoretical synthesis, 327,
339–340, 341

Closed systems, 48–49, 57–58, 59,
62, 104; definition of, 151
Conceptual level (X), 3–4, 52. *See
also* Three-level model
Concrete systems, 1, 3, 15, 47–48;
50, 58–59, 62, 131, 285–322;;
Living Systems Theory, 168–169,
186, 208–209, 210–212, 259–
260, 264, 293; Social Entropy
Theory, 223–227, 244, 327;
theoretical synthesis, 338–340;
three-level model, 55–56, 60–62
Conflict theory, 19–26; autopoiesis,
318–319; concrete explanation,
22–24; criticisms of functionalism,
19–21, 24–25; hypostatization,
24, 35–36, 84, 119–120, 214–
216, 252–253, 318–319, 344–
345; ideology, 19–22, 214–215,
344–345; Living Systems Theory,
214–216; Social Entropy Theory,

www.ingramcontent.com/pod-product-compliance
Lightning Source LLC
Chambersburg PA
CBHW030635270326
41929CB00007B/88